BEG 4033

D1590717

# Advanced Biomedical Technologies

**Series Editor**
*François Ferré*
The Immune Response Corporation

**Editorial Advisory Board**
*Bruce K. Patterson*
Northwestern University

*Kary B. Mullis*
Vyrex Inc.

*Philip L. Felgner*
Vical Inc.

*William N. Drohan*
American Red Cross

*Michael J. Heller*
Nanogen Inc.

*Phillips Kuhl*
Cambridge Healthtech Institute

*Michael Karin*
University of California at San Diego

*Robert E. Sobol*
Sidney Kimmel Cancer Center

*Willem P.C. Stemmer*
Maxygen Inc.

## Forthcoming volumes in the series

*Rational Therapeutic Target in Angiogenesis*
J.M. Pluda and W.W. Li, editors

*Techniques in Localization of Gene Expression*
Bruce K. Patterson, editor

*The Biology of p53*
Michael I. Sherman and Jack A. Roth, editors

*Molecular Genetic Profiling: Applications to Diagnostics and Disease Management*
Lance Fors, editor

# Gene Quantification

François Ferré
*Editor*

Foreword by Edwin M. Southern

Birkhäuser
Boston • Basel • Berlin

François Ferré
The Immune Response Corporation
Carlsbad, CA 92008, U.S.A.

Library of Congress Cataloging In-Publication Data

Gene quantification / François Ferré, editor : foreword by Edwin Southern.
p. cm. – (Advanced biomedical technologies)
Includes biographical reference and index.
ISBN 0-8176-3945-4 (alk. paper). – ISBN 3-7643-3945-4 (alk. paper)
1. Quantitative genetics. 2. Gene amplification. 3. Polymerase chain reaction. I. Ferré, François. II. Series.
QH452.7.G46 1997
572.8'6–dc21 97-21838
CIP

QH
452
.7
.G46
1998

Printed on acid-free paper
© Birkhäuser Boston 1998 Birkhäuser
Copyright is not claimed for works of U.S. Government employees.
All rights reserved. No part of this publication may be reproduced, stored in retrieval system, or transmitted, in any form or by any means, electronic, mechanical, photocopying, recording, or otherwise, without prior permission of the copyright owner.
The use of general descriptive names, trademarks, etc. in this publication even if the former are not especially identified, is not to be taken as a sign that such names, as understood by the Trade Marks and Merchandise Marks Act, may accordingly be used freely by anyone. While the advice information in this book are believed to be true and accurate at the date of going to press, neither the authors nor the editors nor the publisher can accept any legal responsibility for any errors or omissions that may be made. The publisher makes no warranty, express or implied, with respect to the material contained herein.
Permission to photocopy for internal or personal use of specific clients is granted by Birkhäuser Boston for libraries and other users registered with the Copyright Clearance Center (CCC), provided that the base fee of $6.00 per copy, plus $0.20 per page is paid directly to CCC, 222 Rosewood Drive, Danvers, MA 01923, U.S.A. Special requests should be addressed directly to Birkhäuser Boston, 675 Massachusetts Avenue, Cambridge, MA 02139, U.S.A.
ISBN 0-8176-3945-4
ISBN 3-7643-3945-4
Typeset by TypeSmith Graphic Services
Printed and bound by Hamilton Printing Co., Rensselear, New York
Printed in the Unitied States of America
9 8 7 6 5 4 3 2 1

# Contents

# List of Contributors

Michel Bahuau, Laboratoire d'Immunologie des Tumeurs, URA 1484 CNRS et Laboratoire de Génétique Moléculaire, Faculté des Sciences Pharmaceutiques et Biologiques de Paris, 4 Avenue de L'Observatoire, 75006 Paris, France

Dominique Bellet, Laboratoire d'Immunologie des Tumeurs, URA 1484 CNRS et Laboratoire de Génétique Moléculaire, Faculté des Sciences Pharmaceutiques et Biologiques de Paris, 4 Avenue de L'Observatoire, 75006 Paris, France

Ivan Bièche, Laboratoire d'Oncogénétique, Centre René Huguenin, 35 Rue Dailly, 92211 Saint Cloud, France

Elizabeth Bonney, "The Ghost Lab," NIH, NIAID, LCMI, 4 Center Drive, MSC 0420, Bethesda, MD 20891-0420

Beverly S. Chilton, Department of Cell Biology and Biochemistry, Texas Tech University School of Medicine, 3601 Fourth Street, Lubbock, TX 79430

Mark L. Collins, Nanogen Corporation, 10398 Pacific Center Court, San Diego, CA 92121

Peter. J. Dailey, Chiron Corporation, 4560 Horton Street, Emeryville, CA 94608-2916

Abel De La Rosa, Digene Diagnostics, Inc., 2301-B Broadbirch Drive, Silver Spring, MD 20904

Peter A. Doris, Institute of Molecular Medicine, University of Texas Health Sciences Center, 2121 W. Holcombe Boulevard, Houston, TX 77030

Douglas L. Feinstein, Department of Neurology and Neurochemistry, Division of Neurobiology, Cornell University Medical College, 411 East 69 Street, New York, NY 10021

Francois Ferre, The Immune Response Corporation, 5935 Darwin Court, Carlsbad, CA 92008

Elena Galea, Department of Neurology and Neurochemistry, Division of Neurobiology, Cornell University Medical College, 411 East 69 Street, New York, NY 10021

Stuart Geiger, Advanced Bioscience Laboratories, 5510 Nicholson Lane, Kensington, MD 20895

Todd Giles, Genentech, Inc., 460 Point San Bruno Blvd., South San Francisco, CA 94080-4990

Yves Giovangrandi, Laboratoire d'Immunologie des Tumeurs, URA 1484 CNRS et Laboratoire de Génétique Moléculaire, Faculté des Sciences Pharmaceutiques et Biologiques de Paris, 4 Avenue de L'Observatoire, 75006 Paris, France

Frank Gonzales, Gen-Probe Incorporated, 10210 Genetic Center Drive, San Diego, CA 92121

Jukka Hartikka, Cell Biology, Vical Incorporated, 9373 Towne Centre Drive, San Diego, CA 92121

Amanda Hayward-Lester, Department of Cell Biology and Biochemistry, Texas Tech University School of Medicine, 3601 Fourth Street, Lubbock, TX 79430

Chris Heid, Genentech, Inc., 460 Point San Bruno Blvd., South San Francisco, CA 94080-4990

Christine Jacob, INRA, Laboratoire de Biometrie, 78352 Jouy En Josas, France

Janice A. Kolberg, Chiron Corporation, 4560 Horton Street, Emeryville, CA 94608-2916

Jill M. Kolesar, School of Pharmacy, University of Wisconsin-Madison, Madison, WI 53706-1515

John G. Kuhn, Department of Pharmacology, The University of Texas Health Science Center San Antonio, 7703 Floyd Curl Drive, San Antonio, TX 78284-6220

Shirley Kwok, Roche Molecular Systems, 1145 Atlantic Avenue, Suite 100, Alameda, CA 94501

Olivier Lantz, Service d'Hématologie, Université Paris-Sud, Le Kremlin-Bicetre 94270 and INSERM unit 25, Hôpital Necker, Paris, France

Vladimir Lazar, Service de Biologie Oncologique, Institut Gustave Roussy, 39 Rue Camille Desmoulins, 94805 Villejuif Cedex, France

James G. Lazar, Digene Diagnostics, Inc., 2301-B Broadbirch Drive, Silver Spring, MD 20904

Marcia E. Lewis, Chiron Corporation, 4560 Horton Street, Emeryville, CA 94608-2916

Mei-June Liao, Interferon Sciences, Inc., 783 Jersey Ave., New Brunswick, NJ 08901-3660

Attila Lörincz, Digene Diagnostics, Inc., 2301-B Broadbirch Drive, Silver Spring, MD 20904

Marston Manthorpe, Cell Biology, Vical Incorporated, 9373 Towne Centre Drive, San Diego, CA 92121

Lincoln McBride, Department of Physiology, Applied Biosystems, Inc., 850 Lincoln Center Drive, Foster City, CA 94404

Sherrol H. McDonough, Gen-Probe Incorporated, 10210 Genetic Center Drive, San Diego, CA 92121

Mariana G Meijide, Digene Diagnostics, Inc., 2301-B Broadbirch Drive, Silver Spring, MD 20904

P. J. Oefner, Department of Biochemistry, Stanford University, Stanford, CA 94305

Mark Oldham, Department of Physiology, Applied Biosystems, Inc., 850 Lincoln Center Drive, Foster City, CA 94404

Jean Peccoud, TIMC-IMAG, Institut Albert Bonniot, Faculté de Médecine de Grenoble, 38706 La Tronche Cedex, France

Jerald Radich, Fred Hutchinson Cancer Research Center, 1124 Columbia Street, C2S-023, Seattle, WA 98104-1124

Abbas Rashidbaigi, Interferon Sciences, Inc., 783 Jersey Ave., New Brunswick, NJ 08901-3660

Randy Rasmussen, Department of Pathology, University of Utah School of Medicine, Salt Lake City, UT 84132

Kirk Ririe, Idaho Technology, 149 Chestnut Street, Idaho Falls, ID 83402

Joseph Romano, Advanced Bioscience Laboratories, 5510 Nicholson Lane, Kensington, MD 20895

Michael Sawdey, Cell Biology, Vical Incorporated, 9373 Towne Centre Drive, San Diego, CA 92121

Lu-Ping Shen, Chiron Corporation, 4560 Horton Street, Emeryville, CA 94608-2916

Maninder K. Sidhu, Interferon Sciences, Inc., 783 Jersey Ave., New Brunswick, NJ 08901-3660

John Sninsky, Roche Molecular Systems, 1145 Atlantic Avenue, Suite 100, Alameda, CA 94501,

Joanne Spadoro, Roche Diagnostics Systems, Branchburg Township, 1080 US Highway 202 South, Somerville, NJ 08876-3771

Yassine Taoufik, Laboratoire Virus, Neurones et Immunité, Université Paris-Sud, Le Kremlin-Bicêtre 94270 and Service de Microbiologie, Hôpital P. Brousse, Villejuif 94807, France

Ayly Tucker, Genentech, Inc., 460 Point San Bruno Blvd., South San Francisco, CA 94080-4990

Scott Umlauf, T Cell Sciences, Inc., Molecular Immunology Department, 119 Fourth Ave., Needham, MA 02194-2725

P. A. Underhill, Department of Biochemistry, Stanford University, Stanford, CA 94305

Mickey S. Urdea, Chiron Corporation, 4560 Horton Street, Emeryville, CA 94608-2916

H. Lee Vahlsing, Cell Biology, Vical Incorporated, 9373 Towne Centre Drive, San Diego, CA 92121

Paul van de Wiel, Organon Teknika International, Zeedijk 58, B-2300, Turnhout, Belgium

Michel Vidaud, Laboratoire d'Immunologie des Tumeurs, URA 1484 CNRS et Laboratoire de Génétique Moléculaire, Faculté des Sciences Pharmaceutiques et Biologiques de Paris, 4 Avenue de L'Observatoire, 75006 Paris, France

Zhuang Wang, Roche Diagnostics Systems, Branchburg Township, 1080 US Highway 202 South, Somerville, NJ 08876-3771

P. Mickey Williams, Genentech, Inc., 460 Point San Bruno Blvd., South San Francisco, CA 94080-4990

Jane Winer, Genentech, Inc., 460 Point San Bruno Blvd., South San Francisco, CA 94080-4990

Carl Wittwer, Department of Pathology, University of Utah School of Medicine, 50 North Medical Drive, Salt Lake City, UT 84132

Linda J. Wuestehube, Chiron Corporation, 4560 Horton Street, Emeryville, CA 94608-2916

# Foreword

Geneticists and molecular biologists have been interested in quantifying genes and their products for many years and for various reasons (Bishop, 1974). Early molecular methods were based on molecular hybridization, and were devised shortly after Marmur and Doty (1961) first showed that denaturation of the double helix could be reversed – that the process of molecular reassociation was exquisitely sequence dependent. Gillespie and Spiegelman (1965) developed a way of using the method to titrate the number of copies of a probe within a target sequence in which the target sequence was fixed to a membrane support prior to hybridization with the probe – typically a RNA. Thus, this was a precursor to many of the methods still in use, and indeed under development, today. Early examples of the application of these methods included the measurement of the copy numbers in gene families such as the ribosomal genes and the immunoglobulin family. Amplification of genes in tumors and in response to drug treatment was discovered by this method.

In the same period, methods were invented for estimating gene numbers based on the kinetics of the reassociation process – the so-called Cot analysis. This method, which exploits the dependence of the rate of reassociation on the concentration of the two strands, revealed the presence of repeated sequences in the DNA of higher eukaryotes (Britten and Kohne, 1968). An adaptation to RNA, Rot analysis (Melli and Bishop, 1969), was used to measure the abundance of RNAs in a mixed population. A major result of this work was the discovery that the abundance of different mRNAs in any cell type can vary over several orders of magnitude (Hastie and Bishop, 1976). This observation has implications for methods now in use or under development which are based on the analysis of cDNA clone collections made from whole cell populations; without normalization, such collections are likely to contain a few sequences represented many times and a large number of sequences present in low numbers or absent.

Gel transfer methods, though designed for qualitative analysis, (Southern, 1975) were used as a means to measure and compare gene numbers. Related methods for RNA (Alwine, Kemp and Stark, 1977) were applied to the quantitative analysis of gene expression. Methods for applying multiple targets on a single membrane – the dot blot (Kafatos, Jones and Efstratiadis, 1979) – expanded the usefulness of molecular hybridization and with the advent of cloning, this approach could be used to analyze expression levels of individual genes. However, these gene transfer techniques required millions of copies of a given gene for detection and quantitation.

The advent of amplification methods, such as PCR (Mullis and Faloona, 1987), allowed gene quantification to be performed from very small amounts of starting copy numbers. The focus of this book is to describe how amplification techniques can be harnessed to become quantitative methods. These methods have been very successful for individual gene quantification. A case in point is the use of a variety of quantitative methods, described in this book, to measure HIV-1 viral load in HIV-1 infected individuals.

As for all measurements, accuracy and sensitivity are crucial issues, although some applications are more demanding than others; detecting a gene deletion may require sensitivity that will detect a single copy within a whole genome with an accuracy that can distinguish one from two copies. On the other hand, estimating the level of a gene's expression may require only that a high level of expression can be distinguished from a level close to zero.

Recent interest in genome-wide analysis have prompted the development of methods capable of detecting hundreds of genes or gene products simultaneously. These methods capitalized on the development of robotic spotting devices (Nizetic et al., 1991) that initially applied samples to permeable membranes. More recently, to achieve higher density, probes (cDNA or oligonucleotides) have been spotted on glass supports (Schena et al., 1995; Lockhart et al., 1996). Current effort is directed at rendering these methods quantitative. The potential to address biological problems by genetic analysis on a genome-wide scale may revolutionize the life sciences. The engine that is driving the revolution is the set of massively parallel procedures made possible by merging molecular biology with techniques adopted from the semiconductor industry. Developments that will lead to multigene quantitation are principally due to reduction in scale achieved through microfabrication, and to the automation of production methods, of sample preparation, and handling of data production and analysis.

In all the excitement generated by this explosive growth in gene quantification methodology, it is important to reflect on questions defined for us by those with a profound understanding of biological systems. It is important, too, to recognize that many of these new methods have a basis in old ones, and that some of the basic problems we are trying to solve may already have been dealt with. The reader should be aware that some classic papers on gene quantification have publication dates earlier than the abstracting services.

—Edwin M. Southern

Alwine JC; Kemp DJ; Stark GR (1977) Method for detection of specific RNAs in agarose gels by transfer to diazobenzyloxymethyl paper and hybridization with DNA probes. *Proc Natl Acad Sci USA* 1977 Dec; 74(12):5350–4

Bishop JO (1974) The gene numbers game. *Cell* 1974 Jun; 2(2):81–6

Britten RJ; Kohne DE (1968) Repeated sequences in DNA. Hundreds of thousands of copies of DNA sequences have been incorporated into the genomes of higher organisms. *Science* 1968 Aug 9; 161(841):529–40

Gillespie D; Spiegelman S (1965) A quantitative assay for DNA-RNA hybrids with DNA immobilized on a membrane. *J Mol Biol* 1965 Jul; 12(3):829–42

Hastie ND; Bishop JO (1976) The expression of three abundance classes of messenger RNA in mouse tissues. *Cell* 1976 Dec; 9(4 PT 2):761–74

Kafatos FC; Jones CW; Efstratiadis A (1979) Determination of nucleic acid sequence homologies and relative concentrations by a dot hybridization procedure. *Nucleic Acids Res* 1979 Nov 24; 7(6):1541–52

Lockhart DJ; Dong H; Byrne MC; Folletie MT; Gallo MV; Chee MS; Mittman M; Wang C; Kobayashi M; Horton H; Brown E (1996) Expression monitoring by hybridisation to high-density oligonucleotide arrays. *Nature Biotechnol* 1996 Dec 14; 14:1675–83.

Marmur J; Doty P (1961) Thermal renaturation of deoxyribonucleic acids. *J Mol Biol* 1961; 3:585–94

Melli M; Bishop JO (1969) Hybridization between rat liver DNA and complementary RNA. *J Mol Biol* 1969 Feb 28; 40(1):117–36

Mullis KB; Faloona FA (1987) Specific synthesis of DNA in vitro via a polymerase-catalyzed chain reaction. *Meth Enzymol* 1987; 1455:335–50

Nizetic D; Zehetner G; Monaco AP; Gellen L; Young BD; Lehrach H (1991) Construction, arraying, and high-density screening of large insect libraries of human chromosomes X and 21: their potential use as reference libraries. *Proc Natl Acad Sci USA* 1991 Apr 15; 88(8):3233–7

Schena M; Shalon D; Davis RW; Brown PO (1995) Quantitative monitoring of gene expression patterns with a complementary DNA microarray. *Science* 1995 Oct 20; 270(5235):467–70

Southern EM (1975) Detection of specific sequences among DNA fragments separated by gel electrophoresis. *J Mol Biol* 1975 Nov 5; 98(3):503–17

# Key Issues, Challenges, and Future Opportunities in Gene Quantification

François Ferré

Perhaps it should be called an "end of the millennium" phenomenon, but it seems as if the world is entangled in a quantitative frenzy. From the unresolved calculation of the rate of growth of the human population, the never-ending speculation on the amount of food that will be necessary to feed humanity, and the quantification of ozone loss in the Southern hemisphere, to the estimation of the number of square miles necessary to maintain a viable rain forest ecosystem or the enumeration of the panda bears still enjoying the bamboo forests of China, the world is feverishly quantifying. . . . At the nanoscale level, scientists are busy too. Their interest in molecular quantitation is best exemplified by the recent and stupendous crafting of a nanoscale functional abacus, the beads of which are composed of spherical carbon molecules, or buckyballs, that can be moved along a steplike edge on a copper surface (Service, 1996).

Yet another example is the focus of this book: gene quantification. . . . Why a book on gene quantification at this time? First, it is important once in a while to take a snapshot of a rapidly evolving research field in order to reflect upon it. What is the current utility of gene quantification? What are the key requirements or ideal features of quantitative methods? Which quantitative method should be applied when? The aim of this book is to address these questions by presenting an array of different quantitative methods that have broad applicability, yet share a common thread, enabling an in-depth coverage of the issues. These methods were developed to tackle the major challenge in gene quantification, which is to determine quantitatively the level of nucleic acid, based on minute amounts present in a sample. Thus, the common theme to all gene quantification methods presented in this book is the fact that quantification requires amplification. Indeed, these quantitative methods are based either on controlled gene amplification or on signal amplification. Numerous examples of both approaches will be described.

Second, the timing of this book is right because important methodological breakthroughs, such as “real-time” quantitation using kinetic PCR or ultrasensitive branched DNA assays, have recently been described. These technological developments not only allow for faster and more robust quantification, but also permit new paradigms to emerge. For example, the concept of accurate quantitation using PCR without a standard was unthinkable a few years ago. However, as described by Peccoud and Jacob in this book, through the application of mathematical models to the kinetics of PCR amplification, it could soon become a reality.

Since PCR-based techniques still dominate the landscape of gene quantification methods, they will be amply represented in this manual. Indeed, the versatility of PCR and its uncanny ability to amplify and quantify nucleic acids from minute amounts of material make it a key player in the gene quantification field. Two different quantitative formats are extensively discussed in this manual. Most quantitative methods reported throughout the literature are based on the competitive PCR format. In this format, an internal control, close in composition to the target nucleic acid, competes with the latter for reagents (such as common primers) in the same reaction tube. Competitive PCR represents the prototype for the so-called end-point quantitative methods, in which quantification is based on the amount of amplified material (amplicon) obtained at the last amplification cycle. This book includes numerous chapters on the competitive PCR approach (see chapters by Kwok and Sninsky, Wang and Spadoro, Hayward-Lester et al., Kolesar and Kuhn, Lazar et al., Sidhu et al., Radich, and Feinstein and Galea). Major improvements in the sensitivity of detection methods permitted the emergence of kinetic PCR as a strong contender to the competitive PCR format. In this approach, the quantitative information is derived from the kinetics of amplification and thus requires the quantification of the amplicon at a number of cycles (see chapters by McBride et al., Peccoud and Jacob, Wittwer et al., Lantz et al., and Williams et al.). A quick discussion on the advantages and limitations of both methods is presented below.

Two quantitative methods based on isothermal amplification of RNA, NASBA, and Transcription-Mediated Amplification (TMA) have been extensively used in clinical settings and represent sensitive and practical alternatives to PCR (see chapters by Romano et al. and Gonzales and McDonough).

This manual also provides a number of chapters on signal amplification methods and their applications. The branched-DNA (bDNA) technology, primarily developed to quantitate viral entities, is now poised to investigate gene expression, since it can quantitate down to 50 copies of mRNA (see chapters by Collins et al. and Kolberg et al.). The bDNA assay has been instrumental in establishing that nucleic acid quantitation can be rendered accurate when proper algorithms for standard validation are used. The Hybrid Capture method represents another practical method that uses signal amplification to quantify the genomes of a number of viral pathogens, such as HPV, CMV, and HBV (see chapter by Lörincz et al.).

Numerous chapters in this book address the issue of quantitative power for their respective methods. In this introduction, my goal is to set the stage by presenting a number of general issues pertaining to gene quantification.

## The Principles of Gene Quantification

In this section, a few issues and definitions relevant to the description of quantitative assays will be presented. One such issue involves semiquantitative assays. The use of semiquantitation is still very popular today in research settings, but I would argue that there are good reasons why the terminology "semiquantitative assay" should be banned from the gene quantitation lexicon. By definition, semiquantitative means "constituting or involving

less than quantitative precision" (Webster's dictionary). This definition implies that, to extract any quantitative information from a semiquantitative assay, extensive statistical analysis has to be performed. However, paradoxically, semiquantitative assays almost invariably refer to analyses that only contain a few replicates and thus preclude any statistical interpretation. Thus, without proper statistics, apparent reported differences between samples obtained using semiquantitative assays have to be questioned. Furthermore, the semiquantitation terminology has already been banned from clinical settings because semiquantitative assays cannot be validated. Finally, in all situations, "semiquantitation" can be replaced by "relative quantitation" for which, as discussed below, levels of precision and reproducibility in the assay can be defined.

Another important issue in the gene quantitation lexicon is the concept of accuracy, or "the degree of conformity of a measure to a standard or a true value" (Webster's dictionary). Unfortunately, the duality of the definition (to a standard or true value) has brought confusion to the field of gene quantitation. Indeed, from the definition, it is still possible to claim accuracy even if the standard used for quantitation is not exact. Thus, to express accuracy in the sense of true value, scientists (including the author) have been using another description, absolute quantification. Although Absolute applies very nicely to vodka, it does not appear to be a good fit for gene quantification, and thus the use of this terminology should be discouraged. Consider, for example, quantitation at the level of one or a few copy(ies) per sample. To account for the stochastic partition of low copy numbers (Poisson's law), statistical analysis of a significant series of replicates from the same sample needs to be performed. Thus, at this low level, a probability rather than an absolute number defines the true copy number value for a given sample.

To avoid confusion in the field, "accuracy" should only be used in the sense of true value. In other words, accuracy should mean exactness. When quantitating using a standard for which accuracy is not demonstrated, "relative quantitation" terminology should be used in conjunction with precision. Numerous methods can be developed to demonstrate the accuracy of a standard. Interestingly, the chapter written by Wang and Spadoro in this book is entirely dedicated to the validation of PCR standards using statistical analysis. In addition, Collins and co-workers described in their chapter the independent quantitative methods that have been applied for the validation of bDNA standards.

A quantitative assay is also largely defined by its dynamic range, in which the quantitated amount of amplified target or signal is proportional to the initial amount of target molecules. In fact, dynamic range, also known as linear range or range of proportionality, represents the assay's window in which quantitation can be performed. Until recently, most assays, such as quantitative PCR methods, had a rather narrow linear range of 2–3 logs (for

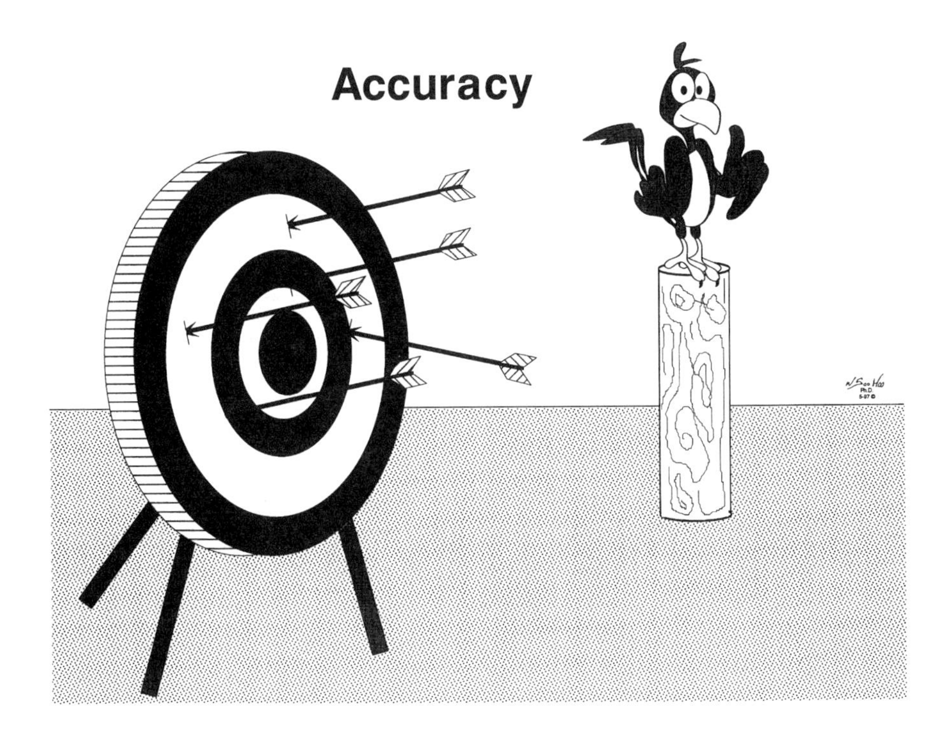


review, see Ferré, 1992; Ferré et al., 1995). Highly sensitive detection methods, such as chemiluminescence, allow the linear range of these end-point quantitative methods to be increased to 4 logs (Ferré et al., 1996). Finally, for recent quantitative assays such as kinetic PCR and bDNA, the quantitative range can be extended to 6 logs, which is powerful enough to quantitate in one reaction any relevant target concentration (for example, from 100 to $10^8$ molecules). Indeed, for the vast majority of applications 4–5 logs of proportionality is sufficient. The lower end of the dynamic range also defines the limit of quantification. In most scenarios, this end of the range is different from the molecular detection limit, which is defined as the lowest level of target that can be confidently distinguished from zero.

To assess the quantitative capability of an assay, precision and reproducibility should be measured within the linear range. Precision represents the degree of agreement among individual test results obtained by repeatedly applying the method to multiple sampling of a homogeneous sample. To establish the precision of an assay, the intra-assay variability needs to be thoroughly investigated, meaning that different operators should perform the assay at different times and, if possible, at different sites. The tighter the

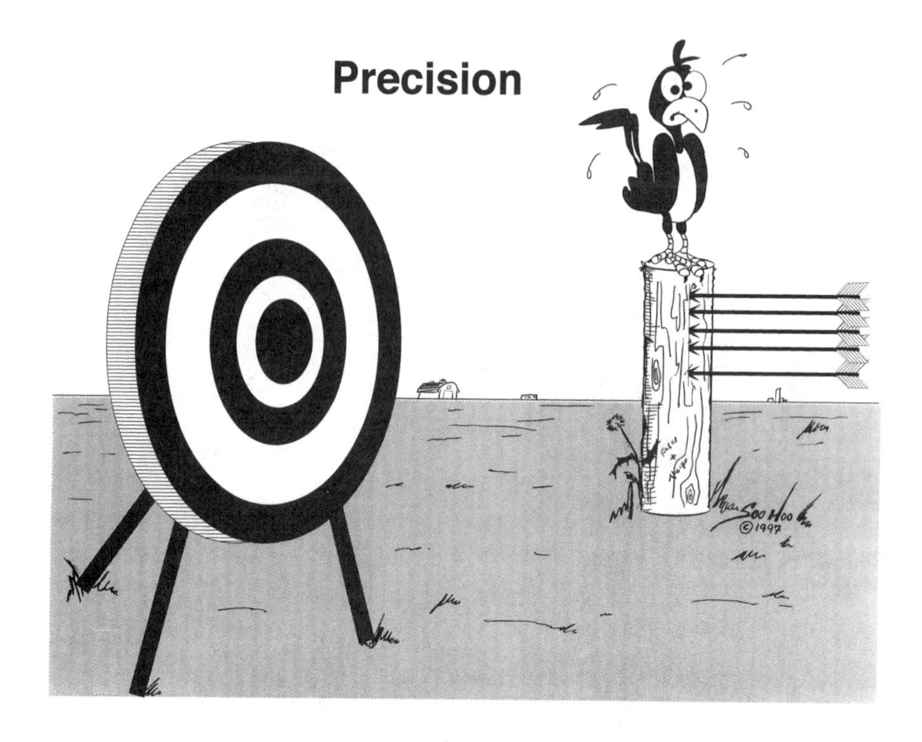


**Reproducibility**

assay's precision, the greater the capacity of the assay to differentiate between two samples. Reproducibility represents the interassay variability, and its assessment is crucial, especially when it is necessary to compare samples run in different assays over time. Achieving high reproducibility requires a tight control on assay reagents and standards, and the failure to exercise such control may lead to a systematic shift in quantitation with each new lot of materials.

## Do You Really Need Accuracy, or Do You Mean Precision?

The applications of gene quantification can be essentially divided into two major categories: biological research and clinical research. The biological research category mainly refers to the analysis of gene expression in its broadest definition, whereas clinical research implies that the focus is on the quantification of species such as viruses, bacterial pathogens, or cancer cells. Although this classification is far from being perfect, it is relevant to the field of quantitation for the following reason: for most applications in biological research, performing relative quantitation is sufficient, whereas accurate quantification is essential in clinical research.

For one thing, achieving accuracy is extremely difficult. This important point can be illustrated with an example from the HIV literature. The need for accurate quantitation of HIV-1 virus for a better understanding of its pathogenesis and for the monitoring of anti-HIV therapies is very clear. Numerous commercial quantitative tests such as Roche's PCR, Chiron's bDNA, and Organon's NASBA have been developed and thoroughly validated as performing accurate HIV viral load quantification. Although these three assays demonstrate equal reliability, differences in the calculated HIV-1 copy numbers have been observed (Schuurman et al., 1996). For example, studies using reconstituted HIV-1 sample have shown that, in general, the bDNA and the NASBA results differed by 0.45 log (about threefold). These results clearly demonstrate the difficulty in achieving accuracy.

In fact, two major prerequisites should be fulfilled in order to claim accuracy: a dedicated effort to guarantee standard accuracy *and* a confirmation of the results by means of other accurate methods. This is an important issue. False claims of accuracy can have very serious consequences. Again, the HIV field can serve as a case in point. In the early days of HIV research, the consensus was that at any stage of the disease, viral load was extremely low. The tools used at the time, such as Southern technology and PCR, were mostly capable of relative quantitation and were certainly not validated for accuracy. Yet, copy numbers were published and taken at face value by the scientific community. The HIV research community jumped to quantitative conclusions that underestimated the viral load by 3–4 orders of magnitude.

This underestimation had a huge impact on HIV research. Suddenly, HIV could no longer be the cause of AIDS—mycoplasma was. Although somewhat exaggerated, this example still illustrates how much damage false claims of accuracy can inflict on an entire field of research.

Another important lesson learned from the previous examples is that accuracy should be sought after only when needed. As previously mentioned, most biological research applications do not require accurate quantification. By far, the most common application of quantification in biological research is the comparison of gene expression levels. For example, quantitative methods have been used to study differential expression of splicing variants (Fasco et al., 1997), of low-abundance mRNA species (St Amand et al., 1996), and of genes from superfamilies such as the T cell receptor family (Uhrberg and Wernet, 1996; Boyer et al., 1993). Relative quantification is perfectly suited for these types of studies. In fact, in most cases, knowing the exact number of mRNA copies would not provide further useful information. It seems that accurate quantitation is important for quantification of species such as viruses (HIV, HCV, etc.) and cancer cells in residual disease, but such quantitation is of limited value for the measurement of biological activity. This is due in part to the fact that there is no established stoichiometry between RNA expression and protein synthesis. Indeed, mRNA levels mostly represent a crude measurement of gene expression. Ultimately, it is the protein expression that needs to be quantitated in order to fully understand the complexity of gene interactions, but the tools to achieve this aim, at the level of sensitivity and specificity seen in the nucleic acid world, are still missing in the protein world. The last chapter of this book, written by Manthorpe and co-workers, describes a method to quantitate, at the protein level, the impact that the modification of regulatory elements, such as promoters/enhancers, has on gene expression.

Unfortunately, the fact that relative quantification is perceived as poor quantitative information has led to the false impression that it is essential to publish copy numbers to increase the credibility of the results. Furthermore, since the fact that the reported copy numbers are only relative to a standard (nonvalidated for accuracy) is often omitted, the reader is led to the conclusion that such numbers are accurate. Finally, an increasing danger in reporting inaccurate copy numbers is that, as previously discussed, the tools now exist to carry out accurate quantitation. Thus, from the literature, it could become very difficult to interpret the true nature, i.e., relative or accurate, of a reported copy number for a given gene.

Is there a simple solution to this problem? For one thing, if copy numbers are reported with a standard that has not been validated for accuracy, it should be made quite clear that these values are relative to the standard. A potentially better solution is to report the relative quantitation results as arbitrary units (AU). Indeed, because high-throughput methods capable of quantitating thousands of target genes simultaneously will become available in the near future, huge quantitative databases will be created in which quantitative information will have to be stored as relative values or AUs. Finally, it is important to recognize that, indeed, methods capable of relative quantitation can provide extremely valuable quantitative information. The quantitative power of such methods will be directly proportional to the extent of their dynamic ranges and to the tightness of their precision.

## From Competitive PCR to Kinetic PCR

Hundreds of recently published protocols for RNA/DNA quantification are based on competitive PCR (for recent review, see Ferré et al., 1996; Zimmermann and Mannhalter, 1996). In the competitive approach, a synthetic RNA/DNA control, also called an internal standard is co-amplified with the target nucleic acid in the same tube. Ideally, both target and internal standard will co-amplify with equivalent efficiency and will be distinguishable after amplification. To this end, most competitive methods use internal standards that are very close in composition to the target and share the same primer sequences. Quantification is then performed by comparing the PCR signal of the target amplicon with the PCR signal obtained with known concentrations of the internal standard. Calibration curves, relating the logarithm of the ratio of both PCR products to the logarithm of the initial amount of competitor, are then constructed (log–log calibration curves). If both target and competitor are amplified with equivalent efficiency, the amount of initial target nucleic acid can be obtained from the point on the curve where target and standard values are equal.

The popularity of competitive PCR is linked to the fact that it does not require sophisticated instrumentation. In addition, numerous protocols for better and faster methods of constructing an internal standard have been published (for review, see Zimmermann and Mannhalter, 1996), thus increasing the flexibility of the approach. In fact, there is such a plethora of methods for internal standard synthesis that, for a novice in the field, the difficulty resides in choosing the proper method! In competitive PCR, quantitation can be achieved without ultrasensitive detection methods. In numerous protocols, PCR is extended beyond the exponential phase, implying that a large number of amplicons are produced. Amplified products can then be quantitated after separation by HPLC (see chapter by Hayward-Lester et al.) or by simple gel electrophoresis using ethidium bromide (EtBr). Alternately, SYBR Green 1 dye, which is 1–2 orders of magnitude more sensitive than EtBr, can be used (Schneeberger et al., 1995). Finally, another advantage to the competitive approach is linked to the fact that the addition of an internal standard increases precision (for review, see Ferré et al., 1994, 1996; Zimmermann and Mannhalter, 1996). In fact, optimal precision may be obtained when the internal standard is added to the sample prior to extraction (Moller and Jansson, 1997).

In a recent review (Ferré et al., 1996), we discussed at length some of the limitations of the competitive PCR approach. In this chapter, I will only make a few additional comments. To achieve accuracy, it is imperative that both amplicons are amplified with equivalent efficiency. This is why most competitors differ only slightly from the target. Interestingly, Steve Sommer's group recently published the information that a single point mutation downstream of a primer can inhibit PCR amplification (Liu et al., 1997). This implies that an internal standard that differs from the target by only a single base change outside the primer sequence could still have an amplification efficiency quite different from that of a very similar target. Concern over limitations caused by different amplification efficiencies may have led Reyes-Engel and co-workers to develop the ultimate competition assay, in which the target competes with itself (Reyes-Engel et al., 1996). The procedure, referred to as additive PCR, might have merit, but it surely seems somewhat extreme. To summarize, great attention should be given to demonstrating the equivalency of amplification efficiency between target and *any* internal standard.

Another limitation of the competitive format is the need for co-amplification of four or five dilutions of the internal standard in order to encompass the expected range of target concentration. To overcome this limitation, a number of different competitors (three or four) can be added at different concentrations to the same reaction tube (Zimmermann et al., 1996; Vener et al., 1996). These competitors contain the primer sequence of the target, but can be differentiated from the target and from each other by length and/or internal sequences. With this multiplexing competitive method, one tube is sufficient to generate a regression curve for quantification. Interestingly, the multiplexing of internal standards has been a major component of the NASBA technology for the accurate quantitation of HIV-1 (see chapter by Romano et al.; van Gemen et al., 1994) and has only recently been incorporated in the competitive PCR format.

It has long been suspected that PCR amplification behaves somewhat erratically at the end of the exponential phase, as it enters the plateau phase, hence the development of methods that use an internal standard to control for this behavior. But this erratic behavior of PCR at plateau also posed the following questions: What is happening in the earlier cycles or during the exponential phase? Does the amplification efficiency fluctuate significantly as well? To answer these questions, new PCR instrumentations were needed.

Higuchi and co-workers developed a system in which the PCR products can be detected in "real time," meaning that the accumulation of PCR products could be visualized at each cycle using an intercalating dye, such as ethidium bromide, an ultraviolet source, and a computer-controlled cooled CCD camera (Higuchi et al., 1992, 1993). This PCR format is also referred to as kinetic PCR. To increase the specificity of the "real-time" detection, specific and fluorescent probes can be added to PCR. Two types of probes are currently used in the "real-time" PCR format, the Taqman probe system (see chapters by McBride et al. and Wittwer et al.), or hydrolysis probe, for which the fluorescent signal is dependent on probe hydrolysis, and the hybridization probe, for which the emission of fluorescence is dependent on hybridization (see chapter by Wittwer et al.). Both systems are extremely sensitive and allow detection below 100 nucleic acid targets. Today, sophisticated instrumentations have been developed to capitalize on these new PCR approaches (see chapters by McBride et al. and Wittwer et al.). A number of remarkable features characterize these "real-time" PCR systems: (i) Because quantification is performed during amplification, the linear range of the assay (5–6 logs) is dramatically increased, as compared to end-point quantitation (2–3 logs); (ii) The fact that there is no need to open the PCR tube after amplification greatly diminishes carryover problems and increases the speed of the quantification process; (iii) It seems that indeed in the earlier cycles, during the exponential phase, PCR is less erratic than during the plateau phase. Since quantitation can be performed during these early cycles, the level of precision of kinetic PCR can be substantially increased, as compared to end-point quantitation; (iv) Finally, the real-time format also allows for within-cycle monitoring. This new feature can be conveniently exploited to distinguish and quantitate two amplicons on the basis of their differences in melting kinetics within a given cycle (see chapter by Wittwer et al.).

Although very powerful, kinetic PCR methods have limitations as well. Sophisticated instrumentation, such as the 7700 system from Perkin-Elmer or the LightCycler from Idaho Technology, as well as the fluorescent probes are expensive. In addition, the selection of a hydrolysis probe is complex and should take into account the fact that mismatches at the probe binding site could compromise the enzymatic cleavage and therefore the quantitation. Finally, for multiplexing quantitation, algorithms able to resolve the spectra of multiple fluorescent dyes need to be developed. Kinetic PCR can be performed without such sophisticated systems, as reported in this book by Lantz and co-workers. In their approach, kinetic PCR is coupled with ELISA for quantitation. Although cheaper, this ELISA approach is cumbersome and is unable to provide a number of the interesting features of kinetic PCR that are listed above, such as the ability to monitor the reaction within a given cycle or the elimination of post-PCR manipulations.

## Toolbox of the Future

This is truly an exciting time for the field of gene quantification. The broad range of tools available today allows for comparative quantitation that should lead to very robust quantitative analyses. In fact, validated and accurate methods can be used to validate other approaches to gene quantitation, such as *in situ* hybridization and *in situ* PCR, which are notoriously more difficult to handle and for which accurate standards may not be readily available. For example, Haase and co-workers have assessed the HIV-1 RNA load in the tonsils of asymptomatic HIV-1 infected individuals using the bDNA assay and *in situ* hybridization and have found that there was a great concordance between the data sets (Haase et al., 1996). Since the bDNA assay has a high level of accuracy, this comparison helps validate the computerized quantitative image analysis performed on the tonsils sections by *in situ* hybridization.

The combining of accurate methods will also be essential to decipher complex pathogenesis such as HIV-1 disease. Again, using bDNA to quantitate HIV-1 RNA in plasma and using quantitative image analysis of lymph node sections, Ashley Haase's group has concluded that the HIV-1 plasma load accurately mirrors the production of virus at its site of generation in the lymph nodes (Haase et al., 1996). As for accuracy, the credibility of relative quantification can be greatly boosted by comparing the results obtained by a number of validated assays. For example, the relative abundance of a number of cytokine mRNAs can today be estimated by QC-PCR, kinetic PCR, and bDNA (see chapter by Kolberg et al.), not to mention the use of commercially available ribonuclease protection assays, when the amount of starting material is sufficient ($\mu$g range).

In addition to robustness, the future promises faster gene quantification. As described by Wittwer and colleagues in their chapter, quantitative analysis of a given sample can already be completed in less than 15 minutes with rapid cycle PCR, thus providing tremendous flexibility to gene analyses in research settings. In addition, because of its speed, this type of system could potentially find its way to the doctor's office, where the analysis of a given gene could be performed on a patient sample before the visit ends.

The field of gene quantification is increasingly moving toward the simultaneous analysis of numerous genes. In addition to its exquisite ability to amplify specific nucleic acid sequences, PCR has been harnessed for whole-genome amplification. In fact, whole-genomic DNA from a small number of cells can be amplified more than a hundredfold using methods such as degenerate oligonucleotide primed-PCR, or DOP-PCR. To be useful in most applications, including gene quantification, such whole-genome amplification needs to be unbiased, meaning that all sequences in the genome are equally represented after cycling. Interestingly, Cheung and Nelson recently reported that DOP-PCR appears to allow such unbiased amplification from as little as 100 genomes (Cheung and Nelson, 1996). This type of analysis has also been conducted for gene expression. Bonnet and co-workers have demonstrated that proportionality of input molecules to output signals can be maintained when amplifying a total mRNA population (Hamoui et al., 1994). Although it is not clear at this point if relative abundance is maintained for all mRNA species after DOP-PCR, it seems reasonable to predict, on the basis of the DNA work, that refining the DOP-PCR technique will allow complete coverage of total mRNA populations to be obtained from a handful of cells.

The power of this whole-genome approach to quantitation becomes truly remarkable when combined with expression-monitoring devices, such as high-density arrays containing tens of thousands of synthetic oligonucleotides (Lockhart et al., 1996). Indeed, it is technically feasible today to synthesize and read arrays containing as many as 400,000 probes in an area of 1.6 $cm^2$. Such a device could potentially quantitate, through hybridization, tens of thousands of RNAs or cDNAs simultaneously. There is no doubt that the combination of whole-genome quantitative PCR and high-density arrays will represent the ultimate tool for gene quantification. In parallel with this massive effort in developing instrumentation for gene analysis, powerful software is currently being developed that will help the scientist explore the incredibly vast amount of quantitative information that will soon be available. In fact, interactive, multidimensional graphics software is being created that will allow researchers to compare the expression of specific genes between tissue types by quantitating their relative abundance (Hood, 1997). Maybe it is time after all to bid farewell to the abacus. . . .

## *Acknowledgments*

I would like to thank Dr. Will Soo Hoo for his great cartoons.

## REFERENCES

Boyer V, Smith LR, Ferre F, Pezzoli P, Trauger RJ, Jensen FC, and Carlo DJ (1993): T cell receptor V$\beta$ repertoire in HIV-1 infected individuals: lack of evidence for selective V$\beta$ deletion. *Clin Exp Immunol* 92:437–41.

Cheung VG and Nelson SF (1996): Whole genome amplification using a degenerate oligonucleotide primer allows hundreds of genotypes to be performed on less than one nanogram of genomic DNA. *Proc Natl Acad Sci USA* 93:14676–9.

Fasco MJ (1997): Quantitation of estrogen receptor mRNA and its alternatively spliced mRNAs in breast tumor cells and tissues. *Anal Biochem* 245:167–78.

Ferré F (1992): Quantitative or semi-quantitative PCR: reality versus myth. *PCR Meth Appl* 2:1–9.

Ferré F, Marchese A, Pezzoli P, Griffin S, Buxton E, and Boyer V (1994): Quantitative PCR: an overview. In: *The Polymerase Chain Reaction*, Mullis KB, Ferre F, Gibbs R (eds). Boston: Birkhauser, pp 67–88.

Ferré F, Pezzoli P, Buxton E, Duffy C, Marchese A, and Daigle A (1995): Quantification of RNA targets using the polymerase chain reaction. In: *Molecular Methods for Virus Detection*, Wiedbrauk D, Farkas DH (eds). San Diego: Academic Press, pp 193–218.

Ferré F, Pezzoli P, and Buxton E (1996): Quantitation of RNA transcripts using RT-PCR. In: *A Laboratory Guide to RNA*, Krieg PA (ed). New York: John Wiley & Sons, pp 175–90.

Haase AT, Henry K, Zupancic M, Sedgewick G, Faust RA, Melroe H, Cavert W, Gebhard K, Staskus K, Zhang ZQ, Dailey PJ, Balfour HH, Erice A, and Perelson AS (1996): Quantitative image analysis of HIV-1 infection in lymphoid tissue. *Science* 274:985–9.

Hamoui S, Benedetto JP, Garret M, and Bonnet J (1994): Quantitation of mRNA species by RT-PCR on total mRNA population. *PCR Methods Appl* 4:160–6.

Higuchi R, Dollinger G, Walsh PS, and Griffith R (1992): Simultaneous amplification and detection of specific DNA sequences: *Bio/Technology* 10:413–17.

Higuchi R, Fockler C, Dollinger G, and Watson R (1993): Kinetic PCR analysis: real time monitoring of DNA amplification reactions. *Bio/Technology* 11:1026–30.

Hood L (1997): Genetic approaches to cancer. Keynote address at the Keystone Symposium on Cellular Immunology and the Immunotherapy of Cancer III, Copper Mountain, Colorado, February 1–7, 1997.

Liu Q, Thorland EC, and Sommer SS (1997): Inhibition of PCR amplification by a point mutation downstream of a primer. *Bio/Techniques* 22:292–300.

Lockhart DJ, Dong H, Byrne MC, Follettie MT, Gallo MV, Chee MS, Mittmann M, Wang C, Kobayashi M, Horton H, and Brown EL (1996): Expression monitoring by hybridization to high-density oligonucleotide arrays. *Nature Biotech* 14:1675–80.

Moller A and Jansson JK (1997): Quantification of genetically tagged cyanobacteria in Baltic Sea sediment by competitive PCR. *Bio/Techniques* 22:512–18.

Reyes-Engel A, Garcia-Villanova J, Dieguez-Lucena JL, Fernandez-Arcas N, and Ruiz-Galdon M (1996): New approach to mRNA quantification: additive RT-PCR. *Bio/Techniques* 21:202–4.

Schneeberger C, Speiser P, Kury F, and Zeillinger (1995): Quantitative detection of reverse-transcriptase PCR products by means of a novel and sensitive DNA stain. *PCR Meth Appl* 4:234–8.

Schuurman R, Descamps D, Weverling GJ, Kaye S, Tijnagel J, Williams I, van Leeuwen R, Tedder R, Boucher CAB, Brun-Vezinet F, and Loveday C (1996): Multicenter comparison of three commercial methods for quantification of human immunodeficiency virus type 1 RNA in plasma. *J Clin Microbiol* 34:3016–22.

Service RF (1996): Tiny abacus points to new devices. *Science* 274:1079–80.

St Amand D, Pottage C, Henry P, and Fahnestock M (1996): Method for quantitation of low-abundance nerve growth factor mRNA expression in human nervous tissue using competitive reverse transcription polymerase chain reaction. *DNA Cell Biol* 15:415–22.

Uhrberg M and Wernet P (1996): Genetic influence on the shaping of the human T-cell receptor repertoire: quantitative assessment by competitive polymerase chain reaction. *Scand J Immunol* 44:173–8.

van Gemen B, van Beuningen R, Nabbe A, van Strijp D, Jurriaans S, Lens P, and Kievits T (1996): A one-tube quantitative HIV-1 RNA NASBA nucleic acid amplification assay using electrochemiluminescent (ECL) labelled probes. *J Virol Methods* 49:157–68.

Vener T, Axelsson M, Albert J, Uhlen M, and Lundeberg J (1996): Quantification of HIV-1 using multiple competitors in a single-tube assay. *Bio/Techniques* 21:248–55.

Zimmermann K and Mannhalter JW (1996): Technical aspects of quantitative competitive PCR. *Bio/Techniques* 21:268–79.

Zimmermann K Schogl D, Plaimauer B, and Mannhalter JW (1996): Quantitative multiple competitive PCR of HIV-1 DNA in a single reaction tube. *Bio/Techniques* 21:480–4.

# PART I: METHODS/TECHNOLOGY ISSUES

## A. GENE QUANTITATION BASED ON PCR AMPLIFICATION

# Present and Future Detection Formats for PCR Quantitation of Nucleic Acids

Shirley Kwok and John Sninsky

We developed a method for quantification of viral RNA in plasma that is based on reverse transcription (RT) of the viral RNA to cDNA and amplification of the DNA with the polymerase chain reaction (PCR) (Mulder et al., 1994). Salient features of this assay include: a) use of a single buffer and single enzyme to catalyze both RT and PCR (Myers and Gelfand, 1991; Young et al., 1993), b) incorporation of a quantitation standard to monitor sample recovery, sample inhibition, and reaction variabilities, c) incorporation of dUTP/uracil-N-glycosylase (UNG) for carryover prevention (Kwok et al., 1992; Longo et al., 1990), d) use of a colorimetric microwell ELISA-type format for product detection, and e) a detection range of 400–750,000 copies/ml with the equivalent of a 25-$\mu$l plasma input.

Different approaches to viral quantitation have been described. Among these, RT-PCR and bDNA (Urdea et al., 1993) have been most widely used, followed by NASBA (van Gemen et al., 1992). The following differences between the assays are noteworthy. First, RT-PCR provides the greatest sensitivity. The 400-copy/ml sensitivity achieved with the current AMPLICOR HIV-1 MONITOR™ test (Roche Diagnostics Systems, Somerville, NJ) represents the minimum sensitivity. If 1 ml of plasma is used, as in the bDNA assay, a sensitivity of ~10–20 RNA copies/ml can be achieved (Mulder et al., 1997). Second, whereas several different internal calibrators are required for the NASBA assay, only a single level of an internal quantitation standard is required with the AMPLICOR HIV-1 MONITOR™ test; bDNA does not use an internal control. Third, carryover prevention measures are provided only by the AMPLICOR HIV-1 MONITOR™ test and are not available for either bDNA or NASBA.

The quantitation standard (QS) is a key component of the AMPLICOR HIV-1 MONITOR™ test. The QS is a synthetic RNA transcript that contains the identical primer binding sites as the target RNA but has a different probe sequence to allow differentiation from the true target.

The QS is added into the lysed plasma sample at a known copy number and is used to monitor the recovery of RNA from the specimen and to normalize any effects of inhibition or reaction variabilities.

Primers for amplification are selected based on sequences that are highly conserved among known isolates. For targets that exhibit extensive heterogeneity, such as HIV-1, additional strategies were used in the design of primers and probes. For example, the incorporation of a 3′ terminal T has been shown to result in better toleration of mismatches at that position than any other base (Kwok et al., 1990). Although internal mismatches are less detrimental to PCR, the number, the position, and the sequence context will affect how well mismatches are tolerated (Christopherson et al., 1997). In hypervariable base positions, modified bases such as 5 methylcytosine (Hoheisel et al., 1990) and inosine (Martin and Castro, 1985) have been successfully used to improve mismatch tolerance. In RNA targets, mutations in the RT primer binding region will have a greater effect as it affects both RT and PCR. Once the cDNA is synthesized and the first round of PCR is completed, the amplicons will be completely homologous to the primer. Thus, the mismatches have an effect only early in the amplification.

## Materials and Methods

### *Sample collection*

Blood should be collected in tubes containing EDTA or ACD as the anticoagulant. Serum samples can be used, but the recovery of virus from serum samples is slightly reduced. The recovery of viral RNA from samples collected in EDTA is slightly better than it is from those collected in ACD (unpublished data). Heparinized blood cannot be used with the sample preparation procedure that uses guanidium alcohol precipitation, because heparin inhibits DNA polymerase activity. Plasma should be separated from whole blood within 6 hours of collection by centrifugation at 800–1600 ×g for 20 minutes. The plasma can be stored at room temperature for up to 1 day, at 2°C–8°C for up to 5 days, or can be frozen at –20°C. For long-term storage, –70°C is recommended.

### *Sample preparation*

Two hundred microliters of plasma are lysed with 600 $\mu$l of a guanidium—based buffer containing the QS RNA. The samples are precipitated with 800 $\mu$l of isopropanol at room temperature, washed with 70% ethanol, and resuspended in 400 μl of specimen diluent. The extracted samples are stored at 4°C until amplification.

*Reverse transcription and PCR*

In contrast to conventional RT and PCR, which uses two different enzymes and two different buffer systems, the RT-PCR system described here uses a single buffer and a single enzyme, and is carried out in a single tube. The enzyme used in this reaction, r*Tth* (*Thermus thermophilus*) DNA polymerase, exhibits both RT and DNA polymerase activity in the presence of manganese (Myers and Gelfand, 1991; Young et al., 1993).

For HIV, 50 μl of the extracted sample is added to 50 μl of a 2× bicine—buffered master mix that contains KOAc, biotinylated primers SK462 and SK431, dNTPs (dU, dT, dA, dG, dCTP), KCl, UNG, and r*Tth* DNA polymerase (Roche Diagnostic Systems, Somerville NJ). Prior to thermalcycling, the reactions are heated to 50°C for 2 min to allow UNG to cleave any dUP—containing products that may have been inadvertently carried over from a previous reaction, and to cleave any extensions that may have occurred during PCR set-up. The reactions are subsequently incubated at 60°C for 30 min for reverse transcription.

Following reverse transcription, PCR amplification is initiated with four cycles of 95°C for 10 sec, 55°C for 10 sec, and 72°C for 10 sec, followed by 26 cycles of 90°C for 10 sec, 60°C for 10 sec, and 72°C for 10 sec. At the end of thermalcycling, the samples are held at 72°C for 15 min and then denatured with 100 μl of a denaturation solution.

*Colorimetric detection*

Amplified products are detected in microwell plates coated with BSA-conjugated probes: one specific for the target and one specific for the QS. Each plate consists of 96 wells in a 12 × 8 format. Each vertical strip contains six wells coated with the HIV probe and two wells coated with the QS probe. Dilutions of the amplified products are necessary in order to expand the dynamic range of the plate assay. Serial fivefold dilutions of the unknown samples and controls are prepared using a denaturation buffer. For HIV, the undiluted amplicon plus five serial five fold dilutions are analyzed in the HIV-specific wells, and the undiluted amplicon and one dilution of the amplicon are analyzed in the QS wells. The plates are then incubated at 37°C for 1 h with gentle mixing to allow hybridization of the biotinylated products to the respective probes. Unbound products are removed by washing the plates five times with wash buffer using an automated plate washer (Bio-Tek Instruments, Inc., Winooski, VT). Avidin—horseradish peroxidase is added to the plates and allowed to incubate for 15 min at 37°C. To remove unbound conjugate, each plate is washed five more times with wash buffer. A chromogenic substrate, tetramethylbenzidine, and peroxide are added to each plate, and color development proceeds for 10 min in the dark. The color reaction is stopped with 100 $\mu$l of stop solution, and the plates are read at 450 nm.

### *Determination of copy number*

The highest dilution that gives an $A_{450}$ between 0.3 and 2.0 on the HIV-1 specific wells and the highest dilution that gives the same on the QS–specific wells are selected and used to calculate the number of HIV-1 copies/ml in the sample as follows:

$$\text{HIV copies/ml} = \frac{\text{Total HIV OD}}{\text{Total QS OD}} \times \text{Input QS copies/PCR} \times 40^{*}$$

* The factor 40 is used to convert copies per PCR (in 25 $\mu$l plasma) to copies per milliliter of plasma.

### *Construction of the quantitation standard*

The quantitation standard was engineered to mimic the true target. The quantitation standard is an RNA transcript generated from a plasmid that contains an SP6 promoter binding site and harbors a sequence that contains the primer binding sites of the target of interest, but has a different probe binding sequence (Mulder et al., 1994). When reverse transcribed and amplified, it generates a product of the same length and base composition as the target. The efficiency of amplification of the QS is identical to the target sequence and is used to correct for any sample or reaction variabilities.

### *Characterization of the quantitation standards*

Since the quantification of the target copies depends on the QS, the accurate determination of the copy number of the QS is critical. The approach we use in quantifying the target number of the quantitative standards is described in the chapter by Wang and Spadoro. The number of copies of the quantitation standard added to a PCR is selected to represent a sample at the low end of the dynamic range, since inefficiencies in recovery or inhibitions are most evident at these levels. The quantification of low numbers of template (less than 100 copies in a PCR) cannot be determined using conventional technologies. Although quantitation of a high–copy stock solution ($\mu$g of DNA or $10^{14}$ molecules) can be readily performed by OD readings and phosphate analysis, the large dilution required to get to the 100–copy level will affect the reliability of the determinations.

Our method for determining the copy number of the standards is based on Poisson analysis of the low–copy (less than 2) target molecules/unit volume. In this procedure, the nucleic acid is quantified by spectrophotometry, and the number of copies calculated. An intermediate solution that contains 5 to 10 times more target than the desired concentration is prepared. This solution is then further diluted to approximately one molecule/PCR. Eighty replicate amplifications are performed by high–cycle PCR, and the products are detected on the microwell plate. The copy number of this solution is determined using the formula:

$$C = \frac{-\ln(\text{number of negative readings})}{80} .$$

The copies determined in the diluted stock are used to determine the concentration of the intermediate stock. The standards are then prepared from the intermediate stock.

### *Controls*

In addition to the QS that is added to each sample, positive and negative controls are included in each run. Two positives representing a high and a low HIV copy number are typically analyzed. These controls are used to ensure that the assay is performing properly.

### *Unexpected results*

a) Low QS signal. A QS value of <0.2 O.D. units indicates that either the processed sample was inhibitory to the amplification, or the RNA was not recovered during sample preparation. The sample should be re-extracted and the amplification/detection repeated. Care should be taken to ensure that the precipitated RNA pellets are not disturbed. Visualizing the pellet after the first alcohol precipitation may be difficult. To help locate the pellet, place an orientation mark on each tube and place the tubes in the microcentrifuge with the orientation mark facing outward. The pellet will align with the orientation mark. Since residual alcohol is inhibitory, it is important to remove as much of the 70% ethanol as possible.

b) High QS values. If the QS values at the two dilutions tested are >2.0 O.D. units, then an error has occurred and the assay needs to be repeated.

*Quantitative capability: precision and reproducibility*

To evaluate the precision of the assay, 24 samples were processed and analyzed daily for 10 days by each of two operators (AMPLICOR HIV-1 MONITOR™ product insert). The 24 samples consisted of four replicates of six samples: a negative, a high, and a low dilution of an RNA transcript and a high, medium, and low dilution of a virus stock. A total of 80 replicates of each sample was analyzed. The coefficient of variation (CV) for the within–run and total precision were similar and ranged from 29.8% to 45.3% for the RNA transcripts and for the high and medium viral dilutions. The performance of the low–virus dilution was less precise and had within-run CV of 98.2% and total precision CV of 104.8% due to four aberrant high results.

The reproducibility of the assay can be illustrated by the results of the Delta Trial. As part of the Delta Trial evaluation of the HIV RNA quantitation methods, three independent laboratories analyzed 21 plasma samples from HIV-1 infected patients, using the AMPLICOR HIV-1 MONITOR™ test. Each sample was analyzed in duplicate in separate runs by each laboratory, resulting in six independent determinations for eachsample. The average CV was 31%, 23%, and 22% for the three laboratories. The CV for the combined results from all three sites was 43%, which is only slightly greater than the CV for the individual sites, demonstrating good interlaboratory reproducibility.

*High–sensitivity sample preparation method*

The dramatic and sustained reduction of HIV plasma RNA to levels below the detection limit of the AMPLICOR HIV MONITOR™ test has resulted in the need for a more sensitive assay. We have developed an alternative sample preparation procedure that can be used to concentrate the virus in plasma and to increase the plasma equivalent per analysis (Mulder et al., 1997). With the AMPLICOR HIV MONITOR™ test, the equivalent of 25 $\mu$l of plasma is analyzed in each amplification. By inputting the equivalent of 450–500 $\mu$l of plasma, a 20 copy/ml or 20–fold increase in sensitivity can be achieved.

The new sample preparation procedure uses high–speed centrifugation to pellet the viruses. After the plasma is removed, the virions are lysed in a solution containing nonionic detergent, RNAsin, and the internal quantitation standard. The extracted sample is added directly to an amplification reaction without further processing.

This sample preparation method offers several advantages. First, the procedure is simple to use and requires little hands–on time. As many as 96 samples can be handled in 2.5 hours with two centrifuges (24 heads each), or 25 samples can be handled in 1.5 hours. Second, the procedure provides flexibility in the amount of samples that can be analyzed. For high sensitivity, 0.5 to 1.0 ml of plasma can be spun and lysed, and the entire content analyzed by RT-PCR. Where high sensitivity is not required, a smaller volume of plasma can be used. Processing 200 $\mu$l of plasma and lysing the virions with the same volume of lysis solution provides sufficient material to perform four amplifications at 50 $\mu$l each. Third, the procedure can be used on plasmas collected in either EDTA, ACD, or heparinized tubes. Fourth, the sample preparation method is compatible with RT-PCR as carried out with the AMPLICOR HIV-1 MONITOR™ test. The dynamic range of the assay is identical to the MONITOR™ test if the same amount of plasma is used. If a concentration step is implemented, the dynamic range is expected to shift from 400–750,000 copies/ml to 20–40,000 copies/ml.

### *Application*

The quantitative assay described here (either prototype or kit) has been used to monitor plasma HIV RNA levels in patients in several clinical trials, including ACTG 116B/117 (Welles et al., 1996), ACTG 116A (Coombs et al., 1996), ACTG 175, ACTG 229, VACSP 298 (O'Brien et al., 1996), and Delta Trials. The results from these trials indicated that HIV RNA levels can predict the risk of clinical progression to AIDS, and that HIV RNA quantitation can be used to measure the effects of antiretroviral therapy. This assay was used as one of the surrogate marker tests in the accelerated approval process for the protease inhibitors: INVIRASE™ (saquinavir), NORVIR™ (ritonavir), and CRIXIVAN® (indinavir), and for the reversetranscriptase inhibitor EPIVIR™ (3TC). AMPLICOR HIV-1 MONITOR™ was approved by the Food and Drug Administration for the measurement of viral load in June 1996 and represents the first nucleic–acid–based quantitative assay to be approved.

## Future Directions

### *Automation*

The necessity to dilute the amplicons in order to achieve the desired dynamic range of the assay is labor intensive. Complete automation of amplification and detection will be possible with the COBAS AMPLICOR™ (Roche Diagnostic Systems). The COBAS AMPLICOR™ consists of four instruments in one: thermalcycler, incubator, washer, and detector. An XYZ arm transfers the reagents and tubes from one station to another and has the capability to prepare the dilutions required for quantification.

The COBAS AMPLICOR™ provides several advantages. First, the automation of amplification and detection allows for a complete walk-away system once the reactions are set up. Consequently, the hands-on time required is dramatically reduced. Second, the incorporation of four instruments into one eliminates the need for multiple instruments and consolidates work space. Third, the precision of the assay is improved because the dilutions prepared with the COBAS are more precise than dilutions prepared manually.

### *Kinetic PCR*

In the AMPLICOR MONITOR™ and COBAS AMPLICOR™ formats, amplified products are quantified post-PCR. With kinetic PCR, real-time detection of product accumulation eliminates the need for post PCR manipulation. Two different kinetic PCR formats have been developed. The first is based on the intercalation of a dye into the amplification reactions (Higuchi et al., 1992; Higuchi et al., 1993), and the second is based on the 5′ nuclease activity of *Taq* or *Tth* DNA polymerase (Holland et al., 1991). In the dye intercalation approach, ethidium bromide is added to the reactions, and the amount of dye intercalated into the newly synthesized product is measured at each cycle by exposing UV light onto the reactions, and measuring the amount of fluorescence emitted. The advantages of the ethidium bromide approach are the following: a) it is inexpensive and readily available, b) the time to result is fast, c) the method is amenable to large–scale screening, d) it does not require a probe, and e) the assays can be quantitative.

Quantitation is possible since the amount of fluorescence emitted is directly proportional to the amount of product synthesized. The larger the target input, the sooner the fluorescent signal appears. By plotting the cycle number at which a sample reaches an "arbitrary fluorescence level" (AFL) vs. the initial copy input, a standard curve can be generated. Using this relationship, one can determine the copy number of an unknown by simply extrapolating from the standard curve.

At present, the ethidium bromide works best when the number of input target copies is at least 1000 copies/PCR. To achieve lower copy sensitivity, exquisite specificity of the reaction is required. "Hot Start" PCR procedures (Chou et al., 1992) that can improve specificity of the reactions include the use of dUTP/UNG (Kwok et al., 1992), AmpliWax (Perkin-Elmer), and modified enzymes such as AmpliTaq Gold (Birch et al., 1996). In addition, nonspecificity due to primer-dimer formation can be significantly reduced through primer design and reaction optimization. Since probes are not used, amplification products from multiplex reactions cannot be differentiated. However, in some settings, the specific identification of each pathogen is not necessary. For example, blood donations that are infected with any of a number of pathogens are automatically rejected. Therefore, multiplex amplification with primers that amplify the targeted pathogens can be used to screen the donations. Once an infected unit is identified, the samples can be sent back for additional testing to identify the specific pathogen.

The second kinetic PCR approach takes advantage of the inherent 5′ to 3′ exonuclease activity of *Taq* and the *Tth* DNA polymerase activity. It was first described by Holland et al. in 1991. Over the years, significant advances in the technology have been made. Most notably, the original assays used radiolabeled probes and required the physical separation of the cleaved product from the full-length probe by gel electrophoresis. Today, the probes are labeled with two fluorophores such that the fluorescence of the reporter molecule is quenched by the second fluorophore through the process of fluorescence energy transfer (Livak et al., 1995). The fluorescent-labeled probe is added directly to the PCR master mix. As amplification proceeds, the probe is cleaved by the polymerase, and the reporter molecule is separated from the "quencher" molecule and its fluorescence measured. The amount of fluorescence generated is directly proportional to the amount of PCR product generated. This technology presently has the capability of detecting 50–100 copies of target per amplification.

Despite significant advances in this technology, a number of technical issues remain. First, the identification of multiple fluorophore/quencher pairs that can be readily resolved will be critical for the co-amplification and detection of multiple targets in a single reaction. Second, instrumentation that can readily resolve overlapping spectra will be needed. Although instruments are currently available for "real-time" fluorescent measurements, further development will be required to detect and resolve multiple fluorophores. Third, mismatches at the probe binding site affect efficient cleavage by the enzymes, which can lead to an underestimation of the target copy number. Before this procedure can be broadly used on heterogeneous targets, methods to minimize the effects of mismatches at the probe binding region will need to be identified.

Despite these challenges, we expect continued improvements in the PCR-based technologies. Rapid, higher-throughput systems that are completely automated are, and will continue to be, the focus of future technology development.

## REFERENCES

Birch DE, Kolmodin L, Laird W, McKinney N, Wong J, Young KY, Zangenberg GA, and Zoccoli MA (1996): Simplified hot start PCR. *Nature* 381:445–6.

Chou Q, Russell M, Birch DE, Raymond J, and Bloch W (1992): Prevention of pre-PCR mis-priming and primer dimerization improves low-copy-number amplifications. *Nucleic Acids Res* 20:1717–23.

Christopherson C, Sninsky J, and Kwok S (1997): The effects of internal primer-template mismatches on RT-PCR: HIV-1 model studies. *Nucleic Acids Res* 25:654–8.

Coombs RW, Wells SL, Hooper C, Recihelderfer PS, D'Aquila RT, Japour AJ, Johnson VA, Kuritzkes DR, Richman DD, Kwok S, Todd J, Jackson JB, DeGruttola V, Crumpacker CS, and Kahn J (1996): Association of plasma human immunodeficiency virus type-1 RNA level with risk of clinical progression in patients with advanced infection. *J Infect Dis* 174:704–12.

Higuchi R, Dollinger G, Walsh PS, and Griffith R (1992): Simultaneous amplification and detection of specific DNA sequences. *Bio/Technology* 10:413–17.

Higuchi R, Fockler C, Dollinger G, and Watson R (1993): Kinetic PCR analysis: real-time monitoring of DNA amplification reactions. *Bio/Technology* 11:1026–30.

Hoheisel JD, Craig AG, and Lehrach H (1990): Effect of 5-bromo- and 5-methyldeoxycytosine on duplex stability and discrimination of the *NotI* octadeoxynucleotide. *J Biol Chem* 265:16656–60.

Holland PA, Abramson RD, Watson R, and Gelfand DH (1991): Detection of specific polymerase chain reaction product by utilizing the 5′ → 3′ exonuclease activity of Thermus aquatius DNA polymerase. *Proc Natl Acad Sci USA* 88:7276–80.

Kwok S, Kellogg DE, McKinney N, Spasic D, Goda L, Levenson C, and Sninsky JJ (1990): Effects of primer–template mismatches on the polymerase chain reaction: Human immunodeficiency virus type 1 model studies. *Nucleic Acids Res* 18:999.

Kwok S, Kinnard S, Spadoro J, and Sninsky JJ (1992): Enhancement of PCR specificity by uracil-N-glycosylase. *VIII Int Conf AIDS Abstract* p. A2388 67.

Livak KJ, Flood SJA, Marmaro J, Giusti W, and Deetz K (1995): Oligonucleotides with fluorescent dyes at opposite ends provide a quenched probe system useful for detecting PCR product and nucleic acid hybridization. *PCR Meth and Appl* 4:357–62.

Longo MC, Berninger MS, and Hartley JL (1990): Use of uracil DNA glycosylase to control carry-over contamination in polymerase chain reactions. *Gene* 93:125–8.

Martin FH and Castor MM (1985): Base pairing involving deoxyinosine: implications for probe design. *Nucleic Acids Res* 13:8927–38.

Meyers TW and Gelfand DH (1991): Reverse transcription and DNA amplification by a Thermus thermophilus DNA polymerase. *Biochemistry* 30:7661–6.

Mulder J, McKinney N, Christopherson C, Sninsky J, Greenfield L, and Kwok S (1994): Rapid and simple PCR assay for quantitation of human immunodeficiency virus type 1 RNA in Plasma: Application to acute retroviral infection. *J Clin Microbiol* 292–300.

Mulder J, Resnick R, Saget B, Scheibel S, Herman S, Payne H, Harrigan R, and Kwok S (1997): A rapid and simple extraction method for HIV-1 plasma RNA: enhanced sensitivity. *J Clin Microbiol* (in press).

O'Brien WA, Hartigan PM, Martin D, Esinhart J, Hill A, Benoit S, Rubin M, Simberkoff MS, Hamilton JD, and The Veterans Affairs Cooperative Study Group on AIDS (1996): *N Engl J Med* 334:426–31.

Urdea MS, Wilber JC, Yehiazarian T, Todd JA, Kern DG, Fong S, Besemer D, Hoo B, Sheridan P, Kokka R, Neuwald P, and Pachl CA (1993): Direct and quantitative detection of HIV-1 RNA in human plasma with a branched DNA signal amplification assay. *AIDS* 7:S11–S14.

van Gemen B, Kievits T, Schuukkink R, et al. (1992): Quantification of HIV-1 RNA in plasma using NASBA during HIV-1 primary infection. *J Virol Meth* 43:177–88.

Welles SL, Jackson JB, Yen-Lieberman B, Demeter L, Japour AJ, Smeaton LM, Johnson VA, Kuritzkes DR, D'Aquila RT, Reichelderfer PS, Richman DD, Reichman R, Fischl M, Dolin R, Coombs RW, Kahn JO, McLaren C, Todd J, Kwok S, and Crumpacker C (1996): Prognostic value of plasma HIV-1 RNA levels in patients with advanced HIV-1 disease and with little or no prior zidovudine therapy. *J Infect Dis* 174:696–703.

Young KKYY, Resnick RM, and Myers TW (1993): Detection of hepatitis C virus RNA by a combined reverse transcription-polymerase chain reaction assay. *J Clin Microbiol* 31:332–886.

# Determination of Target Copy Number of Quantitative Standards Used in PCR–Based Diagnostic Assays

Zhuang Wang and Joanne Spadoro

## Introduction

PCR diagnostic assays are extremely sensitive and generally capable of detecting a small number of specific molecules. The standards used in such highly sensitive assays must contain an extremely low concentration of template, usually in the range of 0.1–100 copies per microliter, or $10^{-10}$–$10^{-14}$ micrograms per microliter, assuming that the size of templates is in the range of 300–3000 base pairs. Concentrations within this range are much lower than the detection limit of all conventional methods for nucleic acid quantitation. Most researchers therefore quantitate the nucleic acid target in a much more concentrated solution (in the magnitude of $10^{-2}$–$10^{-3}$ micrograms per microliter) using conventional methods such as UV spectrophotometry. This solution is then diluted to a PCR standard containing an appropriate number of targets. A major disadvantage of this approach is that the target can only be quantitated in the more concentrated form of the standard, but not in the final reagent. This approach, although commonly used in research laboratories, is not suitable for the manufacture and quality control of diagnostic products, which require more stringent accuracy, reproducibility, and stability.

An analytical system based on Poisson Distribution therefore was developed for the quantitative detection of target copy number. Using this system, target copy number is determined by dilution of the target solution followed by PCR amplification under conditions capable of achieving single–copy sensitivity and with a high number of replicates. The average target number is then calculated by Poisson Distribution. This method allows one to determine the absolute copy number of the diluted standard at the time of manufacture and provides an analytical method by which to monitor the stability of stored reagent.

## Poisson Analysis: The Theory

Poisson Distribution describes the probability that an event will take place when the frequency of that event occurring is very low. According to the Poisson Distribution, if the event takes place at an average frequency $\mu$, then the probability that a single event $x$ ($P_x$) will occur is

$$P_x = \frac{\mu^x e^{-\mu}}{x!} \quad (1)$$

in which $e$ is the base of the natural logarithm (2.71828 . . .). The value of $\mu$ must be small, usually less than 5, and $x$ must be zero or an integer.

In a PCR standard, the concentration of the target molecule is very low. The number of target molecules in a unit volume is always an integer or zero. The relationship between the concentration (the average number of target molecules in a unit volume of solution) and the probability that a unit volume of solution contains a certain number of molecules can therefore be described by Poisson Distribution.

Assume that a solution contains an *average* copy number of target molecules ($C$) per unit volume. Let $N$ represent the *actual* number of molecules in a unit volume. The value of $N$ is always a small integer and does not always equal $C$. According to the Poisson Distribution, the probability that a unit volume contains $N$ copies ($P_N$) is calculated using the equation

$$P_N = \frac{C^N}{N! \cdot e^c} \quad . \quad (2)$$

For example, the probability that a unit volume contains two copies is

$$P_2 = \frac{C^2}{2! e^C} \quad ,$$

and the probability that a unit volume contains 0 copies is

$$P_0 = \frac{C^0}{0!e^C} \quad . \tag{3}$$

Equation (2) is further illustrated in Figure 1 and Table 1.

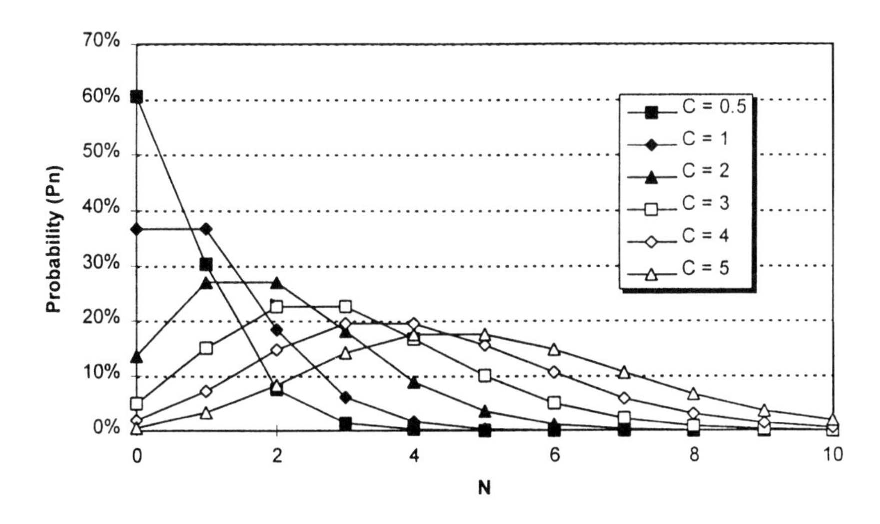


FIGURE 1. Illustration of Poisson Analysis. The relationship between the average copy number and the copy number in a given fraction.

TABLE 1. Poisson Distribution: Relationship between $P_N$ and average copy (C).

| | Probability of a Unit Volume Containing $N$ Copies | | | | | |
|---|---|---|---|---|---|---|
| | $N = 0$ | $N = 1$ | $N = 2$ | $N = 3$ | $N = 4$ | $N = 5$ |
| $C = 0.5$ | 60.7% | 30.3% | 7.6% | 1.3% | 0.2% | 0.0% |
| $C = 1$ | 36.8% | 36.8% | 18.4% | 6.1% | 1.5% | 0.3% |
| $C = 2$ | 13.5% | 27.1% | 27.1% | 18.0% | 9.0% | 3.6% |
| $C = 3$ | 5.0% | 14.9% | 22.4% | 22.4% | 16.8% | 10.1% |
| $C = 4$ | 1.8% | 7.3% | 14.7% | 19.5% | 19.5% | 15.6% |
| $C = 5$ | 0.7% | 3.4% | 8.4% | 14.0% | 17.5% | 17.5% |

Since $C^0 = 1$ and $0! = 1$, Equation (3) can be simplified as

$$P_0 = \frac{1}{e^C}$$

that is,

$$C = -\ln P_0 \quad . \tag{4}$$

Equation (4) establishes the relationship between the concentration of a PCR standard ($C$) and the probability that a unit volume of this solution contains no target ($P_0$), therefore providing an experimental approach by which to determine the copy number of a PCR standard.

The value of $P_0$ can be determined using a highly sensitive amplification and detection system to repeatedly test the standard. This system must achieve single–copy analytical sensitivity. Under these conditions, the value of $P_0$ is equal to the negative rate, that is, the ratio between the number of negative results and the number of total tests:

$$P_0 = \frac{N_0}{N_T} \quad . \tag{5}$$

The average copy number of the PCR standard can then be calculated as

$$C = -\ln\left(\frac{N_0}{N_T}\right) \quad , \tag{6}$$

in which $N_0$ is the number of the negative results, and $N_T$ is the total test number.

## Signal Generating Unit (SGU)

"Copy number" is a commonly used term to describe the quantity of target molecules. Strictly speaking, copy number is a conceptual term rather than an experimental term because its value cannot be measured directly and confirmed by experimentation.

In microbiology, *Colony Forming Unit* (CFU) and *Plaque Forming Unit* (PFU), instead of copy number of bacteria or phage, are used as the measurement of the titer in experimental procedures. We introduce similar terminology, *Signal Generating Unit* (SGU), to measure the concentration of template available for PCR amplification.

A signal generating unit is defined as the smallest unit that generates a positive signal by PCR amplification. On the physical level, an SGU is a particle containing at least one amplifiable molecule. Under ideal conditions, an SGU is equal to a single copy of target molecule.

The copy number determined by Poisson Analysis is expressed as SGUs.

## Poisson Analysis: The Application

With Poisson Analysis, as with other statistical analyses, a sufficient number of replicate tests is required to achieve reasonably good accuracy. The PCR-microwell plate detection format developed by Roche Molecular Systems provides the ability to perform a large number of tests and the sensitivity necessary to perform a Poisson Analysis. An analytical system was developed based upon this format. PCR standards for HIV-1, HTLV-I and -II, HCV, HBV, CMV, *M. tuberculosis,* and *C. trachomatis* have been prepared and characterized using this system.

The process of manufacturing PCR standards is illustrated in Figure 2. The concentrated nucleic acid target is quantitated by UV spectrophotometry, the number of molecules is calculated, and an intermediate solution containing 5–10 times more target than the desired concentration is made by proper dilution. The intermediate solution is then further diluted to approximately one molecule per amplification volume (5–50 $\mu$l, depending upon the product), and referred to as Solution A. The concentration of Solution A is then determined using Poisson Analysis. The concentration of the intermediate solution is calculated. The final product is then diluted accordingly.

A typical Poisson Analysis consists of 96 replicate PCR amplification reactions in a GeneAmp PCR System 9600 thermocycler, using a high–cycle amplification profile (usually 40 or more cycles). The amplification products are examined using the microwell detection format. In addition to 80 replicate tests of Solution A ($N_T = 80$), 12 negative controls and four positive controls are included in the assay. The negative controls are placed at different areas of the amplification tray and microwell plate to monitor

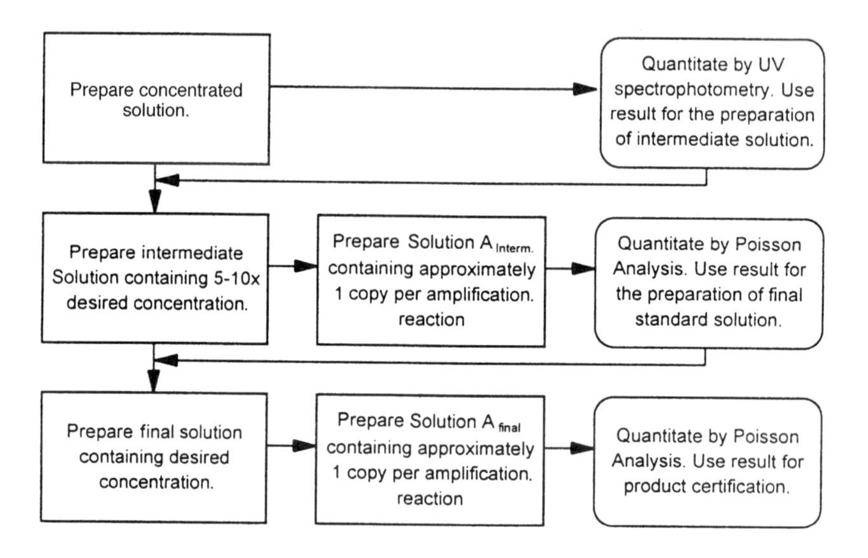


FIGURE 2. The manufacturing process of standards for PCR diagnostic assays.

any possible airborne or reagent contamination. The positive controls, each containing approximately 20 copies of target molecules, are used as reagent controls. The average SGU number of Solution A is calculated according to Equation (6).

The concentration of the intermediate solution is calculated on the basis of the concentration of Solution A. The standard is prepared from the intermediate solution, and the concentration of the standard is confirmed by the same method. In some cases, replicate Poisson Analyses are performed to improve the accuracy of the results.

## Demonstration of Single-Copy Sensitivity

It should be emphasized that Equation (6) is valid only under assay conditions sensitive enough to detect a single target molecule. Only then is $P_0$ equal to the negative rate, a value that can be determined by experimentation. In our analytical system, the sensitivity of the assay is monitored by signal distribution analysis.

Since the number of targets in a unit volume ($N$) is always an integer, the absorbance values generated by these targets should be discontinuous or "quantum." It therefore is predictable that a clear gap should be observed between the negative readings (from the reactions containing no target) and the positive readings (from the reactions containing at least one molecule of target) when single–copy sensitivity is achieved. The higher the sensitivity of the assay, the wider the gap observed for the absorbance values.

The experimental results shown in Figure 3 support this hypothesis. A solution containing an HIV-1 control plasmid was assayed four times using an analytical system with 29, 32, 35, and 40 cycles of PCR amplification, respectively. Eighty replicates were performed each time, and the distribution of the absorbance values was analyzed. Figure 3A shows that the absorbance values generated in the 29-cycle PCR amplification reaction are low and that no gap between the absorbance values generated by the positive and the negative samples is observed. The lack of clarity between positive and negative absorbance values occurs because single–copy sensitivity was not achieved at 29 cycles of amplification. Using a cut-off value calculated on the basis of a commonly used formula

$$Cut\text{–}off = A_0 + 3 \times SD_0$$

where $A_0$ is the average negative reading and $SD_0$ is the standard deviation of the negative readings, a result of 0.43 SGU per amplification volume was suggested. With 32 cycles, a gap is observed between absorbance values generated with the positive and the negative samples. However, two readings fell into the gap, making the calculation ambiguous (Figure 3B). A result of 1.12 or 1.20 SGU per amplification volume could be obtained, depending upon whether the two readings were considered positive or negative. With 35 cycles, a clear gap was observed, and the calculated copy number was 1.015 SGU per amplification volume. At 40 cycles, a large gap in absorbance values was achieved, yielding a result of 0.948 SGU per amplification volume.

These results demonstrate that single-copy analytical sensitivity is essential for accurate copy number determination and that the signal distribution analysis is an effective way to monitor assay sensitivity. At lower cycle numbers, single–copy sensitivity is not achieved, resulting in an ambiguous and/or incorrect copy number determination.

When the single-copy sensitivity is achieved, as it was in the 35-cycle and 40-cycle assays, the copy number remains unchanged while the signal strength still increases with the increase of cycle number, as illustrated in Figure 3(C) and Figure 3(D).

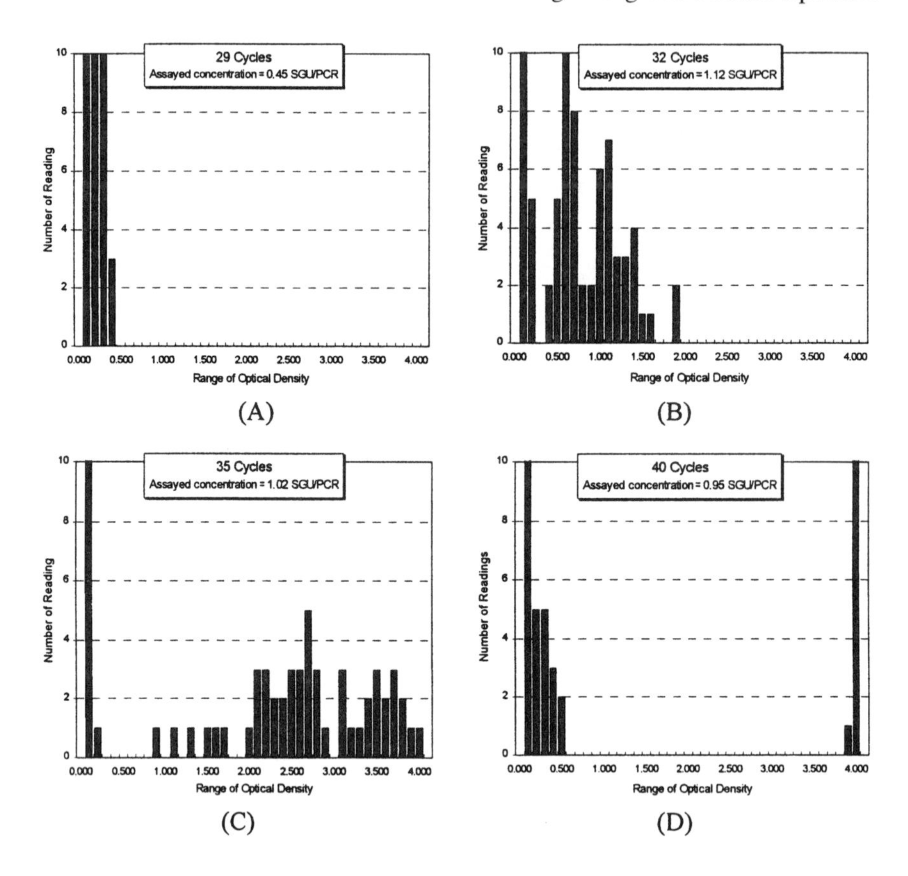


FIGURE 3. The effect of assay sensitivity on copy number determination. When the sensitivity of the assay is low, as shown in (A), the concentration of the target is undersetimated.

## The Accuracy of the Assay

When the copy number of a PCR standard is determined by Poisson Analysis, the accuracy of the result depends on two major factors: the number of replicates in the test ($N_T$) and the concentration of Solution A.

### *Effect of test number*

Test number is the most important factor that affects the accuracy of the copy number. The following mathematical analysis provides a quantitative assessment of how the test number affects the accuracy of the assays.

The resolution of a quantitative method is defined as the difference between the two nearest values that the method can generate. The resolution of Poisson Analysis can be expressed as

$$Resolution = C - C_a \tag{7}$$

where $C$ is a copy number calculated according to Equation (6), and $C_a$ is the copy number nearest to $C$ that can be calculated. The smaller the value of $(C - C_a)$, the better the resolution, meaning that this quantitation method is capable of discriminating between two very close values. The accuracy of the method is a function of the resolution.

According to Equation (6), $C_a$ can be expressed as

$$C_a = -\ln\left(\frac{N_{0,a}}{N_T}\right), \tag{8}$$

where $N_{0,a}$ is the number of negative results used for calculating $C_a$. The value of resolution $(C - C_a)$ can therefore be calculated as

$$C - C_a = -\ln\left(\frac{N_0}{N_T}\right) - \left[-\ln\left(\frac{N_{0,a}}{N_T}\right)\right]. \tag{9}$$

Since both $N_0$ and $N_{0,a}$ are integers, the smallest difference between $N_0$ and $N_{0,a}$ is 1; we therefore have

$$N_{0,a} = N_0 + 1 .$$

According to Equation (5), the number of negative results ($N_0$) can be expressed as

$$N_0 = P_0 \cdot N_T \qquad (0 < P_0 < 1) .$$

Similarly,

$$N_{0,a} = P_0 \cdot N_T + 1 \qquad (0 < P_0 < 1)$$

Substitute $N_0$ and $N_{0,a}$ in Equation (9). The resolution is then expressed as

$$C - C_a = -\ln\left(\frac{P_0 \cdot N_T}{N_T}\right) - \left[-\ln\left(\frac{P_0 \cdot N_T + 1}{N_T}\right)\right] .$$

This equation can be converted to

$$C - C_a = \ln\left(1 + \frac{1}{P_0 \cdot N_T}\right) . \tag{10}$$

According to Equation (10), when $P_0$ remains constant, the greater the value of $N_T$, the smaller the value of $C - C_a$. That is, the better the resolution.

TABLE 2. Effect of test number ($N_T$) on the resolution of the assay.

| $N_T = 8$ | | | $N_T = 80$ | | |
|---|---|---|---|---|---|
| $N_0$ | $C$ | $C - C_a$ | $N_0$ | $C$ | $C - C_a$ |
| 1 | 2.079 | 0.693 | 26 | 1.124 | 0.038 |
| 2 | 1.386 | 0.405 | 27 | 1.086 | 0.036 |
| 3 | 0.981 | 0.288 | 28 | 1.050 | 0.035 |
| 4 | 0.693 | 0.223 | 29 | 1.015 | 0.034 |
| 5 | 0.470 | 0.182 | 30 | 0.981 | 0.033 |
| 6 | 0.288 | 0.154 | 31 | 0.948 | 0.032 |

Table 2 further illustrates the effect of test number by comparing the resolution of the assay when $N_T = 80$ and $N_T = 8$. As shown in Table 2, the assay resolution is substantially increased with a tenfold increase in the number of replicates.

For improved accuracy, standards can be quantitated by multiple Poisson Analyses to increase the total number of replicate tests to several hundred. Mathematical analysis indicates that when three Poisson Analyses ($n = 240$) are employed, the obtained result is within ±15% of the true value, with 90% confidence. When six Poisson Analyses ($n = 480$) are used, the result is within ±10% of the true value, with 90% confidence. These mathematical analyses have been confirmed experimentally and by computer simulation.

### *Effect of the negative rate*

In our studies, we are also interested in the value of relative resolution, which is the ratio between the resolution and the result, expressed as

$$\text{Relative resolution} = \frac{C - C_a}{C} \times 100\% \quad . \tag{11}$$

According to Equation (6) and Equation (10), Equation (11) is converted to

$$\text{Relative resolution} = \frac{\ln\left(1 + \dfrac{1}{P_0 \cdot N_T}\right)}{\ln P_0} \times 100\% \quad . \tag{12}$$

Figure 4 shows the results from a mathematical analysis based on Equation (9). These results illustrate the relationship between the negative rate ($P_0$) and the resolution when $N_T = 80$. The optimal resolution is achieved when the number of negatives is in the range of 6–60. This analysis indicates that in order to achieve the optimal resolution in a Poisson Analysis, the concentration of Solution A should be in the range of 0.3–2.5 SGU per PCR volume.

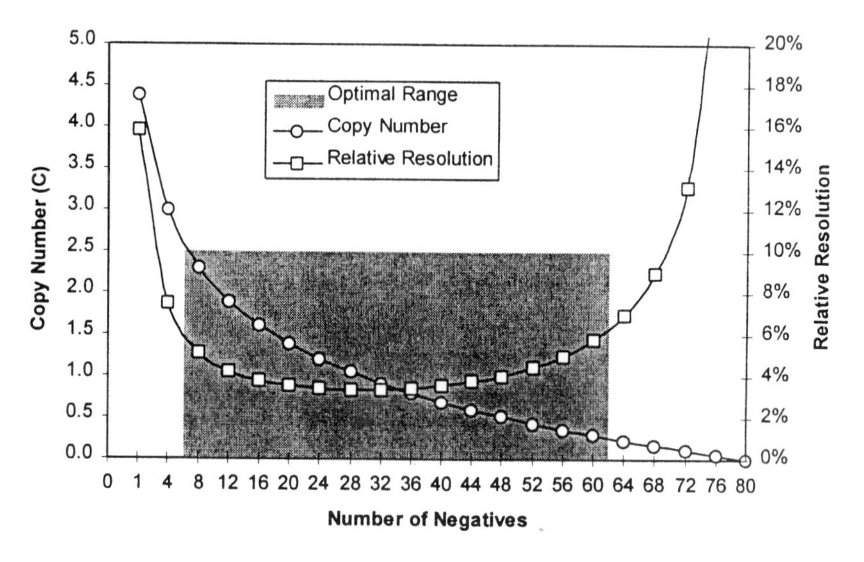


FIGURE 4. Effect of negative rate ($P_0$) on the relative resolution ($N_T = 80$). The shaded area shows the range where the optimal resolution can be achieved.

## Other Factors Affecting the Accuracy of the Assay

The homogeneity of Solution A is an important factor for accurate copy number determination by Poisson Analysis. The target molecules in Solution A must disperse in the form of a monomer for correct determination. Any molecular aggregation or polymerization will significantly affect accurate quantitation. For example, if supercoiled DNA is used as a target, the copy number will usually be underestimated, because supercoiled DNA usually contains concatemers.

It has been observed that at low temperature and in the presence of $Mg^{++}$, the SGU number of a plasmid DNA solution rapidly decreases. Heating of the solution restores the SGU number. This observation suggests the occurrence of reversible aggregation under these conditions.

## Potential Applications in the Research Community

For many research projects, the required accuracy for the quantitative standard is not as high as that for diagnostic products. Poisson Analysis with fewer replicates can then be used.

TABLE 3. Achievable accuracy of Poisson Analysis with low replicate numbers.

| | Number of Replicates | | | |
|---|---|---|---|---|
| Level of Confidence | 12 | 24 | 36 | 48 |
| 60% | ±32% | ±23% | ±18% | ±16% |
| 90% | ±62% | ±44% | ±36% | ±31% |

Table 3 lists the achievable accuracy with fewer replicates, providing that the copy number is about 1 copy per PCR volume, that is, approximately 1/3 of the samples give negative results.

Besides the microwell plate assay system described above, any DNA detection systems capable of clearly discriminating the weakest positive results from the negative results can be used.

## Conclusion

The application of Poisson Distribution Analysis to the determination of low copy number in standards has been described. This method allows one to accurately assign copy number values to dilute standards. Additionally, this method allows one to assess the long-term stability of the standard reagent in a manner independent of functional testing.

## REFERENCES

Colton T (1974): *Statistics in Medicine.* Boston: Little, Brown and Company.

Ott L (1977): *An Introduction to Statistical Methods and Data Analysis.* Belmont: Duxbury Press.

# Quantification of Specific Nucleic Acids, Regulated RNA Processing, and Genomic Polymorphisms Using Reversed-Phase HPLC

A. Hayward-Lester, B.S. Chilton, P.A. Underhill, P.J. Oefner, and P.A. Doris

## Introduction

The ability to make accurate quantitative measurements of specific nucleic acid molecules is becoming increasingly significant in biomedical science. One important clinical application is the quantification of viral nucleic acids in patients' serum and tissues to assess viral burden for determination of disease course and efficacy of treatment. In this case, accuracy and precision of measurement are of primary importance, as decisions concerning selection of medication and dose may depend entirely on perceived change in viral load in an otherwise asymptomatic patient. A second utility in basic biomedical research is the quantification of specific mRNA molecules in functionally characterized cell types. These measurements are useful because they provide a prediction of protein abundance in cell samples in which proteins cannot be quantified directly. These studies often require microdissection to separate the tissues of interest from neighboring tissues. Again, accurate quantification is important for the effective comparison of the expression of multiple genes in multiple tissues and at different laboratories, often under conditions of physiological or pharmacological manipulation.

Traditional methods of gene quantitation include Northern blots and RNAse protection assays. While these provide levels of accuracy and precision acceptable for most comparative studies, the resulting quantitation is relative and may generate estimates that do not accurately reflect the template copy number. In addition, although there have been improvements in sensitivity with the development of RNAse protection assays, relatively large amounts of RNA are still required, and these assays therefore lack the sensitivity necessary for making quantitative measurements in microscopic tissue samples.

In order to make accurate measurements of this kind, some form of signal amplification is required. Two approaches to amplification are currently in use. Reverse-transcription polymerase chain reaction (RT-PCR) and nucleic acid side-chain branching assay (NASBA) both employ nucleic acid template amplification (van Gemen et al., 1994). Another approach is to amplify the detection signal of a specific template, an approach used in the branched DNA assay (Dewar et al., 1994). We have combined RT-PCR, using a synthetic RNA internal competitor with a novel ion-pair reversed-phase HPLC (IP-RP-HPLC) product purification and detection system for quantification of $Na^+$, $K^+$-ATPase (NKA) mRNA expression in microdissected tissue samples. Competitive RT-PCR is ideal for measuring low copy-number sequences. The level of gene expression in a sample can be accurately estimated from the amount of an internal control added to the reaction, if the internal control is properly calibrated. When combined with IP-RP-HPLC, it allows measurement of specific gene expression that is both accurate and precise, with a coefficient of variation (CV) of less than 10%.

## *HPLC analysis of PCR products*

HPLC offers a number of advantages over other available methods for the analysis of competitive RT-PCR reactions. Reaction products can be analyzed immediately with no further processing and no requirement to incorporate radiolabels, or make, load, and run gels. Errors due to inefficient or variable gel blotting or autoradiographs outside the linear detection range are therefore avoided. The use of refrigerated HPLC autosamplers allows 100 or more PCR reactions to be analyzed per day. Other techniques that provide similar rapid separation of double-stranded DNA have been developed. The most advanced of these is capillary zone electrophoresis (CZE). However, HPLC offers some important advantages over CZE (Nathakarnkitkool et al., 1992). PCR reaction products may require further processing (dialysis or ultrafiltration) prior to CZE. Gel scanner techniques provide sensitivity equal to that obtained by HPLC with use of fluorescent primers (Oefner et al., 1994b). However, the equipment is three or more times as expensive, does not provide the automation or flexibility available in HPLC, and requires much greater sample handling.

Efficient and rapid analysis by HPLC of biopolymers requires chromatographic conditions significantly different from those commonly applied to the separation of small molecules. The main reason is that the low diffusivity of large biomolecules necessitates the use of low flow-rates and shallow salt or organic solvent gradients, even when using relatively wide-pore packings (30–40 $\mu$m). One approach to HPLC of nucleic acids employs very small nonporous particles (1.5–2.5 μm) to circumvent some of these problems (Kalghatgi and Horvath, 1987; Kato et al., 1987; Unger et al., 1986). The smaller the particles, the more surface area and capacity. The advantages of such particles are numerous: (a) rapid mass transfer, because permeation is not hindered by diffusion through pores, (b) no size-exclusion effects, (c) short analysis times, (d) biopolymers retain their biological activity because of the short analysis times, (e) higher peak concentrations, due to less diffusion, result in higher mass sensitivity, (f) quantitative recovery, and (g) fast regeneration and equilibration of the column because of the absence of pores. A drawback of nonporous beads is the relatively small surface area compared with that of porous particles, which results in low loading capacity, typically in the lower microgram range. This loading capacity, however, is more than sufficient for most applications in molecular biology.

Nonporous particles allow the rapid separation of both proteins and nucleic acids, as confirmed independently by Kato (Kato et al., 1989) and Oefner (Huber et al., 1995; Oefner et al., 1994a; Oefner et al., 1994b). However, these investigators differ with regard to the matrix used and, particularly, with regard to the functional groups attached to the beads. Kato creates a solid anion-exchanger by covalently attaching DEAE groups to methacrylate beads. A drawback of this approach is that no size-dependent separation of double-stranded DNA is obtained, although a recent modification has reduced this problem, to a certain extent, by replacing the sodium chloride gradient with a tetramethylammonium bromide gradient. However, anion-exchangers are far more difficult to handle than alkylated stationary phases and require longer equilibration times, i.e., time of separation may be short, but up to 15 minutes may be required to re-equilibrate the system to starting conditions, resulting in fewer possible runs per day.

More recently, Oefner and co-workers demonstrated that highly efficient DNA separations can be obtained on alkylated poly-(styrene/divinylbenzene) (PS-DVB) particles using an ion-pairing reagent, typically tri- or tetraalkylammonium salts, in the mobile phase (Huber et al., 1995; Oefner et al., 1994a; Oefner et al., 1994b). The ion-pairing reagent converts the stationary phase into a dynamic anion-exchanger. The number of methylene groups in the alkyl chains of the ion-pairing reagent also determines the extent to which the surface of the beads retains its hydrophobic properties, which in turn exerts a significant effect on the mode of separation. Short alkyl chains (number of methylene groups per alkyl chain) result in incomplete coverage, and the stationary phase partially retains its hydrophobic or reversed-phase properties. Longer alkyl chains create complete coverage, and anion-exchange becomes the predominant mechanism of separation (Oefner et al., 1994b). This also explains why the separation of single-stranded oligonucleotides is so dependent on the ion-pairing reagent chosen. Triethylammonium ions, for instance, cover the stationary phase only partially, and therefore separation is not only governed by size, but also by base composition. Tetrabutylammonium ions, on the other hand, result in complete coverage of the stationary phase; consequently, separation of single-stranded oligonucleotides is almost strictly size-dependent (Oefner et al., 1994b).

The reverse situation is observed in the separation of double-stranded DNA. Tri- and tetraalkylammonium salts with short alkyl chains are able to ion-pair specifically with AT-base pairs. This suppresses their preferential interaction with the dynamic anion-exchanger and enables a degree of size–dependent separation otherwise only observed in denaturing polyacrylamide slab gel electrophoresis (Huber et al., 1995). The preferential interaction of AT-base pairs with the anion-exchange stationary phase is also the reason that size-dependent separations cannot be obtained using a NaCl gradient (Kato et al., 1989). Tetraalkylammonium ions with longer alkyl chains do not fit into the major groove of the double helix and so are unable to ion-pair with double-stranded DNA. As a consequence, when ion–pairing agents with longer alkyl chains are used, separation becomes governed to a large degree by the AT-content of the DNA fragments, instead of by their size (Oefner et al., 1994b). A further advantage of the PS-DVB phase is the chemical and physical stability of the matrix. Several thousand runs can be carried out on a single column.

The advantages conferred by the PS-DVB stationary phase combined with an ion-pairing mobile phase for double-stranded DNA separation led us to examine its suitability for ion-pair reversed-phase (IP-RP-HPLC) analysis of competitive RT-PCR reaction products. In addition to those already mentioned, several benefits are offered by the IP-RP-HPLC technology. The sensitivity and extensive linear dynamic range of UV detectors make detection of reaction products simple and accurate. The column offers excellent resolution of DNA molecules (Figure 1). In addition, quantitative PCR techniques have consistently required the generation of titration curves, resulting in large numbers of samples. The potential ability to automate sample loading and analysis, using HPLC autoloaders and computerized data acquisition and analysis software, provides major advances in productivity and efficiency over other methods of PCR product separation and quantitation.

### *IP-RP-HPLC heteroduplex detection: an essential property for gene quantification, mutation detection, and analysis of regulated mRNA splicing*

One of the most important advantages that IP-RP-HPLC offers over other nucleic acid separation techniques is the ability to detect and resolve

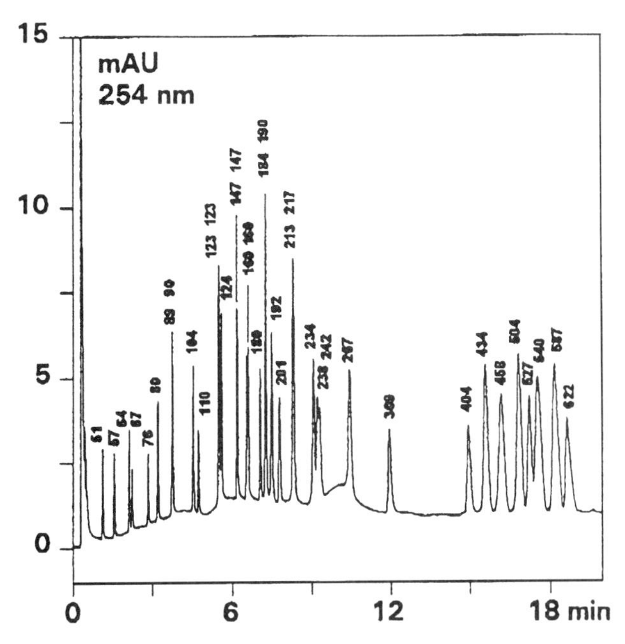


FIGURE 1. Chromatogram of pBR322-HaeIII digest (0.75 $\mu$g) and pBR322-MspI digest (0.4 $\mu$g) on PS-DVB-C18 HPLC column (50 × 4.6 mm I.D.). Fragment sizes are labeled. Elution profile was modified to optimize separation of the large number of fragments present. Mobile phase composition: A = 0.1 M TEAA, pH 7.0, B = 0.1 M TEAA pH 7.0, 25% acetonitrile. Gradient: linear 37–55% B in 6 min; 55–65% B in 14 min; flow rate 1 ml/min; temp. = 50°C; detection UV 254 nm.

heteroduplex molecules that may form by hybridization of partially homologous DNA (Hayward-Lester et al., 1996b; Hayward-Lester et al., 1995; Oefner and Underhill, 1995; Underhill et al., 1996). The generation of such products can perturb the mathematical relationships between native and competitor product ratios. The formation of heteroduplexes is likely in any competitive PCR reaction that approaches or enters the plateau phase. However, we have shown that these heteroduplexes are generally not resolved in nondenaturing gel electrophoresis. Analysis of competitive RT-PCR reaction products by HPLC (Figure 2) demonstrates that, in addition to the two expected homoduplex product peaks, a third peak is evident that elutes immediately prior to the smaller of the two products and is most prominent when individual reactions produce significant amounts of both homoduplex products, rather than when one product

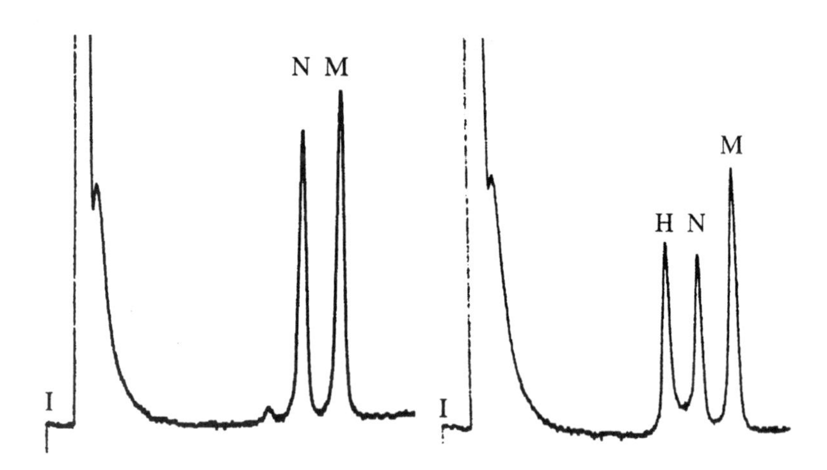


FIGURE 2. Chromatograms generated by IP-RP-HPLC analysis of RT-PCR products, illustrating the formation of the heteroduplex products. The left-hand chromatogram was generated by proportionately mixing products from reactions generating native product alone or mutant product alone. 10 $\mu$l of this mixture was injected onto the column, and the two expected peaks were resolved. The right-hand chromatogram was produced by analysis of a 10 $\mu$l aliquot of the same mixture, heated to 97°C then cooled to 4°C, resulting in the appearance of a third peak representing the formation of heteroduplex products. For HPLC, Solvent A was 0.1 M TEAA, pH 7. Solvent B was 0.1 M TEAA, pH 7 with 25% acetonitrile. The products are resolved in a gradient of 50% B to 57% B over 4 min.

predominates. Further analysis of the additional peak revealed that it represents the heteroduplex products. Separate RT-PCR reactions with native and mutant templates were performed. When the reaction products were mixed together and analyzed by HPLC, only the two homoduplex peaks were evident. When an aliquot of this mixture was heat-denatured and then cooled slowly to room temperature, the third peak evident from the competition reactions was again revealed. The relative size of this peak is influenced by cooling rate and salt concentration, and the peak can be selectively eliminated by S1 nuclease treatment (Hayward-Lester et al., 1996b). Partial denaturation of the homologous portion of the double-stranded heteroduplex molecule occurs at the peripheries of the region of nonhomology. Retention time on the column is determined primarily by the interaction of ion-pairing reagent with complementary paired bases. Because the heteroduplex molecule contains fewer paired bases, it elutes prior to the homoduplex molecules.

## Accurate Quantification of Sodium, Potassium ATPase mRNA Using Single-Tube Competitive RT-PCR and IP-RP-HPLC

Sodium, potassium ATPase (NKA) is a multisubunit protein with all catalytic functions currently ascribed to the alpha subunit. Consequently, alterations in the regulation or function of the alpha subunit might be expected to affect the ion-transport function of the pump. Four mammalian alpha isoforms (termed alpha 1, 2, 3, and 4), encoded by separate genes, have been identified, cloned, and sequenced. The expression of these isoforms has been demonstrated to be highly tissue- and developmentally-specific (Orlowski and Lingrel, 1988; Schafer, 1994; Schneider et al., 1988; Shamraj and Lingrel, 1994). Studies characterizing the different alpha isoforms have demonstrated that they possess different biochemical properties (Jewell and Lingrel, 1991; Sweadner, 1989). The kidney is a heterogeneous organ, whose basic functional unit, the nephron, can be divided into at least 12 morphologically and functionally distinct segments (Berry, 1982). Sodium–ion–handling properties differ between segments; therefore, nephrons may be expected to exhibit a distinct pattern of expression and regulation of NKA. Sodium balance could be influenced by differential expression of the NKA alpha isoforms in various nephron segments. Studies examining sodium pump activity along the length of the nephron imply the presence of more than one NKA isoform in nephron tissue (Doucet and Barlet, 1986). However, examinations to date of NKA alpha isoform protein and mRNA expression in rat kidney have yielded conflicting results. Some groups have reported the presence of more than one isoform in various nephron segments (Clapp et al., 1994). Others have only detected the expression of alpha 1 RNA (McDonough et al., 1994). The ability to make accurate quantitative measurements of NKA gene expression is a major advantage. The possible functional significance of qualitative detection by RT-PCR of a specific message is difficult to assess. Leaky transcription could result in the conclusion that a gene may be both expressed and functional, simply because PCR amplification yields a specific nonquantitative signal. Accurate quantification not only allows meaningful comparison of the expression of one isoform in various nephron segments (and other tissues), but also permits comparison between isoforms within a tissue and between animals in different physiological states. Therefore, we used competitive RT-PCR combined with IP-RP-HPLC to develop an assay to make quantitative measurements of NKA alpha isoform gene expression in microscopic tissue samples. Development of such a system requires the consideration and fulfillment of theoretical ideals associated with competitive nucleic-acid amplification. These mathematical constraints will now be described.

## *Theoretical questions concerning the reliability of the competitive RT-PCR approach*

Competitive PCR approaches provide the prospect of both accurate and precise quantification. However, the exponential nature of the amplification reaction imposes two primary constraints, which must be met in order for measurements to be accurate and precise. First, the PCR amplification efficiency of native and competitor inputs must be identical. Small differences in efficiency of amplification caused by temperature, buffer composition, volume, or presence of contaminants are easily magnified during exponential increases due to replication in repetitive cycles. Second, the decline in amplification efficiency that occurs as reactions proceed to plateau (a necessary constraint for the quantification of low abundance signals) must affect both native and competitor inputs identically. Finally, accuracy requires the use of a properly calibrated internal standard. The effect of variations in efficiency ($E$) can lead to dramatic variations in product yield, as can be seen from the equation below:

$$N = N_0\left(1 + E\right)^n \tag{1}$$

where $N$ is the final amount of reaction product, $N_0$ is the initial amount of DNA in the reaction, $n$ is the number of cycles, and $E$ is the efficiency of the reaction. Thus, if variations in reaction efficiency of only 10% occur between two similar samples, changes of product amount of more than 700% after 35 reaction cycles will result.

Competitive PCR seeks to overcome this problem. An internal mutant is designed to be amplified by the same primer pair as the gene of interest, but to be distinguishable from the native product on subsequent analysis of PCR products. If the initial competitor concentration is known or is always constant between samples, it can be used as a quantitative reference. We chose to utilize an internal RNA competitor that differs in amplified sequence from the native molecule only by a small insertion or deletion, allowing the two products to be resolved by their relative length. Although the deletion and insertion mutant strategy using competitor RNA in the RT reaction step is a major advance in permitting precise determination of gene expression levels, it is not without some theoretical problems. The most obvious of these are: (a) that the processivity rate of Taq polymerase (about

50 bases/sec) may be sufficiently low that during the 1–min elongation step in our PCR reaction, shorter DNAs are amplified with greater efficiency than longer ones, and (b) that there may be differences in the efficiency of reverse transcriptase that bias native or mutant signal generation. To understand the significance of these factors requires a mathematical approach to PCR amplification. Raeymaekers has devised such an approach, and our methods provide an opportunity to test the mathematical model he has developed (Raeymaekers, 1993). This model is described briefly below.

If the initial unknown amount of a gene $U$ in a competitive RT-PCR reaction is $U_o$ and that of its specific (mutant) competitor RNA is $C_o$ and we subject these to $n$ reaction cycles in which the efficiency of amplification is $E_u$ and $E_c$ for the unknown and competitor, respectively, then from Equation (1) we can describe the amount of reaction products at the end of $n$ cycles by:

$$U_n = U_o \cdot \left(1 + E_u\right)^n \tag{2}$$

$$C_n = C_o \cdot \left(1 + E_c\right)^n \, . \tag{3}$$

Making a ratio of Equation (2) and Equation (3) and taking the logarithm gives:

$$\log\left(U_n / C_n\right) = \log U_o - \log C_o + n \cdot \log\left[\left(1 + E_u\right) / \left(1 + E_c\right)\right] . \tag{4}$$

A basic assumption of competitive RT-PCR is that $E_u$ and $E_c$ remain equal throughout the reaction. In this case, Equation (4) reduces to:

$$\log\left(U_n / C_n\right) = \log U_o - \log C_o \, . \tag{5}$$

In calculating competitive RT-PCR reactions to obtain the unknown amount of a gene ($U_o$) present in a sample, a plot is made that relates $\log(U_n/C_n)$ to $\log C_o$, the known amount of starting competitor RNA. This allows $U_o$, the initial amount of gene $U$, to be calculated.

Equation (5) indicates that such a plot will form a straight line having a slope of –1, or of 1 if $\log(C_n/U_n)$ is plotted. Most competitive PCR work has paid attention to the property of linearity in this relationship, but less attention has been applied to the slope, and numerous reports contain data in which the slope requirement is not met (Gilliland et al., 1990; Siebert and Larrick, 1992; Zachar et al., 1993).

Raeymaekers has modeled the effects of various factors that might influence the accuracy of competitive RT-PCR quantification (Raeymaekers, 1993). He has pointed out that the basic assumption of $E_u = E_c$ may be preserved, even if the actual values of $E_u$ and $E_c$ decline as PCR reactions progress. This decline in efficiency could be due to product accumulation, thermal denaturation of DNA polymerase, and reduction in primer concentration, and possibly to other factors. He has estimated the effect of such a situation on accuracy of quantification and shown that both linearity and slope conditions can be preserved, while accurate (but not relative) quantification can be lost. Raeymaekers also proposes that other theoretical factors may invalidate accurate quantification.

We used RT-PCR in conjunction with competitive IP-RP-HPLC to ascertain whether or not our system meets the theoretical requirements addressed in Raeymaekers' work. A concern that presents itself when an internal competitor is used in quantitative RT-PCR is that the discrepancy in size or altered sequence composition between the native and mutant templates may result in disparate values for their amplification efficiencies ($E_u$ and $E_c$). Our first objective was therefore to examine transcription and amplification efficiency in our competitive RT-PCR reactions. Second, we wished to determine the accuracy of our quantitation, using predetermined RNA and DNA inputs. Our third aim was to assess the precision of the measurement system and quantitative accuracy over a wide range of input RNA, including RNA extracted from microdissected rat nephron segments. Finally, given that our assay met the above criteria, we wished to determine whether or not equivalent measurements could be achieved using single–tube assays rather than multiple titrations.

## *Methods*

RNA was transcribed from the T7 RNA polymerase promoter of a pGem4Z construct containing a 396 bp rat alpha 1 sodium, potassium-ATPase (A1NKA) partial cDNA. Two mutant RNAs (541 bp insertion mutant and 382 bp deletion mutant) were transcribed from similar constructs containing the same 396 bp native sequence interrupted by a 145 bp insertion or a 14 bp deletion, respectively. RNA yield was determined by UV absorbance, and then transcribed RNA was diluted in a solution of yeast total RNA to reduce binding to plastic surfaces and to minimize degradation by RNAse. Subsequent RNA dilutions were made in the same yeast RNA buffer. RNA mixtures were reverse transcribed at 42°C for 25 min in a Peltier effect thermal cycler (MJ Research, Watertown, MA). Reaction volumes (10 μl with a 20 μl wax overlay) comprised Perkin Elmer reagents including MMLV reverse transcriptase and random hexameric primers. After heating the reactions to 99°C and cooling to 5°C, 40 μl of PCR reaction mastermix, containing Amplitaq DNA polymerase (1.25 U/40 μl) and 0.3 μM of each alpha 1 primer, was added to each tube. Standard cycling parameters were as follows: 2 min at 95°C, followed by 36 cycles comprising 50 sec at 94°C, 60 sec at 56°C, and 70 sec at 72°C, with a final extension for 5 min at 72°C. Samples were rapidly cooled and held at 0°C for analysis.

Primers were designed to span an intron, so as to distinguish cDNA amplification from possible contaminating genomic DNA amplification and to avoid self-complementarity. The forward primer was an 18mer with sequence: 5′-CCCTAGTTCCCGCCTCTC, the reverse primer was a 21mer: 5′-TGGTCGTCCATAGACACTTCC. Reaction products were analyzed by agarose gel electrophoresis (15 μl aliquots), using a 5% Metaphor gel (FMC, Rockland, ME) and IP-RP-HPLC (10 μl aliquots) (DNASep column, Sarasep Inc, Santa Clara, CA) (Huber et al., 1993; Oefner et al., 1994a). HPLC was performed without labeling or further treatment of reaction products. Aliquots of reaction products were injected sequentially onto the column at approximately 6.5-min intervals. The elution system was a gradient of acetonitrile in 0.1 M triethylammonium acetate (TEAA), pH7, 1 ml/min. The amount of each product was determined by on-line UV absorbance detection (254 nm).

*Results*

Reaction products from an NKA competitive RT-PCR titration series were analyzed by agarose gel electrophoresis and IP-RP-HPLC (Figure 3). Each reaction tube initially contained the same amount of native rat brain total RNA, with different known levels of a 541 bp insertion mutant synthetic RNA competitor. Gel analysis with ethidium bromide staining revealed two principal reaction product bands, the relative intensities of which reflected RNA input. The migration distances of these bands corresponded to that expected of the native and mutant homoduplex products. IP-RP-HPLC demonstrates the native and mutant products together with a third product peak, eluting immediately prior to the smaller (native) product. Our previous observations suggested that this peak constituted heteroduplexes that form between homologous portions of the native and mutant products. This assumption was supported by the observation that the proportion of this product was greatest in reactions in which the

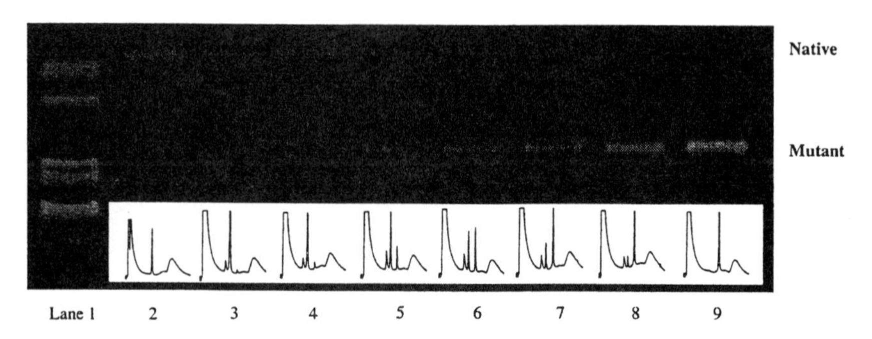


FIGURE 3. Analysis by both HPLC and agarose gel electrophoresis of products from a single RT-PCR titration series. Each reaction in the titration contained 25 ng rat brain total RNA input with varying mutant input. On the gel, lane 1 is a pUC18 HaeIII ladder, lane 2 contains RT-PCR products of native RNA only, lanes 3–8 contain competitive reaction products in which native RNA was subject to RT-PCR with 0.25, 0.5, 1, 2, 4, and 8 pg mutant respectively, and lane 9 contains RT-PCR products of 4 pg mutant RNA only. Reprinted with permission of Cold Spring Harbor Press from *Genome Research* 5:494–499, 1995.

amounts of native and mutant products were similar. In addition, at both outer limits of the titration series, this third peak was larger than the least prominent homoduplex peak. Even though the formation of heteroduplexes is less thermodynamically favored than formation of homoduplexes, application of probability theory indicates that within the plateau phase of the reaction, this pattern of heteroduplex peak formation would be expected. We analyzed the ethidium-bromide-stained gel by densitometric scanning, and compared the subsequent regression result with that provided by IP-RP-HPLC product analysis. This resulted in an estimate of molecular number approximately 30% of that estimated by the HPLC data analysis. This indicates that the heteroduplex molecules are comigrating during electrophoresis with the insertion mutant product. It is generally not possible to separate heteroduplex products from the homoduplex products using nondenaturing electrophoresis. This presents the danger not only that heteroduplex molecules might be excluded from regression analysis, but that they might be erroneously included with one of the homoduplex products. Thus, employing agarose gel electro–phoresis to analyze these competitive RT-PCR products can result in incorrect estimation of native and mutant product formation with a subsequent skew in quantification.

*Data analysis*

The competitive RT-PCR quantification method requires analysis of the ratio of reaction products (Gilliland et al., 1990). The formation of heteroduplex molecules in a reaction would subtract equally from the numerator and denominator of this ratio only when the reaction products are present in exactly equal amounts. Therefore, for correct quantification, it is imperative that the portion of the heteroduplex peak constituting native and mutant reaction products be reallocated to the homoduplex product peak values. A spreadsheet was developed that calculated and reapportioned the absorbance due to heteroduplex molecules, according to the size of each component of the heteroduplex. The spreadsheet was further used to correct for the discrepancy in UV absorbance due to product size, and then to perform regression analysis of the logarithm of the mutant competitor input against the log of the corrected ratios of reaction products. As discussed above, such a plot should generate a straight line with a slope of unity if the PCR amplification efficiencies of the native and mutant inputs remain identical throughout the reaction. Titrations that consistently conformed to slope and linearity properties constraining competitive reactions were obtained (mean slope and $R^2$ for these reactions were $1.01 \pm 0.01$ and $0.99 \pm 0.005$, respectively, mean $\pm$ SEM, $n = 18$).

*Mutant design and amplification efficiency*

Selection of mutant for competitive RT-PCR is a critical component of the strategy. Our 541 bp insertion mutant was engineered to produce a PCR product that shared only 63% homology with the native product, so as to minimize the likelihood of heteroduplex formation. However, heteroduplex formation freely occurred in reactions that were allowed to proceed beyond exponential phase. The production of mutants that differ by the presence of a single base alteration so as to provide a specific restriction endonuclease cleavage site would be expected to generate an even greater proportion of heteroduplex products. In this case, error due to the inability of the heteroduplex to be cleaved will occur. Error can also be caused by the extra postreaction manipulation required for restriction digestion.

Although our regression analyses were robust in their compliance with theoretical requirements, we wished to ensure that a significant difference in length between the native and insertion mutant products does not result in an inequality in their PCR amplification efficiencies. It is evident from Raeymaekers' models that linear regressions with slopes approaching unity could still be obtained if initial relative amplification efficiencies differed, as long as their subsequent rates of decline were equal. See Equation (5) above.

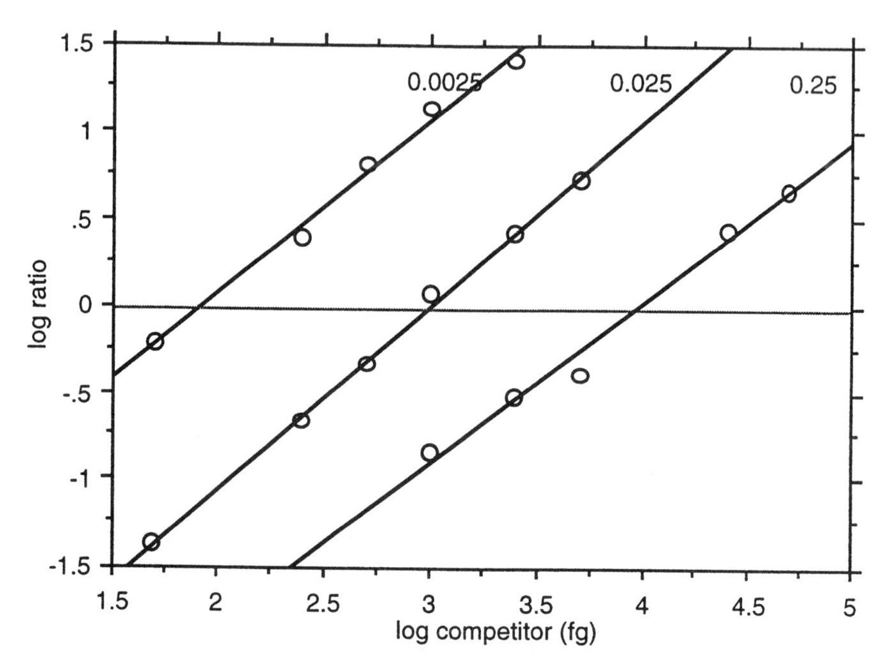


FIGURE 4. Regression plots from three RT-PCR titration reactions with different starting levels of rat brain total RNA. From left to right, plots generated by reactions initially containing 0.0025 $\mu$g, 0.025 $\mu$g, and 0.25 $\mu$g brain total RNA are shown superimposed to illustrate their relationship. Reprinted with permission of Eaton Press from *Bio/Techniques* 20:250–257, 1996.

We examined this by diluting native total rat brain input RNA over three orders of magnitude and performing competitive titrations at each level of input. The results revealed preservation of slope and linearity attributes of the regression analysis and reproducible quantification at each of the levels of input (Figure 4). With a difference in input level, the reactions might be expected to enter the plateau phase after different numbers of cycles; thus the effect of a discrepancy would be compounded with increasing cycle number and become evident. Our results therefore imply that the initial amplification efficiencies of native and mutant sequence are equal and demonstrate precision of measurement over four orders of magnitude of input.

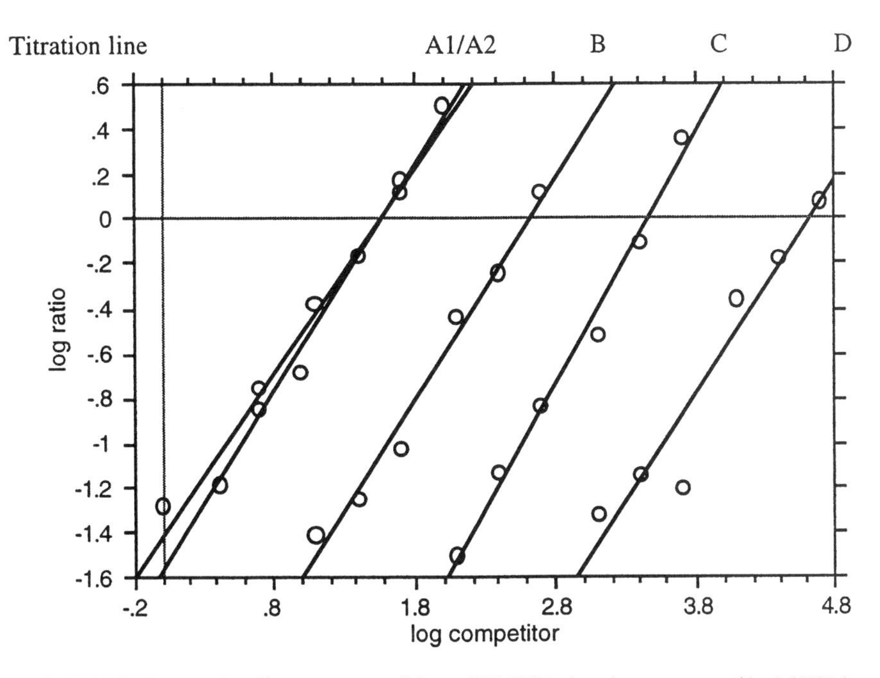


FIGURE 5. Regression lines generated from RT-PCR titrations to quantify A1NKA molecules at four known input levels. Each titration comprised a fixed initial number of native molecules per tube (A, $1.13 \times 10^4$; B, $1.13 \times 10^5$; C, $1.13 \times 10^6$; and D, $1.13 \times 10^7$) with appropriate varying competitive mutant RNA input. For each regression, 10 $\mu$l aliquots of reaction products were analyzed by HPLC. The logarithm of the ratio of the two products was plotted against the logarithm of the initial amount of mutant RNA in each tube. The initial amount of A1NKA native RNA can be estimated by extrapolation from the point on the line at which the log product ratio = 0. The intersection of the titration lines with log ratio = 0 shifts appropriately with change in input level. The slopes of the lines were shown by analysis of covariance to be indistinguishable from one another. In addition, slope and $R^2$ values did not differ significantly from unity, demonstrating feasibility of quantification over at least four orders of magnitude. Reprinted with permission of Cold Spring Harbor Press from *Genome Research* 5:494–499, 1995.

### *Accurate quantitation*

Our primary objective was to develop a method of quantification that was accurate (able to correctly measure the number of input native mRNA molecules) (Ferré, 1992). In order to determine if IP-RP-HPLC analysis provided these conditions, we performed quantitative titrations using known amounts of synthetic native RNA (over four orders of magnitude of native input) in competitions with known amounts of the 541 bp insertion mutant RNA (Figure 5). The results again demonstrated excellent preservation of regression slope and linearity properties for a wide range of RNA input. Estimation of native RNA molecule input number, however, revealed a consistent discrepancy when compared with the actual input. This difference was estimated at 3.79 ± 0.20 fold (all values are mean ± SEM, $n = 18$). This value was independent of input level, cycle number, various cycling parameters, and reaction conditions. The consistency of the estimation discrepancy is strong evidence for a difference in RT efficiency between the two RNA templates. To test this hypothesis we repeated the experiments using known native and insertion mutant plasmid DNA. In this case the titrations accurately estimated the known amount ($15.5 \times 10^3$ molecules), of input DNA (estimate $16.8 \times 10^3 \pm 1.4 \times 10^3$ molecules), and the slope and linearity properties were preserved as before (mean slope and $R^2$ were 1.02 ± 0.03 and 0.99 ± 0.003, respectively). Thus a constant difference in RT efficiency between native and mutant RNA templates permits accurate quantification of gene expression, as long as this constant is determined and the appropriate correction factor applied.

Recent studies in our laboratory have examined the possibility of reducing or removing such discrepancies in RT efficiency. We have found that increasing the temperature at which RT is performed reduces, but does not overcome, the discrepancy. Construction of new competitors with smaller mutations is helpful in some instances. A correlation was observed between the formation or loss (in insertion and deletion competitors, respectively) of stem-loop structures predicted (by the RNA modeling program, *m*fold) to form in the region of the mutation. Interestingly, a mutation that resulted in predicted loss of a stemloop resulted in a competitor with increased RT efficiency compared to the native, a mutation predicted to have no effect on stem-loop structure resulted in a competitor that had identical RT efficiency to its native template, and a mutation predicted to produce new stem-loop structure had lower RT efficiency than its corresponding native template. Such correlations are interesting, but probably reflect an overly simplified view of the source of RT efficiency difference (Hinojos et al., unpublished observations). This work has important implications for accuracy and the ability to make comparisons between results from laboratories using different competitive RT-PCR systems.

The ready formation of heteroduplexes as reactions proceed beyond the exponential phase and the need to identify and quantitate them in order to achieve accuracy of quantification require that competitive mutants be designed to permit clear heteroduplex resolution. Within the constraints of IP-RP-HPLC resolution, this requires that mutants must be sufficiently different in size so that the two homoduplex products can be separated, and of sufficient nonhomology so that the resulting heteroduplexes are also adequately resolved from both homoduplex products. The PS-DVB column permits resolution of products differing by as few as ten base pairs. Consequently, we constructed a second competitive mutant, this time differing by only 14 base pairs (internal deletion) from the native amplicon. We then used RNA synthesized from the new mutant construct to repeat the quantitative accuracy studies, estimating known input amounts of synthetic native RNA. In this case the difference in the estimated amount of native RNA input compared with the known amount was 0.95 ± 0.06 fold (all values are mean ± SEM, $n = 12$). Slope and $R^2$ values obtained in these titrations were 0.98 ± 0.032 and 0.99 ± 0.005, respectively. Optimizing the design of the competitive mutant makes accurate quantification without a correction factor possible.

*Determination of precision*

In addition to assessing accuracy measurements, we also determined the precision of quantification with this quantification system. Precision was examined by making repeated measurements of alpha 1 expression in the same sample of total RNA by competitive RT-PCR and HPLC and consequently determining the experimental variance. Using RNA from a tissue in which the NKA alpha 1 mRNA copy number is relatively high (brain), we observed a coefficient of variation of 8.3% in samples ($n = 5$) in which gene expression was $8.82 \times 10^7 \pm 0.33 \times 10^7$ molecules of A1NKA/$\mu$g total RNA. Precision was also estimated for a low abundance source by analyzing NKA alpha 1 expression in RNA extracted from microdissected nephron segments. In this case we observed a CV of 17.8% in samples ($n = 6$) in which the level of expression was $16.8 \times 10^2 \pm 1.22 \times 10^2$ molecules/0.025 mm tissue.

Titration is required in competitive RT-PCR quantification of gene expression to ensure that the system conforms to the theoretical requirements for slope and linearity discussed above. However, once the components of a competitive system are shown to meet mathematical ideals, theoretically titration is no longer required. We have demonstrated for our NKA alpha 1 system that the amplification efficiencies of native and mutant templates are equal. Under these circumstances, the final ratio of products must directly reflect the initial ratio of native and mutant templates. Analysis of the products of a single competitive reaction containing a known number of molecules of mutant competitor should suffice for determination of the initial concentration of native RNA (as long as the approximate concentration of native mRNA is known and the detection system has sufficient linear range). We tested this supposition

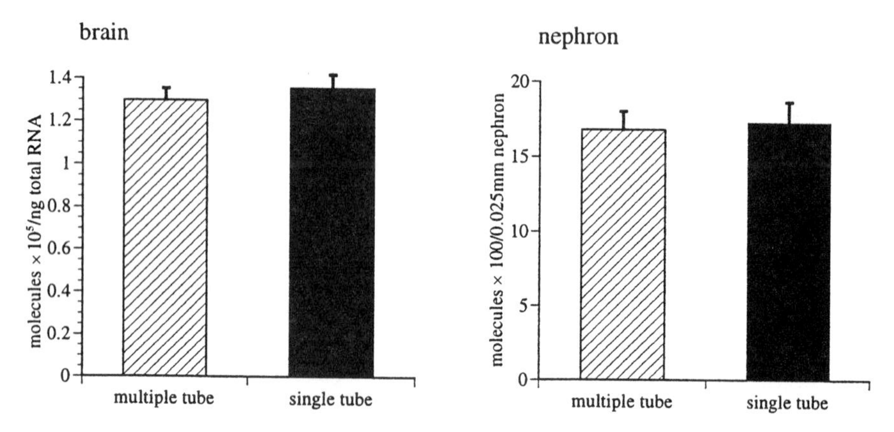


FIGURE 6. Estimation of precision of measurement of A1NKA RNA comparing multiple-tube titration with single-tube quantification for high-abundance (brain) and low-abundance (nephron) samples. For each tissue, multiple measurements were made from one RNA sample, analysis of reaction products was by HPLC, and the results are shown as mean ± SEM. Reprinted with permission of Cold Spring Harbor Press from *Genome Research* 5:494–499, 1995.

by comparing the result of quantification by titration with the result obtained by single–tube assays using the same sample. Further, we compared the precision of single–tube quantification with that of titration. Comparison was made at both high and low initial signal abundance by using brain and nephron segment RNAs. Results were essentially identical in both systems (Figure 6), and revealed that under these circumstances titration is not required.

## *Discussion*

The present work demonstrates the accurate and precise measurement of gene expression. It confirms the utility of IP-RP-HPLC as a rapid and sensitive means of analyzing reaction products. It also reveals that the detecting and quantifying of heteroduplexes formed in competition reactions is important to the accuracy of the system. Using this system, we demonstrate that all theoretical requirements for competitive RT-PCR quantification are satisfied. In addition, accuracy and precision of measurements are excellent, and once acceptable performance parameters for a specific native/mutant template pair have been established, each time-consuming, expensive titration series may be replaced by a single competitive RT-PCR reaction.

Heteroduplex detection and subsequent quantification constitute a significant advantage of this quantitation system over others. We have demonstrated that heteroduplex formation occurs readily when competitive PCR reactions are allowed to proceed beyond the exponential phase (Figure 3). At the commencement of every cycle of competitive PCR, all products are denatured. Any molecules that subsequently fail to bind a primer and undergo extension will reanneal either to their complement (forming a homoduplex molecule) or to the partially homologous competitive mutant complement (resulting in a heteroduplex). A decline in the proportion of template molecules undergoing extension in consecutive cycles constitutes entry into the plateau phase of the reaction. There is no longer exponential increase in product concentration; thus, the presence and relative abundance of heteroduplexes provides a convenient indication of whether or not a particular reaction has proceeded beyond the exponential phase. Subsequent quantification of heteroduplex products by IP-RP-HPLC permits accurate quantification of reactions that have been allowed to proceed beyond the linear phase, resulting in greater product concentration and obviating the necessity for sample-dependent cycle number optimization.

The failure to detect heteroduplex molecules that have formed during competitive PCR may result in inaccurate estimation of nucleic acid concentration. This is of particular significance when assays utilize internal competitors constructed to differ from the native sequence only by the presence or absence of a restriction enzyme site. This is often achieved by the alteration of a single base. Under such circumstances, heteroduplex molecules will fail to be cleaved appropriately by the restriction enzyme, resulting in their erroneous inclusion, during subsequent analysis, with the nondigestible homoduplex products. One solution to this problem is to ensure that the reaction does not proceed beyond the exponential phase (Zhou and Hoffman, 1994; Zhou et al., 1994). This requires careful reaction optimization, with a subsequent possible decrease in assay sensitivity. One way of circumventing this problem may be the use of fluorescent primers, which offer increased sensitivity and a broader dynamic range for detection. Under ideal conditions, it may also be possible to estimate the initial ratio of competitor to native templates mathematically, if the amount of digestible products can be accurately measured (De Kant et al., 1994). However, in either of these circumstances an extra restriction enzyme digestion step is required, with a subsequent increase in cost and analysis time and the introduction of potential new sources of imprecision. Using IP-RP-HPLC analysis, postreaction handling is minimized and quantification beyond the linear range of PCR reactions is feasible and accurate.

## Competitive RT-PCR and IP-RP-HPLC for Analysis and Proportional Quantification of Regulated RNA Splicing

A second utility we have developed for IP-RP-HPLC involves the detection and proportional quantification of alternatively spliced mRNA molecules. RT-PCR is performed using total or poly (A+) RNA extracted from cells or tissues of interest and a pair of primers designed to flank a cDNA sequence believed to be alternatively spliced. If more than one splice variant is expressed, IP-RP-HPLC will reveal the specific homoduplex products and the heteroduplex products, which characteristically elute immediately prior to the smallest homoduplex product. The presence of heteroduplex products confirms that two specific PCR products amplified by the same primer pair share significant homology. RT-PCR and IP-RP-HPLC can then be used to accurately quantitate changes in the relative levels of the two messages between different samples and under conditions of physiological and pharmacological manipulation. This can provide a measure of regulated splicing efficiency. We applied this technology to the study of alternatively spliced uteroglobin promoter-binding proteins in rabbit endometrial tissue.

In the rabbit uterus, uteroglobin (UG) is a progesterone-induced preimplantation secretory protein, which binds progesterone and has immunosuppressive and anti-inflammatory properties (Miele et al., 1994; Miele et al., 1987). It has been a useful model system for studies of progesterone action, because its transcription is stimulated by progesterone (Chandra et al., 1981; Shen et al., 1983; Snead et al., 1981). UG is present in human endometrium (Peri et al., 1994), where like rabbit UG, the highest level of expression occurs in response to progesterone dominance and coincides with the events of embryo implantation. The recent localization of a high-affinity UG-binding protein on human trophoblast cells (Kundu et al., 1996) suggests that UG may play a pivotal role in regulating cellular invasiveness.

The UG gene contains two strong and two weak progesterone receptor binding sites that are located between positions −2.7 kb and −2.3 kb (Jantzen et al., 1987). Gel shift assays, Southwestern blots, and UV-crosslinking *in situ* identify four proteins that bind to the UG promoter (−194/+9). cDNAs for two of the UG promoter binding proteins (RUSH-1$\beta$ and RUSH-1$\alpha$, 95 kDa and 113 kDa, respectively) were cloned by recognition site screening (Hayward-Lester et al., 1996a). RUSH-1$\alpha$ and $\beta$ are products of the RUSH1 gene that result from alternative splicing of a 57-bp exon.

Northern analysis of uterine endometrial poly(A+)RNA revealed a single RUSH band of approximately 5.2 kb. Quantification of autoradiograms by computer-assisted image analysis indicated that progesterone increased the uterine content of message for the RUSH proteins in estrous

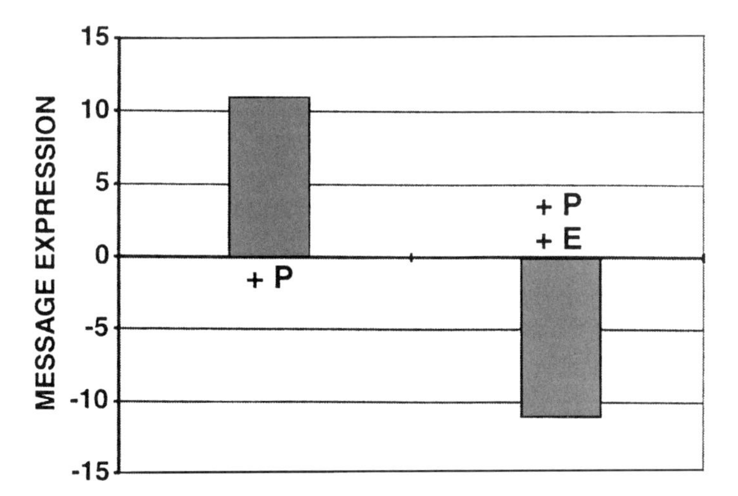


FIGURE 7. Summary of results from Northern blots. Poly($A^+$)RNA was isolated from uterine endometrium of estrous rabbits injected subcutaneously with hormones (*+P* = five injections of 3 mg progesterone/kg/24 hr; *+E* = five injections of 50 μg estradiol benzoate/animal/24 hr) or vehicle (ethanol: corn oil), and killed 24 hr after the last injection. Message is expressed as fold-increase or -decrease over estrous control animals, i.e., the zero base line.

rabbits (Figure 7). However, because the RUSH-1$\alpha$ and RUSH-1$\beta$ proteins result from alternative splicing, the small difference in their mRNA sizes could not be detected by Northern analysis.

In addition to hormonal regulation of net expression of RUSH mRNA, there may also be hormone-dependent differences in the relative levels of expression of the alternatively spliced RUSH mRNAs. To assess this possibility, primers were designed to nucleotide sequences on either side of the 57-bp insert. This results in the amplification of two templates (282 and 225 bp, respectively) which differ only by the presence or absence of the alternatively spliced insert. Amplification of these templates simulates a competitive RT-PCR reaction. Initial ratios of these homologous templates reflect the proportion of the alternatively spliced transcripts, and these ratios should be preserved during amplification. However, heteroduplex formation can occur between homologous amplicons and must be both detected and quantified to provide an accurate estimate of initial ratios.

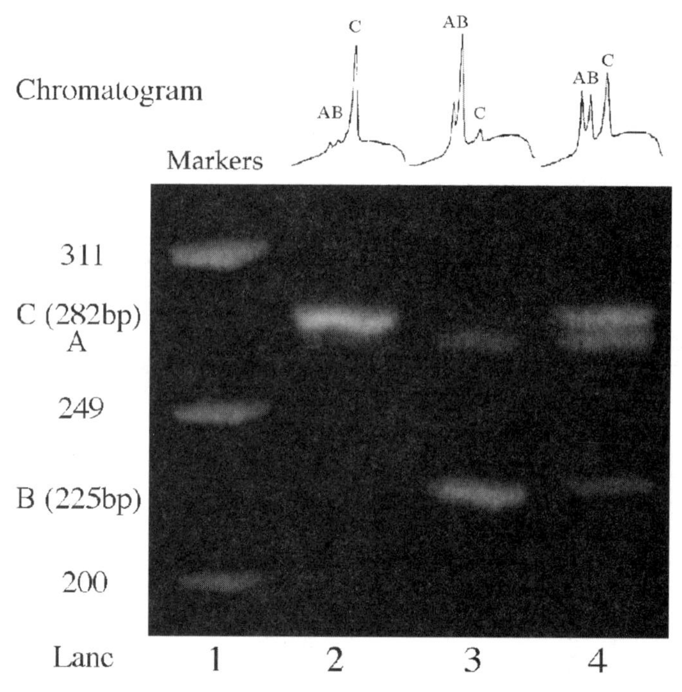


FIGURE 8. Analysis of competitive RT-PCR products by HPLC and electrophoresis. Product A = heteroduplexes formed between RUSH-1$\alpha$ and RUSH-1$\beta$ amplicons. Product B is from RUSH-1$\alpha$ amplification, and product C is from RUSH-1$\beta$ amplification. Lane 1 illustrates øX174/Hinf I molecular size markers, and lanes 2–4 illustrate competitive RT-PCR products with mRNA from animals treated as follows: estrous control (2), estrous + progesterone (3), and estrous + progesterone + estradiol benzoate (4).

Analysis of competitive RT-PCR reaction products by agarose gel electrophoresis and ion-pair reversed-phase HPLC is shown in Figure 8. Metaphor agarose gel electrophoresis demonstrated the two principal reaction products visualized by ethidium bromide fluorescence, the ratios of which changed according to the hormonal condition of the animals. A third reaction product was also detectable by gel electrophoresis. HPLC analysis revealed that it eluted immediately prior to the smaller homoduplex product, which identified it as a heteroduplex from the annealing of the two principal reaction products. A spreadsheet was used to mathematically reallocate the absorbance due to the heteroduplex (A) and to calculate the ratio (B, mRNA for RUSH-1$\alpha$; C, mRNA for RUSH-1$\beta$) of the two reaction products

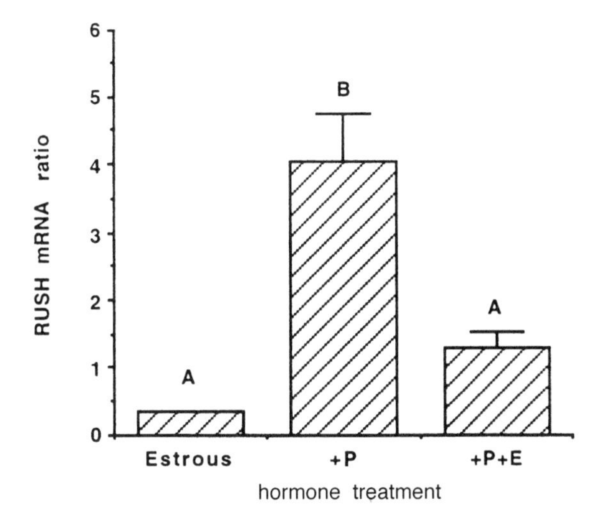


FIGURE 9. Message ratios of RUSH-1$\alpha$ to RUSH-1$\beta$ change in response to hormones. Ratios were calculated from peak areas generated by HPLC analysis of products from competitive RT-PCR with poly(A+)RNA. Values are expressed as mean ± SEM. Data were analyzed by a one-way analysis of variance and Duncan's multiple range test ($P < 0.05$). Values with the same letter designation are not significantly different.

(Figure 9), which is proportional to splicing efficiency. It is important to note that in the absence of an internal PCR control, one cannot compare the amounts of products between individual lanes.

The mRNA for RUSH-1$\beta$ is preferentially expressed in estrous animals. For these animals, the RUSH-1$\alpha$ isoform is a minor component of the transcripts generated from the competitive RT-PCR reaction. The RUSH–1$\alpha$ amplification product was detectable by HPLC, even though it was barely visible by ethidium bromide staining (Figure 8). When estrous rabbits were treated with progesterone, there was a dramatic change in the preference of splice site selection in favor of the RUSH-1$\alpha$ isoform (Figures 8 and 9). The increase ($p < 0.005$) in the RUSH-1$\alpha$ isoform correlates with the overall increase in accumulated message as determined by Northern analysis (Figure 7). However, a change ($p < 0.005$) occurred when estrous animals were treated sequentially with progesterone and estradiol benzoate. Northern data (Figure 7) revealed a decline in RUSH message, and HPLC analysis showed RUSH-1$\beta$ mRNA is once again the predominant isoform (Figures 8 and 9). Transcripts for both messages were detected in lung and liver control tissues, suggesting that RUSH protein isoforms may regulate genes in numerous cell types.

Splicing regulation of RUSH proteins may prove to be a critical adjunct to the regulation of the UG promoter. As a mechanism to regulate gene expression, it allows for changes in protein isoforms without changes in transcriptional activity (Smith et al., 1989). Pre-mRNA splicing occurs in a large macromolecular ribonucleoprotein particle, or spliceosome, whose assembly requires two stable intermediate stages. Formation of the so-called committed complex is the earliest detectable stable precursor to the spliceosome, and it is purported to be the key regulatory step. To date, *in vitro* splicing systems have provided most of the details of the biochemical reactions that are involved in alternative splicing. Mathematical models for splicing events indicate that the ratio of the spliced product to unspliced product is the most informative, sensitive indicator of splicing efficiency (McKeown, 1990, 1992). Therefore, the 8–12-fold progesterone-induced increase in the ratio of the RUSH-1$\alpha$ splice variant to RUSH-1$\beta$ parallels an equivalent change in the relative rate of splicing.

Regardless of changes in degradation rates, more complex splicing patterns, and issues of translatable versus nontranslatable mRNAs, the HPLC PCR product ratio calculation provides an internally controlled method of confirming that progesterone plays a regulatory role in posttranscriptional processing of RUSH message isoforms.

## Heteroduplex Detection in Differential Genome Screening by Denaturing HPLC

IP-RP-HPLC has been shown to allow the successful detection of mismatches in DNA hybrids shorter than approximately 50 base pairs, with the mismatch resulting in the complete denaturation of the hybrid into its two single-stranded components at elevated column temperatures. Perfectly matched hybrids, in contrast, elute as single peaks under identical conditions (Oefner et al., 1994b). Single-base mismatches may still be detected successfully in DNA fragments as large as 1.5 kb by means of partial heat denaturation of heteroduplex molecules (Oefner and Underhill, 1995). In contrast to the application of IP-RP-HPLC to quantitative RT-PCR reactions involving competitors that differ by several base pairs in length from the native template, column temperatures higher than 50°C are required to resolve the heteroduplexes from the homoduplexes, with the latter usually remaining unresolved. The higher temperature is required to increase the number of unpaired bases around the mismatch. The ensuing decrease in paired bases results in the early elution of the heteroduplex species.

In order to ensure that homo- and heteroduplex molecules are present in equimolar ratios, PCR products from diploid templates are heated prior to denaturing HPLC (DHPLC) analysis to 95°C for 3min before cooling the sample gradually from 95 to 65°C over a period of 30min. This will ensure the presence of equal amounts of homo- and heteroduplex species in the sample independently of the extent to which PCR has progressed into the plateau phase.

Key to the detection of such small differences in base composition is the rapid heating of the injected DNA to the temperature at which the separation is carried out, because the stationary phase seems to preserve the conformation in which the DNA molecules get onto the column. When the sample was injected into mobile phase at ambient temperature with the column connected directly to the sample injection port, no heteroduplex signal could be observed (Figure 10). By placing additional tubing in the column oven between the sample loop and the column, a heteroduplex signal of increasing resolution was observed. This indicated that partial denaturation of the sample has to occur before the DNA reaches the stationary phase. Subsequently, it was observed that the tubing between injection port and column could be kept very short, provided that the sample was injected into a preheated mobile phase. This configuration

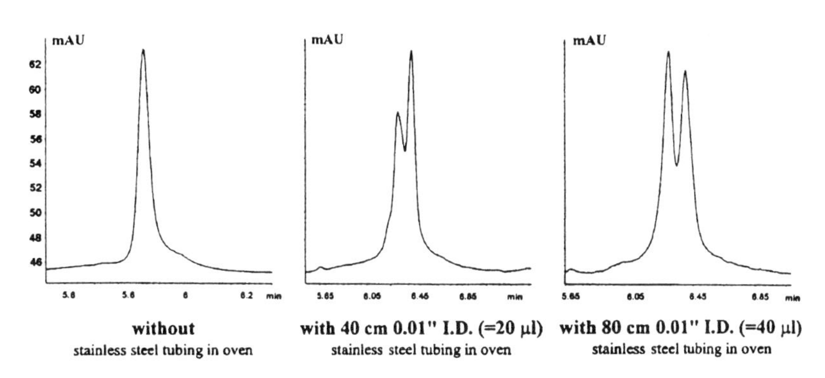


FIGURE 10. Effect of preconditioning on the detection of a single-base mismatch in a 1-kb PCR product. Injected into a mobile phase of ambient temperature, hetero-duplexes could not be resolved successfully from the homoduplexes even when the column was kept at an appropriate temperature. Improvements in resolution were obtained, however, when additional tubing was placed inside the column oven between the sample loop and the column. Column: DNA Sep™, 50×4.6mm I.D., Sarasep Inc. (Santa Clara, CA). Eluant A: 0.1 M TEAA, pH 7.0. Eluant B: 0.1 M TEAA, pH 7.0, 25% acetonitrile. Gradient: 40–66% B in 4min, 66–70% B in 3min. Flow rate: 1 ml/min. Temperature: 56°C. Detection: UV (254nm).

keeps sample dispersion to a minimum. In the automated system currently in use, preheating of the mobile phase is attained in an 80-cm loop of polyetheretherketone (PEEK) tubing encased into a tin alloy block. Since the tin alloy possesses high thermoconductivity, conditioning of the mobile phase is accomplished rapidly whenever the column oven temperature is changed. To minimize the time required for adjusting column temperature, and to ensure excellent temperature uniformity, a forced air circulation system with Peltier cooling is used. This allows the temperature of the mobile phase to be adjusted within 5 min compared to the 30–45 min usually required for a conventional heating block. Such a system, when automated, increases throughput to ~200 injections per day.

The optimum temperature resulting in a bubble of unpaired bases of sufficient size to permit the resolution of homo- and heteroduplexes is determined by the GC content of the fragment and especially of the sequences flanking the actual mismatch. Optimum temperature may range from 51 to 65°C for very AT- and GC-rich sequences, respectively. More importantly, mismatches can usually be detected over a temperature range of at least 2–3°C, before melting results in the complete disintegration of the signal. In practice, the easiest way to determine the optimum temperature at which to screen PCR products of a given sequence for the presence or absence of mismatches is to increase the column temperature gradually, until the PCR product peak begins to shift significantly towards shorter retention times. At this point, the presence of a mismatch will be clearly detected by the appearance of one or two additional peaks eluting immediately before the homoduplex signal (Figure 11). Whether one or two heteroduplex peaks are observed depends on the base composition of the unpaired bases, because single-stranded DNA, in contrast to double-stranded DNA, is retained in IP-RP-HPLC, mainly as a function of its base composition (Oefner et al., 1994b). It has been shown, for instance, that homotetrameric deoxynucleotides elute $G < C < A < T$, with triethylammonium acetate as the ion-pairing reagent, whereas the exchange of the 3′ base of a 22mer oligonucleotide results in an elution order of $C < G < A < T$ (Huber et al., 1996). Depending on the nature of the mismatch, one or two heteroduplex peaks may be observed. In the case of mismatches located within low-melting domains, the onset of partial denaturation even in the homoduplexes may result in the appearance of four peaks, representing the two homo- and heteroduplex species that are present in the sample, respectively.

Among the other factors that might influence the ability of the system to resolve hetero- from homoduplex molecules, increasing concentrations of triethylammonium acetate were found only to increase the temperature required to partially denature the heteroduplex molecules. In accordance

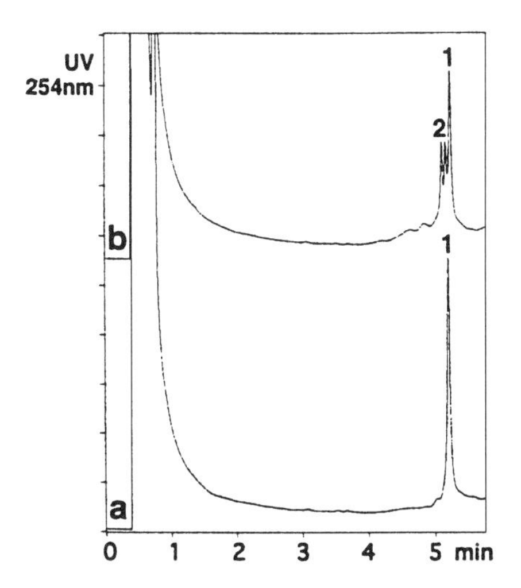


FIGURE 11. Representative chromatograms of homo- (peak 1) and heteroduplexes (peaks 2) created upon reannealing 241-bp PCR products of a male reference with an Asian (a) and an American Indian (b) individual. Heteroduplexes reflect a single-base substitution. Column: DNA Sep™, 50 × 4.6 mm I.D. (Sarasep Inc., Santa Clara, CA). Eluant A: 0.1 M TEAA, pH 7.0. Eluant B: 0.1 M TEAA, pH 7.0, 25% acetonitrile. Gradient: 37–63% B in 5.5 min. Flow rate: 1 ml/min. Temperature: 55°C. Detection: UV (254 nm). Reprinted with permission from *Proc. Natl. Acad. Sci. USA* 93:196–200, 1996.

with a previous study on the effect of pH on the temperature of transition from the native to the denatured state (Privalov and Ptitsyn, 1969), no improvement in resolution was detected at a given temperature over a pH range of 6–9. Theoretically, more acidic pH values will allow the detection of heteroduplex molecules at lower temperatures. Practically, this is not an option because the decreasing dissociation of the phosphate residues of the DNA backbone with decreasing pH abolishes the basis of separation. Higher pH values are feasible, but the absorption of carbon dioxide makes them difficult to maintain. Furthermore, higher pH values would not preferentially affect the transition temperature of GC-rich domains. The latter may pose a problem in the detection of mismatches, when the flanking sequences exhibit significantly lower melting temperatures, because they will denature more readily than the GC-rich domain harboring the mismatch. In such cases, it has proven beneficial to substitute dGTP in the PCR with 7-*deaza*-2′-dGTP, which is known to effectively lower the melting temperature of GC-base pairs (Dierick et al., 1993), hence makingit possible to observe mismatches even within extremely high-melting domains.

Using an HPLC instrument as described, heteroduplexes representing all possible combinations of mismatches as well as 1- to 4-bp insertions or deletions (which had been created using site-directed mutagenesis) were readily detected by on-line UV absorbance monitoring. On the basis of the analysis of 150 known mutations, the sensitivity of DHPLC in detecting single-base changes is estimated to be ~95%. Sensitivity might have been further improved by amplifying the undetected mutations with 7-*deaza*-2′-dGTP instead of dGTP (see above).

One of the first successful applications of this new technology to differential genome screening has been the identification of single-base substitution polymorphisms on the nonrecombining portion of the Y chromosome. Such polymorphisms are highly useful in studies of human evolution, because they represent presumably unique mutational events in human history, hence preserving haplotype information crucial in tracking human origins and migrations. However, base substitution polymorphisms on the Y chromosome are rare. One study reported three single-base substitutions from a survey of 18.3 kb in five individuals (Whitfield et al., 1995), while we have observed only two polymorphic sites in 4235 bp of Y-specific sequence in 20 human males from around the world (Underhill et al., 1996). The scarcity of such polymorphisms renders their detection by conventional slab–gel–electrophoresis-based sequencing extremely laborious and expensive. In such circumstances, DHPLC proves to be most beneficial, because it allows one to exclude rapidly all monomorphic chromosomes. Conventional DNA sequencing can thus be limited to those rare cases in which a heteroduplex signal is observed. Such an instance is shown in Figure 11, which represents a C to T transition, which has been found exclusively in Native Americans, with the frequency of the T allele ranging from >90% in Central and South American populations to ~50% in indigenous North American Navajo and Eskimo populations (Underhill et al., 1996).

Overall, DHPLC offers several significant advantages over other existing techniques for the detection of unknown mutations:

1) The column matrix lasts for thousands of runs, provided that a biocompatible chromatographic system, preferably made out of titanium, is used. The column eliminates the need for preparing electrophoretic gels, and with the cost per analysis being as low as 25 cents, based on an average column lifetime of 5000 injections, tremendous cost savings are achieved.

2) On-line UV detection eliminates gel staining or the use of fluorescent/radioactive labels. The sensitivity of UV detection equals that of ethidium–bromide–stained gels.

3) Using commercially available equipment, the search for mutations and polymorphisms can be completely automated, eliminating such time- and labor-consuming tasks as the loading of gels. Further, PCR products can be loaded directly, eliminating purification of PCR products as is required for various enzymatic methods for mutation detection and capillary–electrophoresis–based techniques.

4) Given the high separation efficiency and the short time required for regenerating the column, up to 200 samples can be analyzed per day on a single instrument.

5) DHPLC detects with equal efficiency single-base changes as well as short deletions and insertions, which often cannot be identified by commonly-used methods for detecting point mutations, such as enzymatic assays or single-strand conformation polymorphism analysis.

## Conclusion

High sensitivity and excellent size-separation properties, combined with remarkable heteroduplex resolution capability make IP-RP-HPLC a powerful and versatile tool for nucleic acid quantitation. It can be utilized to identify and quantify genetic differences at a number of levels, any of which may prove important in the study of disease. First, we have demonstrated the importance of IP-RP-HPLC in the development of a system for precise and accurate gene quantitation in microscopic tissue samples. This provides the ability to study both transcription and viral disease progression in a quantitative manner. Second, we have described its utility in the detection and quantitation of RNA splice-variants, providing a quantitative measure of posttranscriptional regulation. Finally, we have demonstrated the exciting ability of DHPLC to detect point mutations. This constitutes a fast and sensitive method for identifying genetic polymorphisms and provides the means for quantifying the accumulation of genetic change in disease and by evolution.

*Acknowledgments*

This work was supported in part by National Institutes of Health grants DDK45538 (PAD), IPO1-HG00205 (PJO), and HD29457 (BSC). We are grateful for the cooperation of our colleagues Drs. Sandra Sabatini, Aveline Hewetson, and Eduard Grill and for technical assistance provided by Betty Lonis. We wish to thank Dr. Jerry Lingrel (University of Cincinnati, OH) and Dr. Kevin Lynch (University of Virginia, Charlottesville) for supplying cDNAs used in these studies. For frequent updates on HPLC of nucleic acids, readers are referred to our Web site at URL: http://lotka.stanford.edu/dhplc.

## REFERENCES

Berry CA (1982): Heterogeneity of tubular transport processes in the nephron. *Ann Rev Physiol* 44:181–201.

Chandra T, Bullock DW, et al. (1981): Hormonally regulated mammalian gene expression: steady-state level and nucleotide sequence of rabbit uteroglobin mRNA. *DNA* 1:1, 19–26.

Clapp WL, Bowman P, et al. (1994): Segmental localization of mRNAs encoding Na, K-ATPase alpha and beta-subunit isoforms in the rat kidney using RT-PCR, *Kidney Int* 46:627–38.

De Kant E, Rochlitz C, et al. (1994): Gene expression analysis by a competitive and differential PCR with antisense competitors. *Bio/Techniques* 17:5, 934–42.

Dewar R, Highbarger H, et al. (1994): Application of branched DNA signal amplification to monitor human immunodeficiency virus type 1 burden in human plasma. *J Infect Dis* 170:1172–9.

Dierick H, Stul M, et al. (1993): Incorporation of dITP or 7-deaza dGTP during PCR improves sequencing of the product. *Nucleic Acids Res* 21:18, 4427–8.

Doucet A and Barlet C (1986): Evidence for differences in the sensitivity to ouabain of Na,K-ATPase along the nephrons of rabbit kidney. *J Biol Chem* 261:3, 993–5.

Ferré F (1992): Quantitative or semi-quantitative PCR: reality versus myth. *PCR Meth Appl* 2:1, 1–9.

Gilliland G, Perrin S, et al. (1990): Analysis of cytokine mRNA and DNA: detection and quantitation by competitive polymerase chain reaction. *Proc Natl Acad Sci USA* 87:7, 2725–9.

Hayward-Lester A, Hewetson A, et al. (1996a): Cloning, characterization and steroid-dependent posttranscriptional processing of RUSH-1$\alpha$ & $\beta$, two uteroglobin promoter binding proteins. *Mol Endocrinol* 10:1335–49.

Hayward-Lester A, Oefner PJ, et al. (1996b): Rapid quantification of gene expression by competitive RT-PCR and ion-air reversed-phase HPLC. *Bio/Techniques* 20:250–7.

Hayward-Lester A, Oefner PJ, et al. (1995): Accurate and absolute quantitative measurement of gene expression by single tube RT-PCR and HPLC. *Genome Res* 5:494–9.

Huber CG, Oefner PJ, et al. (1993): High-resolution liquid chromatography of oligonucleotides on nonporous alkylated styrene-divinylbenzene copolymers. *Anal Biochem* 212:2, 351–8.

Huber CG, Oefner PJ, et al. (1995): Rapid and accurate sizing of DNA fragments on alkylated nonporous poly(styrene-divinylbenzene) particles. *Anal Chem* 67:3, 578–85.

Huber CG, Stimpfl E, et al. (1996): A comparison of micropellicular anion-exchange and reversed-phase stationary phases for HPLC analysis of oligonucleotides. *LC-GC* 14:2, 114–17.

Jantzen K, Fritton HP, et al. (1987): Partial overlapping of binding sequences for steroid hormone receptors and DNaseI hypersensitive sites in the rabbit uteroglobin gene region. *Nucleic Acids Res* 15:11, 4535–52.

Jewell EA and Lingrel JB (1991): Comparison of the substrate dependence properties of the rat Na, K-ATPase 1, 2, and 3 isoforms expressed in

HeLa cells. *J Biol Chem* 266:25, 16925–30.

Kalghatgi K and Horvath C (1987): Rapid analysis of proteins and peptides by reversed-phase chromatography. *J Chromatogr* 398:335–9.

Kato Y, Kitamura T, et al. (1987): High-performance ion-exchange chromatography of proteins on non-porous ion-exchangers. *J Chromatogr* 398:327–34.

Kato Y, Yamaskai Y, et al. (1989): Separation of DNA restriction fragments by high-performance ion-exchange chromatography on a non-porous ion exchanger. *J Chromatogr* 478:264–8.

Kundu GC, Mantile G, et al. (1996): Recombinant human uteroglobin suppresses cellular invasiveness via a novel class of high affinity cell surface binding site. *Proc Natl Acad Sci USA* 93:2915–19.

McDonough AA, Magyar CE, et al. (1994): Expression of Na+-K+–ATPase alpha- and beta-subunits along rat nephron: Isoform specificity and response to hypokalemia. *Am J Physiol-Cell Physiol* 36:4, C901–8.

McKeown M (1990): Regulation of alternative splicing. *Genet Engin* 12:139–81.

McKeown M (1992): Alternative mRNA splicing. *Ann Rev Cell Biol* 8:133–55.

Miele L, Cordella-Miele E, et al. (1994): Uteroglobin and uteroglobin-like proteins: the uteroglobin family of proteins. *J Endocrinol Invest* 17:8, 679–92.

Miele L, Cordella-Miele E, et al. (1987): Uteroglobin: structure, molecular biology, and new perspectives on its function as a phospholipase A2 inhibitor. *Endocr Rev* 8:4, 474–90.

Nathakarnkitkool S, Oefner PJ, et al. (1992): High-resolution capillary electrophoretic analysis of DNA in free solution. *Electrphoresis* 13:1–2, 18–31.

Oefner PJ, Huber CG, et al. (1994a): High-performance liquid chromatography for routine analysis of hepatitis C virus cDNA/PCR products. *Bio/Techniques* 16:898–908.

Oenfer PJ, Huber CG, et al. (1994b): High resolution liquid chromatography of fluorescent dye labeled nucleic acids. *Anal Biochem* 223:39–46.

Oefner PJ and Underhill PA (1995): Comparative sequencing by denaturing high-performance liquid chromatography (DHPLC). *Am J Hum Genet* 57(Suppl.):1547.

Orlowski J and Lingrel JB (1988): Tissue-specific and developmental regulation of rat Na,K-ATPase catalytic alpha isoform and beta subunit mRNAs. *J Biol Chem* 263:21, 10436–42.

Peri A, Cowan BD, et al. (1994): Expression of Clara cell 10-kD gene in the human endometrium and its relationship to ovarian menstrual cycle. *DNA Cell Biol* 13:5, 495–503.

Privalov PL and Ptitsyn OB (1969): Determination of stability of the DNA double helix in an aqueous medium. *Biopolymers* 8:559–71.

Raeymaekers L (1993): Quantitative PCR: theoretical considerations with practical implications. *Anal Biochem* 214:2, 582–5.

Schafer JA (1994): Salt and water homeostasis — Is it just a matter of good bookkeeping? *J Am Soc Nephrol* 4:12, 1933–50.

Schneider JW, Mercer RW, et al. (1988): Tissue specificity, localization in brain, and cell-free translation of mRNA encoding the A3 isoform of Na+, K+-ATPase. *Proc Natl Acad Sci USA* 85:284–8.

Shamraj OI and Lingrel JB (1994): A putative fourth Na+, K+-ATPase alpha–subunit gene is expressed in testis. *Proc Natl Acad Sci USA* 91:26, 12952–6.

Shen XZ, Tsai MJ, et al. (1983): Hormonal regulation of rabbit uteroglobin gene transcription. *Endocrinology* 112:3, 871–6.

Siebert PD and Larrick JW (1992): Competitive PCR. *Nature* 359:557–8.

Smith CW, Patton JG, et al. (1989): Alternative splicing in the control of gene expression. *Ann Rev Genet* 23:527–77.

Snead R, Day L, et al. (1981): Mosaic structure and mRNA precursors of uteroglobin, a hormone-regulated mammalian gene. *J Biol Chem* 256:22, 11911–16.

Sweadner K (1989): Isozymes of Na/K-ATPase. *Biochim Biophys Acta* 988:185–220.

Underhill PA, Jin L, et al. (1996): A pre-Columbian Y chromosome-specific transition and its implications for human evolutionary history. *Proc Natl Acad Sci USA* 93:196–200.

Unger KK, Jilge G, et al. (1986): Evaluation of advanced silica packings for the separation of biopolymers by high-performance liquid chromatography. *J Chromatogr* 359:61–72.

van Gemen B, van Beuningen R, et al. (1994): A one-tube quantitative HIV-1 RNA NASBA nucleic acid amplification assay using electrochemiluminescent (ECL) labelled probes. *J Virol Meth* 49:157–68.

Whitfield LS, Sulston JE, et al. (1995): Sequence variation of the human Y chromosome. *Nature* 378:6555, 379–80.

Zachar V, Thomas RA, et al. (1993): Absolute quantification of target DNA: a simple competitive PCR for efficient analysis of multiple samples. *Nucleic Acids Res* 21:8, 2017–18.

Zhou J and Hoffman EP (1994): Pathophysiology of sodium channelopathies: studies of sodium channel expression by quantitative multiplex fluorescence polymerase chain reaction. *J Biol Chem* 269:18563–71.

Zhou J Spier SJ, et al. (1994): Pathophysiology of sodium channelopathies: correlation of normal/mutant mRNA ratios with clinical phenotype in dominantly inherited periodic paralysis. *Hum Mol Genet* 3:1599–603.

# Capillary Electrophoresis for Quantitative Genetic Analysis

Jill M. Kolesar and John G. Kuhn

## Introduction

Electrophoresis was introduced by Tiselius in 1937. He received the Nobel Prize for his work that demonstrated that protein solutions migrate in a direction and at a rate determined by their charge and mobility when an electrical current is applied. Slab gel electrophoresis (SGE) is an adaptation of this technique and has been used for decades as the primary tool for purification and analysis of nucleic acids and proteins. Gels are typically made of cross–linked polyacrylamide or agarose, and the separations are carried out in low electric fields. While many slab gel techniques are well established and effective in the analysis of nucleic acids, accurate quantitation and ease of automation remain obstacles.

Nucleic acids, including PCR products, are frequently quantified by isotopic detection in conjunction with slab gel electrophoresis. This technique is able to detect 1–5 pcg of target DNA with an overnight exposure (Brandsma et al., 1980). However, radioactive detection methods have significant disadvantages with respect to safety, stability of the labeled nucleic acids, and automation.

High pressure liquid chromatography (HPLC) was initially studied as a replacement for slab gel electrophoresis (Haupt et al., 1983), although restricted intraparticle diffusion of biopolymers resulted in only limited improvement of resolution and speed.

Alternately, electrophoretic separations can be performed in narrow–bore tubes or capillaries, and capillary electrophoresis (CE) may be viewed simply as the instrumentation of electrophoresis. Originally, capillary electrophoresis utilizing hydroxyethylcellulose and ethidium bromide was shown to have increased resolving power when compared to HPLC for dsDNA; however, the detection level remained in the nanogram range (Oefner et al., 1992). Capillary electrophoresis has since been used successfully to separate and quantitate restriction digest fragments of the erb2 oncogene (Ulfelder et al., 1992) and PCR products of HIV-1 (Piatek et al., 1993), (Schwartz and Ulfelder, 1991). The introduction of capillary electrophoresis with laser–induced fluorescence (CE-LIF) (Srinivasen et al., 1993a), fluorescence detectors, and intercalating dyes (Srinivasen et al., 1993b) improves nonisotopic detectability to the attomole (Schwartz and Ulfelder, 1992) level with sample volumes as small as a few picoliters (Lu et al., 1994). Capillary electrophoresis represents a safe and automatable assay system for quantitative analysis.

## Basic Principles

Separation by electrophoresis is based on differences in solute velocity in an electric field. The velocity ($v$) at which an ion travels is a function of ion charge ($q$), solution viscosity ($h$), the ion radius ($r$), and the applied electrical field ($E$) and is represented by the equation:

$$v = \frac{q}{6phr} \times E$$

Increases in charge and electrical force will increase the velocity of a molecule while increasing molecule size, and solution viscosity will slow the molecule velocity (Karger et al., 1995).

Electroosmotic flow (EOF) is the bulk flow of fluid through the capillary and is controlled by the surface charge on the inside of the capillary. The EOF determines the amount of time the solute remains in the capillary (Tsuda, 1987). The solid surface of the capillary usually contains an excess of negative charge to which counterions will be attracted to balance the charge, forming a double layer and creating what is known as the zeta potential. When an electrical charge is applied to the capillary, the cations from the double layer are attracted to the cathode, and since they are in solution their movement drags the bulk solution in the capillary to the cathode as well. When samples migrate across the capillary by EOF, the driving force is uniformly distributed across the capillary and the flow is uniform (Hjerten, 1985). This is beneficial since it will not contribute to dispersion of solute zones, and peaks will be sharper. Since the pH of a solution is a measure of the ions present, the pH and ionic strength of a solution can alter the EOF. Interventions that alter solution viscosity and integrity of the capillary wall can also alter EOF. Some proteins interact with capillary walls, and coating is an option to decrease this interaction

(Schmalzing et al., 1993). Since DNA is negatively charged and migrates toward the positive pole, capillaries are coated to reverse the EOF, making a positively charged inner wall, which attracts negatively charged species in solution. When the electrical field is applied, negatively charged DNA migrates to the positive pole.

The time required for the sample to migrate from the point of injection to the point of detection is called the migration time (Figure 1). The effective length of the capillary is the length from the injection site to the exit reservoir, and it defines the electric field.

Electrophoretic separation is based on the difference in mobility between the two zones, and the difference required to resolve two zones is dependent on the length of the zones. Dispersion, or the spreading of the solute zone, is the primary determinant of zone length and should be minimized to decrease the zone length and enhance the efficiency of separation. Dispersion is determined mostly by longitudinal diffusion along the capillary, although generation of Joule heat, injection length, and sample adsorption can also contribute, and attempts should be made to minimize these factors (Kuhr and Monnig, 1992).

Resolution of the sample components is the primary goal of separation science. In the case of capillary electrophoresis, the separation is driven mainly by efficiency, or the ability of the particles to migrate through the

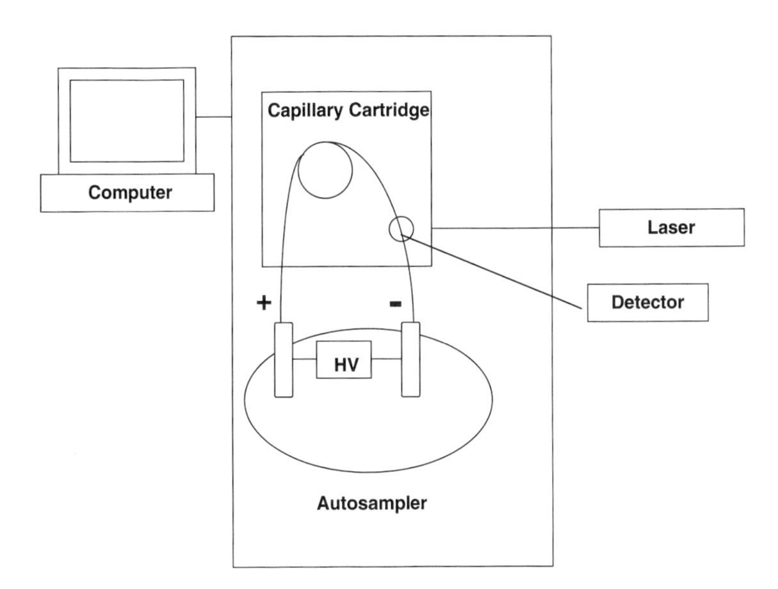


FIGURE 1. Schematic of CE-LIF instrumentation.

matrix, which is in contrast to chromatography where separation is usually driven by selectivity.

*Modes*. High performance capillary electrophoresis (HPCE) is a highly versatile technique with numerous modes of operation, each with different separation mechanisms. Capillary zone electrophoresis (CZE) is the most widely used technique and separates on the basis of charge to mass ratios. CZE is an effective means of separation for small ions, pharmaceuticals, and macromolecules, including proteins and peptides. Both drugs and proteins may be separated in the same mode, making this is a major advantage of CZE. Proteins and peptides may also be separated on the basis of their isoelectric point by capillary isoelectric focusing. Both charged and neutral small molecules and peptides may be separated by micellar electrokinetic capillary electrophoresis based on the molecules' ability to partition into micelles according to hydrophobicity (Landers, 1995).

Capillary gel electrophoresis (CGE) uses either a fixed or immobilized polymerized matrix that acts as a "molecular sieve" within the capillary to separate DNA on the basis of size and charge. Since the mass to charge ratio of DNA remains constant with increasing mass, separations are made on the basis of differences in molecular weight. As charged solutes migrate through the polymer network, they are hindered. Larger molecules are hindered more than smaller ones, allowing for separation based on molecular weight. Initially, cross–linked agarose and polyacrylamide were used for the separation of DNA by capillary electrophoresis. However, polymerization of gels within the capillary is difficult and time–consuming. Polymerization that occurs too rapidly, use of impure chemicals, and solutions that are not degassed can lead to bubble formation and unstable gels. Additionally, these capillaries are very rigid, making hydrodynamic injections impossible and the capillary susceptible to breakage (Karger et al., 1995). Linear polymer solutions are more flexible, and pressure can be used to refill the capillary. Additionally, they are much less susceptible to bubble formation and breakage. Although the polymer structure of the cross–linked gel is much different from that of the linear polymer solution, the mechanism of separation is identical, and the ease of use of the replaceable polymer networks has made these the capillary system of choice for most laboratories (Heller, 1995). CGE is used most frequently for the analysis of nucleic acids and will be the focus of this discussion.

## Instrumentation

The typical CE analysis involves a series of steps, including injection of the sample, application of the separation voltage across the capillary, and migration of the sample through the capillary past the optical window where detection takes place, followed by automated data analysis.

*Injection*. In most modes of capillary electrophoresis, the injection site is the anode or positive pole. The exception is CGE, where the injection site is the cathode. The reversal of polarity is required because DNA is negatively charged and migrates from negative to positive poles.

Since the capillaries are of very narrow bore, the volume injected is correspondingly small. Sample overloading, which broadens sample zones, broadens peak shape, and decreases the efficiency of the separation, should be avoided. The injection plug length is the critical parameter and should be less than 1–2% of the total capillary length. This corresponds to a volume of approximately 1–50 nl (Rossomondo et al., 1994). The actual volume will usually be unknown, but can be calculated, and the quantifiable parameters will be pressure/time for a volume injection or voltage/time for an electrokinetic injection. In general, a 1 s pressure injection will correspond to a 1 nl volume. The small sample volume required is an advantage when a limited volume of sample is available. Samples are injected on the capillary either electrokinetically or more frequently hydrodynamically. Pressure (typically 0.5 psi for 1–10 sec) is applied at the injection end of the capillary. Volume of sample loaded will be a function of the capillary dimensions, the buffer viscosity, the pressure applied, and the time. Injection reproducibility can be better than 1–2% RSD (relative standard deviation) for hydrodynamic injection. Electrokinetic injections are generally made at a field strength 3–5 times lower than that used for separation, and the solute enters the capillary by both migration and EOF. Sample loading is dependent on EOF, sample concentration, and sample mobility as well as on the factors that influence hydrodynamic injection. Electrokinetic injections are not as reproducible as hydrodynamic injections for quantitative analysis (Huang et al., 1988; Kunkel et al., 1996).

*Separation*. Electrophoretic separation is carried out within the capillary. Capillaries are usually constructed of fused silica because it is both chemically and electrically inert, UV-visible transparent, flexible, robust, and (relatively) inexpensive. Capillaries are coated with a layer of polyimide to enhance strength and flexibility. An optical window can be placed within the capillary for detection by removal of the protective coating by either scraping with a razor blade or burning off a few millimeters with an electrical arc. Ready–to–use capillaries are also commercially available from a variety of sources. Capillaries have very narrow bores, with inner diameters (ID) generally 25–75 mm, and with a range of 10–200 mm ID available. Capillary length ranges from 10–100 cm, with effective lengths of 50–75 cm used most frequently. Shorter capillaries require less time for analysis, conditioning, and fraction collection; however, longer capillaries are sometimes required for very complex separations (Schmalzing et al., 1993).

Factors that are important for reproducibility include capillary conditioning and thermostatting the power supply. The capillary must have a constant surface charge to provide reproducible and quantifiable results. Proper conditioning of the capillary is required to maintain the charge. When the sample to be analyzed interacts with the capillary wall, the capillary must often be rinsed with a base, strong acid, organic, or detergent. These washing techniques can also interact with the capillary wall and affect the surface charge. The most reproducible type of conditioning is with buffer alone, and this should be employed whenever possible (Karger et al., 1995). DNA does not adhere to capillaries, and buffer conditioning should always be used.

The large electric fields used in capillary electrophoresis generate Joule heat. Because of the properties of the capillary, high surface area to inner diameter ratio, the capillary can dissipate much of the generated heat. However, temperature changes as small as 1°C can result in a 2–3% change in viscosity, resulting in variability in both the sample injection and migration time (Pariat et al., 1993). Temperature control is a feature available in commercially available units and is usually accomplished by bathing the capillary in either a high–velocity airstream or liquid. Efficiency of either system is similar at up to 5 W/m, while at higher powers, the liquid–cooled system is more effective (Landers, 1995).

Since capillary electrophoresis is carried out at very high voltages (up to 30 keV), it is important that the power source provide stable regulation of voltage (±0.1%) to maintain reproducibility. It is often advisable to ramp the voltage at the beginning of an analytical run to avoid rapid heating and generation of Joule heat, which decreases the reproducibility of the reaction.

*Detection.* The sample migrates through the capillary towards the positively charged anode through the detector. Detection of DNA can be either by UV (with detection limits of $10^{-5}$ or $10^{-6}$ M) or on-column-laser-induced fluorescence (LIF) detection. LIF increases the sensitivity of the system to the atto- or zeptomole range, but requires the addition of fluorescent labels, either directly to the DNA by labeled primers or by inclusion in the buffer. In addition to greatly enhanced sensitivity, fluorescence detection is very specific. Most commonly used fluorescent labels can be excited by an argon laser at 488 nm. Double-stranded DNA can also be complexed with intercalating dyes such as ethidium bromide or thiazole orange (Schwarz et al., 1994).

*Data analysis.* Software programs are available that control the hardware system, making the capillary electrophoresis process completely automatic. Additionally, the software systems collect, integrate, analyze, and report chromatographic sample data.

## Quantitation

To truly quantitate gene expression, the measured concentration must be a reliable gauge of the amount of RNA present in the original sample (Ferré et al., 1994). This determination is usually a two-part process: amplification followed by detection. RNA samples are usually amplified by RT-PCR. An internal standard that amplifies under the same conditions as the target sequence is introduced prior to amplification. The internal standard controls for variability present within the PCR reaction as well as provides a ratio to calculate the initial concentration of the unknown. Detection of PCR products is often done by SGE with autoradiography (Kuypers et al., 1994) or silver–stained polyacrylamide gels (Dieguez-Lucena et al., 1994).

To ensure that peak areas reported on the electropherogram by CE-LIF are a dependable measure of the amount of RT-PCR product present in the sample, the instrument must be calibrated. This is usually done by injecting DNA of known concentration and generating a standard curve. The same solution of standard DNA fragments can be used to calibrate the capillary and determine the molecular weight of fragments based on retention times.

In addition to being an additional step towards the accurate quantitation of gene expression, this technique offers several advantages. Required sample volumes are small (1–5 $\mu$l), sensitivity is high (attomolar), and the hazards associated with isotopic storage, use, and disposal are eliminated (Schwarz and Ulfelder, 1992). This technique is also readily automatable, can be used for fraction collection, and can be validated. This methodology may also represent an advantage over HPLC methodologies, with increased sensitivity and no requirement for sample clean-up.

There are drawbacks to CE-LIF, including expense. An initial equipment investment of the CE hardware, laser, computer, and applications software must be made. Consumable supplies, primarily intercalating dyes and capillaries, are also expensive and fragile.

With existing CE-LIF technology, samples are analyzed individually. Each sample has a 15–45 min run time, while multiple samples can be analyzed simultaneously by SGE. For analysis of a single sample, CE-LIF may be faster, but when multiple analyses are required, SGE may be more time–efficient. The technology termed parallel processing, which allows analysis of multiple samples simultaneously by CE-LIF, is under development.

In addition to requiring a longer time, individual sample processing also means that each sample is analyzed by a separate process, with an increased risk for bias. Of the steps involved in analysis by CE-LIF, injection bias is the most frequent source of error, although this can be minimized by using hydrodynamic injections (Kleparnik et al., 1995). Interassay variation can also be minimized by the addition of a reference standard of known concentration and migration into each of the samples after PCR amplification and before CE-LIF analysis. DNA fragments of similar, but not identical, molecular weight to the unknown sample are suitable for use as reference peaks.

## Applications

The quantitative analysis of RNA is central to the study of gene expression. Northern hybridizations and dot blots are frequently utilized for the analysis of RNA. These techniques generally require relatively large quantities of template RNA, are technically difficult, are only amenable to relative quantification, and are not automatable. RT-PCR with a reaction–specific internal standard linked to CE-LIF has the advantages of being able to detect very low abundance mRNAs, being readily automatable, and producing quantifiable results. This technique has been used successfully to quantitate basic fibroblast growth factor gene expression (Gelfi et al., 1995), NQO1 gene expression (Kolesar et al., 1995a), the polio virus (Rossomondo et al., 1994), gastric H+, K+, ATPase (Stalborn et al., 1994), and the cytochrome p4501A1 subtype (Fasco et al., 1995).

Gene therapy is quickly becoming a reality. The goal of gene therapy is often to replace or augment gene expression in target tissues, and this type of assay may be clinically relevant in the foreseeable future.

## Experimental Conditions and Reagents

*RNA purification.* Total cellular RNA and RNA obtained from human tissue samples, including tumor biopsies and whole blood, can be obtained by standard procedures. Alternately, we have used a commercially available system (Ultraspec II RNA isolation system, Biotecx, Houston, TX). This system isolates total RNA by disruption and homogenization of samples with 14 M solution guanidine salts and urea followed by chloroform extraction. The sample is centrifuged, and the upper aqueous phase containing the RNA is isolated, followed by isopropanol precipitation. The RNA is purified by a proprietary RNATack resin that binds RNA specifically, and eluted with TE (Tris/EDTA, pH 7.4) buffer. The concentration of RNA is quantitated spectrophotometrically (Glasel, 1995).

The entire isolation can be completed in approximately 1 hr, a significant advantage over standard methods. RNA can also be extracted directly from lymphocytes obtained from whole blood. We routinely obtain 10–15 $\mu$g of RNA from 10 ml of whole blood, which compares favorably to standard methods.

Care must be taken with this product to ensure that the lysis buffer is completely in solution and homogenous prior to use. If the lysis step is carried out on ice, yields are improved by as much as 50%.

*Design of internal standard.* We generally design an internal standard by gel purifying the desired PCR product and identifying restriction sites 30–70 base pairs apart that will generate compatible ends. Digestion with the appropriate enzymes and religation generate a DNA fragment that is identical to the target DNA, but 30–70 base pairs smaller. The primer recognition sites and the sequence are identical, and the internal standard should amplify under identical conditions. After gel purification and elution in TE, the internal standard is quantified spectrophotometrically and stored at –20°C. We then titrate the internal standard to determine what concentration is optimal for amplification, usually $10^{-6}$ ng per PCR run.

Using a DNA standard has the benefit of ease of preparation and storage, amplification under identical conditions, and low expense. It does not control for variability present within the RT step. RNA standards can be introduced prior to reverse transcription to control for variability throughout the RT-PCR process.

*RT-PCR.* To improve reproducibility, all RT and PCR steps are done with master mixes that contain all components except the target nucleotides and the Taq polymerase. Since PCR is by nature prone to contamination and false positive results, precautions must be taken to ensure the validity of results. All reagents should be aliquotted into single-use portions; separate pipettes should be set aside for PCR use only. Reactions can be set up in a biological hood with ultraviolet light exposure to surfaces between reactions. Adequate controls (both positive and negative) should be used for all reactions.

RNA aliquots of 1.25 $\mu$g are reverse transcribed into cDNA by adding 5 $\mu$l of 25 mM MgCl2, 5 $\mu$l 10X buffer (250 mM Tris HCl, pH 8.3, 250 mM KCl, 50 mM MgCl2, 2.5 mM spermidine, 50 mM DTT), 2.5 $\mu$l of 10 mM aqueous 2′-deoxynucleoside 5′-triphosphate solution in 1 mM Tris-HCl (pH 7.2) (GibcoBRL, Gaithersburg, MD), 0.625 $\mu$l of recombinant RNAsin ribonuclease inhibitor 20 u/$\mu$l, 1.25 $\mu$l of 5 u/$\mu$l AMV reverse transcriptase, 1.25 $\mu$l of 20 mg/500 $\mu$l oligo(dT)15 primer, and sufficient DEPC–treated water for a final volume of 25 $\mu$l. This gives a standard cDNA solution of

50 ng/μl. Reactions are incubated at 42°C for 15 min, followed by 95°C for 5 min, and then cooled to 4°C. Only 5 μl of this solution is used for the PCR, leaving 20 μl for other analyses.

Each PCR contains 61.3 μl DEPC–treated water, 10 μl of 10X PCR buffer (GibcoBRL, Gaithersburg, MD) (200 mM Tris-HCl, pH 8.4, 500 mM KCl), 3.75 μl of 50 mM MgCl2, 3 μl of 10 mM aqueous 2′-deoxynucleoside 5′-triphosphate solution in 1 mM Tris-HCl (pH 7.2), 5 μl each of each primer 10 pmol/μl, 25–250 ng of cDNA, and internal standard $1 \times 10^{-6}$ ng/5 μl. We often amplify two unknowns (Figure 2) in addition to the internal standard.

One drop of mineral oil is placed over the aqueous layer, and tubes are heated in the thermocycler (MJ Research, Cambridge, MA) at 95°C for 5 min. This will inactivate the reverse transcriptase, prevent nonspecific elongation during the first few cycles, and completely denature the cDNA. Next Taq polymerase (GibcoBRL, Gaithersburg, MD), 1.67 units in 2 μl of DEPC–treated water is added through the mineral oil. Samples are spun in

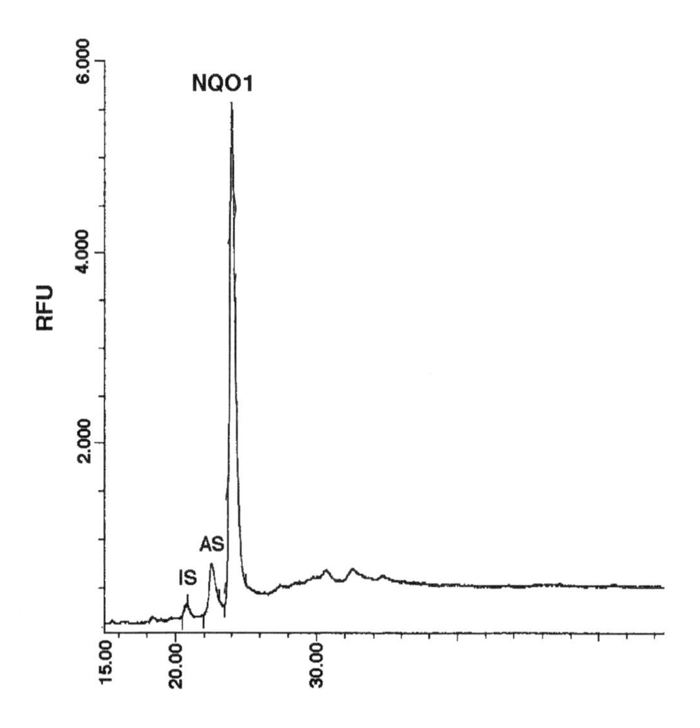


FIGURE 2. Electropherogram of pGEM ladder. pGEM is a solution of DNA fragments ranging from 36 to 2645 bases in a concentration of 1.0 mg/ml. The solution was diluted 1:10 and analyzed as described in “Instrument parameters.”

a tabletop microcentrifuge for 5 sec at maximum speed and placed back in the thermocycler. The PCR cycles are 1 min at 94°C, 30 sec at 55°C, 1.5 min at 72°C for 40 cycles, followed by 72°C for 7 min, then 4°C. The PCR amplification process is no longer linear after a certain number of cycles, depending on the efficiency of the amplification and the abundance of the starting material. For low-abundance transcripts the PCR is no longer linear after 40 cycles.

## Analysis and Quantification by Capillary Electrophoresis with Laser-Induced Fluorescence (CE-LIF)

*Preparation of DNA solution for standard curve.* In our laboratory, we routinely use a commercially available DNA ladder (pGEM, Promega Corp, Madison, WI) to calibrate our equipment. pGEM is a solution of DNA fragments ranging from 36 to 2645 bases in a concentration of 1.0 mg/ml. The DNA standard is aliquotted in 10 *m*l portions and stored at -20°C.

pGEM contains a 222 bp fragment present in a concentration of 129 mcg/ml initially. This solution is diluted 1:10 with DEPC-treated water and injected for 5–20 sec. Rossomondo et al. (1994), using a similar capillary electrophoresis system, determined that a pressure injection at 0.5 psi for 11 sec, created a water plug 1 mm long, and on the basis of this observation, Butler et al. (1994) calculated that a 45-sec injection under these conditions introduces 32 nl into the 100 mm ID capillary. Using very similar conditions, we calculated that the concentration of 222 bp fragment, when diluted 1/10 and injected for 10 sec, is 91.6 pcg/7.1 nl (Kolesar et al., 1995a).

*Instrument parameters.* Separations are performed on the P/ACE 5510 CE system (Beckman, Fullerton, CA) or similar system, with the temperature held constant at 20°C. PCR products are detected with laser-induced fluorescence in the reversed polarity mode (anode at the detector site) at excitation of 488 nm and emission of 520 nm. Samples are introduced hydrodynamically by 10-sec injections at 0.5 psi across a 65 cm by 100 mm ID coated eCAP dsDNA capillary filled with TBE containing replaceable linear polyacrylamide (Beckman, Fullerton, CA). No sample preparation of PCR products is required.

The capillary is conditioned with eCAP ds 1000 Gel buffer containing 60 of LiFlour dsDNA 1000 EnhanCE (thiazole orange) intercalator per 20 ml (Beckman, Fullerton, CA) and rinsed at high pressure for 3 min. Separations are performed under constant voltage at 7.0 kV for 50 min. The capillary is rinsed with gel buffer for 3 min prior to each injection. Data can

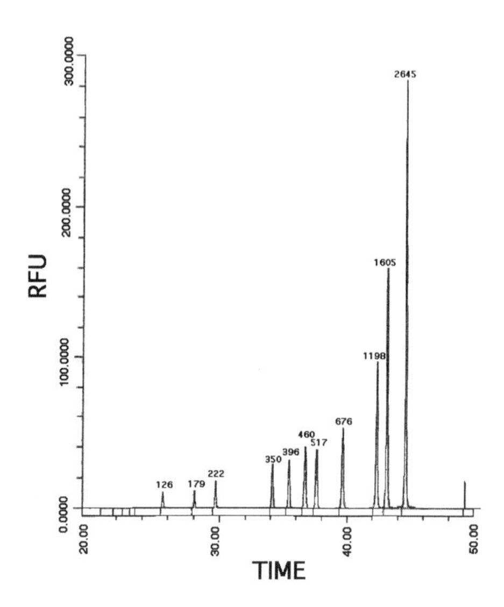


FIGURE 3. Electropherogram of polymerase chain reaction products derived from blood sample to quantitate NQO1 gene expression and alternatively spliced NQO1 (AS). Cellular RNA was isolated from whole blood and amplified by RT-PCR with a reaction–specific internal standard as described in "RT-PCR." Analysis was performed with a P/ACE 5510 CE with coated eCAP dsDNA capillary (100 mm ID) containing replaceable linear polyacrylamide. Detection was achieved by laser–induced fluorescence at 488 nm with a separation potential 7.0 keV. Peak area of the internal standard is compared to the peak area of NQO1 and AS to quantitate gene expression.

be analyzed with the System Gold chromatography data system (Beckman, Fullerton, CA) or other similar system.

*Calibration of CE-LIF.* To generate a standard curve (Figure 3) 10 $\mu$l of diluted DNA solution is placed in a microvial, and six pressure injections at 0.5 psi are made from the vial for 5, 10, 12.5, 15, 17.5, and 20 sec. Separations are performed on the basis of the parameters detailed under instrument parameters.

To determine the linear range, the peak area reported on the electropherogram versus the concentration of the 222 bp fragment injected is plotted. With this assay, a linear relationship between the peak area and concentration exists for the 222 bp fragment over the range 0.5–183 pcg. The concentration of unknown samples can be determined in this range by comparison of the peak areas obtained to the standard curve.

The same solution of standard DNA fragments can be used to determine the molecular weight of fragments on the basis of retention times. Plot the migration time for each peak versus the known molecular weights and determine the linear range. The linear range will not be over the entire range of molecular weights, since intercalating dyes such as thiazole orange can affect migration times at higher (>1000 bp) molecular weights. With this assay, a linear relationship exists between retention time and molecular weights of the standard fragments between 126 bp and 460 bp ($r^2 = 0.996$). This standard can then be used to determine the molecular weights of unknown samples.

*Validation of RT-PCR CE-LIF assay*. To validate the assay, cDNA from two different sources is amplified in triplicate and analyzed on three different days. This generates 18 data points over 3 days. The ratio of the peak area of the internal standard to the peak area of the gene of interest is calculated, and the linear range is determined by plotting the ratio versus the starting amount of cDNA (Figure 2). We have found a linear relationship ($r^2 = 0.991$) over the range of 0–312.5 ng of starting cDNA (the RNA is quantitated spectrophotometrically, and the RT process is assumed to be 100%). The PCR is no longer linear when more than 400 ng of cDNA is amplified. This is consistent with data reported by Rossomondo and colleagues (Rossomondo et al., 1994), who also showed a linear relationship between peak area and starting amount of RNA at low concentrations and a nonlinear relationship when higher concentrations were amplified.

The mean retention time for a reference peak is determined (we commonly analyze the internal standard fragment as well as the target sequence) and used to calculate interday precision (standard deviation [SD] of runs between days × 100/average of within-run means) and intraday precision (SD of runs on same day × 100/mean measured concentration). With this assay, we have interday and intraday migration time precision of <1.0%. This is a measure of the reproducibility of the CE-LIF aspect of the assay.

The same samples can be used to determine the peak area precision. The peak area precision is a measure that represents the reproducibility of both the CE and PCR aspects of this assay. The ratio of the peak area of the gene of interest to internal standard is calculated and compared. Because PCR with an internal standard is competitive, the peak area of one or the other product is not representative. This assay has a peak area intraday precision of 12–15% and interday precision of 10–16%. Strategies to improve assay precision should target the RT-PCR portion and could include automation and use of RNA standards.

The major limitations to accurate quantitation in the assay are: 1) the inability of spectophotometric analysis to accurately predict RNA concentration (Glasel, 1995), 2) the assumption that the reverse transcription step is 100% efficient, and 3) the need for accurate calibration of RNA internal standard. To overcome these limitations a calibrated RNA internal standard could be added prior to reverse transcription.

## Detection of Single Base Substitutions

*Restriction fragment length polymorphism (RFLP).* RFLP analysis is a commonly used technique for detecting mutations and loss of heterozygosity, and for restriction length mapping. RFLP is generally analyzed by running the samples on agarose and determining qualitatively the presence and absence of bands. CE-LIF has been used with RFLP to identify point mutations in NQO1 (Kolesar et al., 1995b) and the oncogene K-ras (Mitchell et al., 1995).

By analyzing the samples with CE-LIF and comparing predigestion samples with postdigestion samples, the actual percentage of digestion with restriction enzymes can be calculated. Approximately 50% digestion by gel electrophoresis is assumed to mean heterozygous mutation or allele presence. Determining the percentage of digestion may help distinguish between heterozygous and homozygous mutation.

*Single-stranded conformational polymorphism (SSCP) and constant denaturing gradient electrophoresis (CDGE).* RFLP analysis is only useful for the detection of specific point mutations that result in the loss or creation of restriction sites. To detect point mutations in any sequence, CE-LIF methods analogous to SSCP and constant denaturing gradient electrophoresis have been developed (Kharpko et al., 1994; Kumar et al., 1995).

*Single-stranded oligonucleotides.* Single–stranded oligodeoxynucleotides are under investigation as a new class of therapeutic agents. These agents can inhibit protein expression by binding to mRNA, halting *in vitro* transcription. Methods to quantitate these molecules utilizing CE-LIF are under development (Reyderman and Stavchansky, 1996).

## Conclusion

Capillary electrophoresis can be viewed as the instrumentation of electrophoresis. In addition to being a further step towards the accurate quantitation of gene expression, this technique offers several advantages over traditional slab gel electrophoresis with autoradiography. Required sample volumes are small (1–5 $\mu$l), sensitivity is high (attomolar), and the hazards associated with isotopic storage, use, and disposal are eliminated. This technique is also readily automatable, can be used for fraction collection, and can be validated. This technique can be applied to the analysis of gene expression, gene therapy, single–stranded oligonucleotide, and point mutations.

# REFERENCES


Brandsma J and Miller G (1980): Nucleic acid spot hybridization: Rapid quantitative screening of lymphoid cell lines for Epstein-Barr virus. *PNAS* 77:6851–5.

Butler JM, McCord BR, and Jung JM (1994): Quantitation of polymerase chain reaction products by capillary electrophoresis using laser fluorescence. *J Chromatogr B* 658:271–80.

Dieguez-Lucena JL, Ruiz-Galdon M, Morell-Ocana M, Garcia-Villanova J, Flores-Polanco FJ, and Reyes-Engel A (1994): Capillary electrophoresis compared with silver staining of polyacrylamide gels for quantification of pcr products. *Cl Chem* 40:493–4.

Fasco MJ, Treanor CP, Spivack S, Figge HL, and Kaminsky LS (1995): Quantitative RNA-polymerase chain reaction -DNA analysis by capillary electrophoresis with laser induced fluorescence. *Anal Biochem* 224:140–7.

Ferré F, Marchese A, and Pezzoli P (1994): Quantitative PCR: An overview. In: *PCR. The Polymerase Chain Reaction*, Mullis KB, Ferré F, Gibbs RA, eds. Boston: Birkhauser.

Gelfi C, Leoncini F, Righetti PG, Cremonsei L, di Blasio AM, Carniti C, and Vignali M (1995): Separation and quantitation of reverse transcriptase polymerase chain reaction fragments of basic fibroblast growth factor by capillary electrophoresis in polymer networks. *Electrophoresis* 16:780–3.

Glasel J (1995): Validity of nucleic acid purities monitored by 260 nm/280 nm absorbance ratios. *Biotechniques* 18:62–3.

Haupt W and Pingoud A (1983): Comparison of several high performance liquid chromatography techniques for the separation of oligodeoxynucleotides according to their chain lengths. *J Chromatogr* 260:419–23.

Heller C (1995): Capillary electrophoresis of proteins and nucleic acids in gels and entangled polymer solutions. *J Chromatogr A* 698:19–31.

Hjerten S (1985): High performance electrophoresis: elimination of electroendosmosis and solute absorption. *J Chromatogr* 347:191–8.

Huang X, Gordon MJ, and Zare RN (1988): Bias in quantitative capillary zone electrophoresis caused by electrokinetic sample injection. *Anal Chem* 60:375–7.

Karger BL, Chu YH, and Foret F (1995): Capillary electrophoresis of proteins and nucleic acids. *Annu Rev Biophys Biomol Struct* 24:579–610.

Khrapko K, Hanekamp JS, Thilly WG, Belenkii A, Foret F, Karger BL (1994): Constant denaturant capillary electrophoresis (CDCE): a high resolution approach to mutational analysis. *Nucleic Acids Res* 22: 364–9.

Kleparnik K, Garner M, and Bocek P (1995): Injection bias of DNA fragments in capillary electrophoresis with sieving. *J Chromatogr A* 698:375–83.

Kolesar JM, Rizzo JD, and Kuhn JG (1995a): Quantitative analysis of NQO1 gene expression by RT-PCR and CE-LIF. *J Capillary Electrophor* 2:287–90.

Kolesar JM, Burris H, and Kuhn JG (1995b): Detection of a point mutation in NQO1 (DT-diaphorase) in a patient with colon cancer. *J Natl Cancer Inst* 87:1022–4.

Kuhr WG and Monnig CA (1992): Fundamental reviews: capillary electrophoresis. *Anal Chem* 62:403R–414R.

Kumar R, Hanekamp JS, Louhelainen, Burvall K, Onfelt A, Hemminki K, and Thilly WG (1995): Separation of transforming amino acid-substituting mutations in codons 12, 13, and 61 of the n-ras gene by constant denaturant capillary electrophoresis (CDCE). *Carcinogenesis* 16:2667–73.

Kunkel A, Degenhardt M, and Watzig H (1996): Precise quantitative results by capillary electrophoresis (CE): instrumental aspects (an update). Proceedings from the 8th International Symposium on High Performance Capillary Electrophoresis 17:140 (abstract).

Kuypers A, Meijerink JP, Smetsers T, Linssen P, and Mensink E (1994): Quantitative analysis of DNA aberrations amplified by competitive polymerase chain reaction using capillary electrophoresis. *J Chromatogr B* 660:271–7.

Lander JP (1995): Clinical capillary electrophoresis. *Clin Chem* 41:495–509.

Lu W, Han DS, Yuan J, and Andrieu JM (1994): Multi-target PCR analysis by capillary electrophoresis and laser induced fluorescence. *Nature* 368: 269–71.

Mitchell CE, Belinsky SA, Lechner JF (1995): Detection and quantitation of mutant K-ras codon 12 restriction fragments by capillary electrophoresis. *Ann Biochem* 224:148–53.

Oefner PJ, Bonn GK, Huber CG, and Nathakarnkitkoll S (1992): Comparative study of capillary zone electrophoresis and high performance liquid chromatography in the analysis of oligonucleotides and DNA. *J Chromatogr* 625:331–40.

Pariat YF, Berka J, Heiger DN, Schmitt T, and Vilenchik M (1993): Separation of DNA fragments by capillary electrophoresis using replaceable linear polyacrylamide matrices. *J Chromatogr A* 652:57–66.

Piatak M, Saag MS, and Yang LC (1993): High levels of HIV in plasma during all stages of infection determined by competitive PCR. *Science* 259:1749–54.

Reyderman L and Stavchansky S (1996) Determination of ss-oligodeoxynucleotides by capillary gel electrophoresis with laser induced fluorescence and on column derivatization. *J Chromatogr B* (in press).

Rossomando EF, White L, and Ulfelder KJ (1994): Capillary electrophoresis: Separation and quantitation of reverse transcriptase polymerase chain reaction products from polio virus. *J Chromatogr B* 656:159–68.

Schmalzing D, Piggee CA, Foret F, Carrilho E, and Karger BL (1993): Characterization and performance of a neutral hydrophilic coating for the capillary electrophoretic separation of biopolymers. *J Chromatogr A* 652:149–59.

Schwartz HE and Ulfelder KJ (1991): Analysis of DNA restriction fragments and polymerase chain reaction products towards detection of the AIDS (HIV-1) virus in blood. *J Chromatogr* 559:267–83.

Schwartz HE and Ulfelder KJ (1992): Capillary electrophoresis with laser induced fluorescence detection of PCR fragments using thiazole orange. *Ann Chem* 64:1737–40.

Schwarz HE, Ulfelder KJ, Chen FTA, and Pentoney SL (1994): The utility of laser induced fluorescence detection in applications of capillary electrophoresis. *J Capillary Electrophor* 1:36–54.

Srinivasan K, Girard JE, and Williams P (1993a): Electrophoretic separations of polymerase chain reaction-amplified DNA fragments in DNA typing using a capillary electrophoresis-laser induced fluorescence system. *J Chromatogr* 652:83–91.

Srinivasan K, Morris SC, and Girard JE (1993b): Enhanced detection of PCR products through use of TOTO and YOYO intercalating dyes with laser induced fluorescence-capillary electrophoresis. *Appl Theoret Electrophor* 3:235–9.

Stalbom BM, Torven A, and Lundberg LG (1994): Application of capillary electrophoresis to the post polymerase chain reaction analysis of rat mRNA for gastric H+, K+-ATPase. *Ann Biochem* 217:91–7.

Tsuda T (1987): Modification of electroosmotic flow with cetyltrimethylammonium bromide in capillary electrophoresis. *J High Resolut Chromatogr* 10:622–4.

Ulfelder KJ, Schwartz HE, Hall JM, and Sunzeri FJ (1992): Restriction fragment length polymorphism analysis of ERB2 oncogene by capillary electrophoresis. *Analyt Biochem* 200:260–7.

# Quantitative PCR Technology

Lincoln McBride, Ken Livak, Mike Lucero, Federico Goodsaid, Dane Carlson, Junko Stevens, Traci Allen, Paul Wyatt,
Daniel Thiel, Peter Honebein, John Shigeura, Tim Woudenberg, Eugene Young, Raymond Lefebvre, Susan Flood, Bruce Goldman, Jeff Lucas, Kevin Bodner, Robert Grossman, Bashar Mullah, Charles Connell, Linda Lee, and Mark Oldham

## Introduction

Higuchi et al. (1992, 1993) pioneered the analysis of PCR kinetics by constructing a system that detects PCR products as they accumulate. This "real-time" system includes the intercalator ethidium bromide in each amplification reaction, an adapted thermal cycler to irradiate the samples with ultraviolet light, and detection of the resulting fluorescence with a computer-controlled cooled CCD camera. Amplification produces increasing amounts of double-stranded DNA, which binds ethidium bromide, resulting in an increase in fluorescence. By plotting the increase in fluorescence versus cycle number, the system produces amplification plots that provide a more complete picture of the PCR process than assaying product accumulation after a fixed number of cycles.

We have developed an instrument system that automates real-time PCR detection using fluorescent probe technology.

### *Fluorogenic probe chemistry for real-time PCR*

Real-time systems for PCR were later improved by probe-based, rather than intercalator-based, PCR product detection. The principal drawback to intercalator-based detection of PCR product accumulation is that both specific and nonspecific products generate signal. An alternative method, the 5′ nuclease assay (Holland et al., 1991; Gelfand et al., 1993) provides a real-time method for detecting only specific amplification products. Holland et al. were the first to demonstrate that cleavage of a target probe during PCR by the 5′ nuclease activity of *Taq* DNA polymerase could be used to detect amplification of the target-specific product. In addition to the components of a typical amplification, reactions included a probe labeled with $^{32}P$ on its 5′ end and blocked at its 3′ end so that it could not act as a primer. During amplification, annealing of the probe to its target sequence generates a substrate that is cleaved by the 5′ nuclease activity of *Taq* DNA polymerase when the enzyme extends from an upstream primer into the region of the probe. This dependence on polymerization ensures that cleavage of the probe occurs only if the target sequence is being amplified. After PCR, Holland and co-workers (1991) measured cleavage of the probe by using thin layer chromatography to separate cleavage fragments from intact probe.

Figure 1 diagrams what happens to a fluorogenic probe (Lee et al., 1993; Livak et al., 1995) during the extension phase of PCR. If the target sequence is present, the probe anneals downstream from one of the primer sites and is cleaved by the 5′ nuclease activity of *Taq* DNA polymerase as this primer is extended. This cleavage of the probe separates the reporter dye from the quencher dye, increasing the reporter dye signal. Cleavage removes the probe from the target strand, allowing primer extension to continue to the end of the template strand. Thus, inclusion of the probe does not inhibit the overall PCR process. Additional reporter dye molecules are cleaved from their respective probes with each cycle, effecting an increase in fluorescence intensity proportional to the amount of amplicon produced.

## 7700 Sequence Detection System

The 7700 Sequence Detection System comprises an analytical instrument, a microcomputer running proprietary software, and reagents to automate fluorescent PCR-based detection and quantification of nucleic acid sequences during PCR. Manufactured by PE Applied Biosystems, the 7700 System integrates thermal cycling and real-time fluorescence detection using closed-vessel detection through optical caps on MicroAmp® tubes.

The 7700 System detects PCR product using fluorescent intercalator binding or fluorogenic DNA probes that are excitable with an argon ion laser and detectable with a cooled CCD camera. With probe-based detection, increases in fluorescence as a function of cycle number are related to increases in PCR product concentration. After thermal cycling is completed, the computer rigorously compares fluorescence versus time data of unknowns with standards run simultaneously. This comparison automatically gives the amount of analyte nucleic acid sequence present in each of

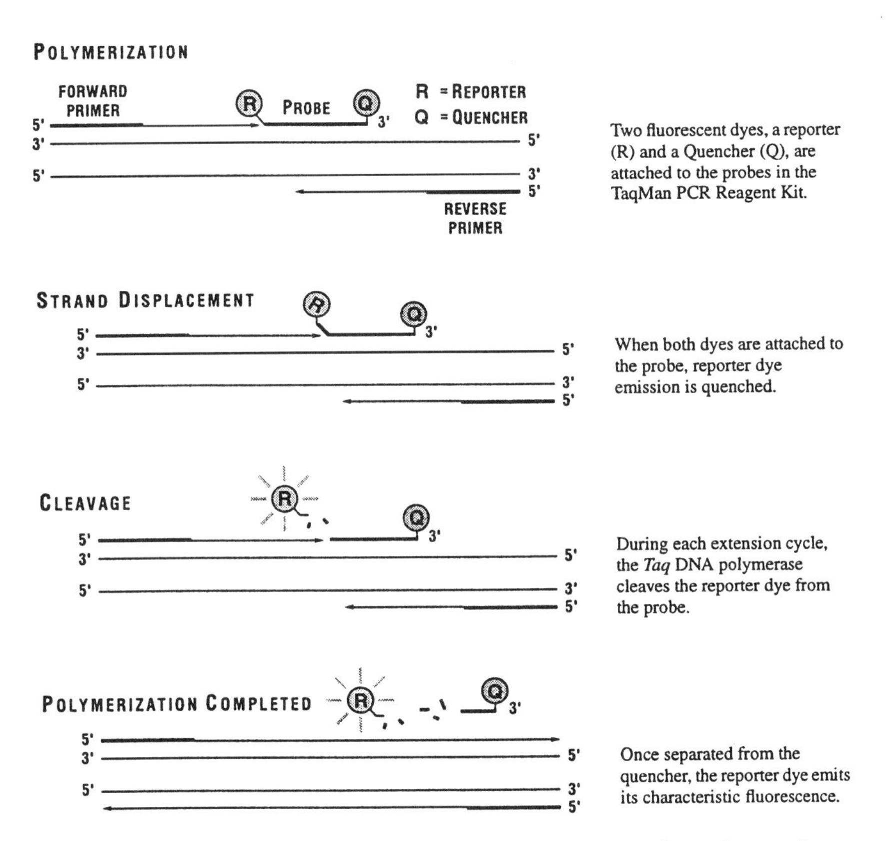


FIGURE 1. Stepwise representation of the forklike-structure-dependent, polymerization-associated, 5′ and 3′ nuclease activity of *Taq* DNA polymerase acting on a fluorogenic probe during one extension phase of PCR (Lyamichev et al., 1993).

up to 96 samples. The closed MicroAmp tubes or trays are then either discarded, archived, or subjected to further analyses.

### *Hardware*

The 7700 System makes use of Perkin-Elmer 9600 thermal cycler technology and adds real-time fluorescence detection. A 488-nm argon ion laser serves as the light source, and a spectrograph interfaced to a CCD camera serves as the detector. The 96 reactions are scanned in real time via an optical multiplexer.

For each sample, the CCD camera collects the emission data between 520 nm and 660 nm once every few seconds. The multiplexer distributes excitation and emission light to and from each of the 96 reactions via 96 optical fibers and 96 lenses above each tube. Both excitation and fluorescence emission light pass to and from the PCR samples through optically clear MicroAmp caps. The system essentially eliminates carryover contamination between samples because it monitors the fluorescent signal in sample tubes with specially designed closed optical caps, eliminating the need to open the sample tubes.

### *Software*

Proprietary software running on a Power Macintosh computer coordinates the 7700 System instrument setup, control, instrument status, data acquisition, and other management functions. The software is also responsible for run-time data display, postrun analysis, and data storage. For real-time applications, the 7700 System supports a single reporter plate setup. For plate–reading applications, the user can choose from single–reporter or allelic discrimination plate setup. Plate setup information for a particular run can be saved and recalled later for other runs.

### *Thermal cycler conditions*

A configuration window for defining the thermal cycler protocol is also provided. The user can specify the data collection window within the thermal cycler protocol. Like the information in the plate setup dialog box, this information can be saved and recalled at a later time.

## Chemistry

The 7700 System performs fluorescence-based analysis inside the PCR tube without the need for any post-PCR purification or analysis. Samples and reagents are simply added to the PCR tube, then the tube is capped and placed on the 7700 System. The 7700 System automatically gives a quantitative answer immediately after the PCR is completed. This approach can be referred to as a "homogeneous closed-tube assay." The assay comprises standard PCR components and a fluorogenic probe.

### *Chemistry and kits*

The principal fluorescent chemistry supported on the 7700 System is the fluorogenic 5′ nuclease assay (Figure 1). In this approach, a fluorescent probe complementary to the desired PCR product sequence is added to the normal PCR cocktail. This probe contains a fluorescent dye (FAM, TET, or JOE) at the 5′ terminus and a fluorescent quencher (TMRA) at the 3′ terminus. Probes labeled in this way are generally greater than 95% nonfluorescent relative to the same probe that lacks the quencher (Lee et al., 1993). When this probe is placed in PCR, the *Taq* DNA polymerase that catalyzes the PCR also cuts any probe that specifically binds to PCR product being generated. The probe is designed so that this cutting occurs between the quencher and the fluorophore. Cutting results in a fluorescence increase above a large fluorescence background (from the excess probe) (Figure 1).

### *Hot Start and AmpliTaq Gold*

For low copy detection, many PCR users ensure properly performing PCR with a method known as "Hot Start," which originally required the use of a wax layer to separate critical PCR components. Current strategies dictate that one or more of the required PCR components be added to the PCR only at the exact time when PCR is started, usually at a temperature greater than 70°C. This eliminates unwanted side reactions that can occur when all the reactants are combined at room temperature just seconds or minutes before thermal cycling commences. However, many researchers found the manual Hot Start technique so cumbersome, time-consuming, or expensive, especially for high-throughput applications, that they avoided Hot Start PCR and risked the chance of decreased specificity and yield.

The 7700 System eliminates the frustrations associated with the manual Hot Start technique by employing a chemical method for Hot Start: AmpliTaq Gold™. Developed by Roche Molecular Systems, AmpliTaq Gold is a modified form of AmpliTaq® DNA Polymerase that is provided in an inactive state. The new enzyme is activated by a pre-PCR heat step at 93–95°C that can be easily programmed into any thermal cycler. The transition from enzyme activation to PCR cycling is seamless and begins without going below optimal primer annealing temperature.

AmpliTaq Gold is the enzyme of choice for any experiment in which yield of product is important. This enzyme provides the sensitivity required for the detection of low copy number targets such as low-level virus particles, foreign DNA in quality control assays, rare genes, viruses, and bacteria. The Gold enzyme successfully amplifies low copy target in the presence of high concentrations of sample DNA, such as genomic DNA.

Because AmpliTaq Gold is inactive at room temperature, high-throughput laboratories can set up the reactions in advance without the fear of increasing the amplification of nonspecific products. The ability to mix all of the reagents in advance reduces the number of pipetting steps and makes the setup of multiple reactions easy. The result is a significant savings in labor with better results.

## Disposables

The 7700 System uses two types of disposables: Optical Tubes and Optical Caps.

### *MicroAmp Optical Tubes*

MicroAmp Optical Tubes undergo quality controlled screening to eliminate tubes with fluorescent background. The tubes are also frosted to minimize any fluorescence from the wells of the cycling block. The MicroAmp Optical Tubes are used with a MicroAmp Tray Retainer Set.

### *MicroAmp Optical Caps*

Unlike traditional caps, MicroAmp Optical Caps have a much thinner dome (0.010in), which allows fluorescent emissions to penetrate through the cap. This feature is what enables the unique closed-tube detection system.

## Quantitation of Starting Copy Number Using PCR

### *Real-time methodology*

The ability to monitor the real-time progress of the PCR completely revolutionizes the way one approaches PCR-based quantitation of DNA and RNA. Reactions are characterized by the point in time during cycling when amplification of a PCR product is first detected, rather than by the amount of PCR product accumulated after a fixed number of cycles. The higher the starting copy number of the nucleic acid target, the sooner a significant increase in fluorescence is observed.

The parameter $C_T$ (threshold cycle) is defined as the fractional cycle number at which the reporter fluorescence generated by cleavage of the probe passes a fixed threshold above baseline. The default baseline setting is 10 times the standard deviation of noise in cycles 3–15.

As shown by Higuchi and co-workers (1993), a plot of the log of initial target copy number for a set of standards versus $C_T$ is a straight line. Quantitation of the amount of target in unknown samples is accomplished by measuring $C_T$ and using the standard curve to determine starting copy number. Figure 2 shows an example in which dilutions of a known amount of RNA were analyzed by RT–PCR on the 7700 detector to determine $C_T$ values and prepare a standard curve. The entire process of calculating $C_T$s, preparing a standard curve, and determining starting copy number for unknowns is performed by the software of the 7700 system.

a)

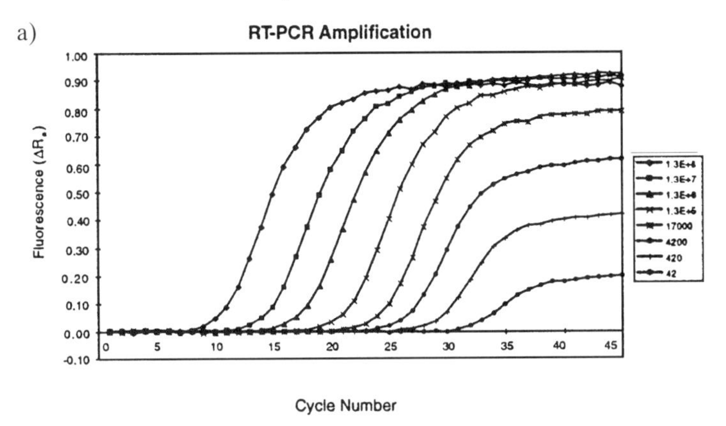


b)

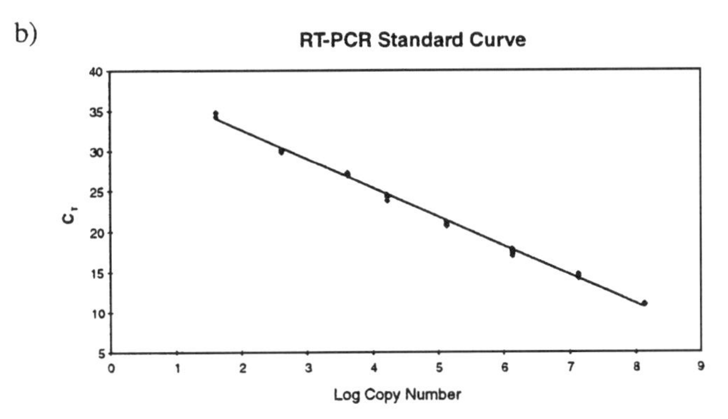


FIGURE 2. RT–PCR amplification of RNA over six orders of magnitude. A segment of human GAPDH cDNA was cloned and used as template for in vitro transcription with T7 RNA polymerase. Dilutions of a known amount of this synthetic GAPDH transcript were the templates for RT–PCR using the components of the EZ r*Tth* RNA PCR Kit (Perkin-Elmer).

(a) Amplification plots for reactions with starting RNA copy number ranging from 42 to $1.3 \times 10^8$. Cycle number is plotted versus change in normalized reporter signal ($\Delta R_n$). Four replicates for each copy number were performed, but the data for only one are shown here.

(b) Standard curve plotting log starting copy number versus threshold cycle ($C_T$). In this case, the data from all four replicates at each copy number are plotted.

### *Effect of limiting reagents*

The early cycles of PCR are characterized by an exponential increase in target amplification. As reaction components become limiting, the rate of target amplification decreases until a plateau is reached and there is little or no net increase in PCR product. The sensitive fluorescence detection of the 7700 allows the threshold cycle to be observed when PCR amplification is still in the exponential phase. This is the main reason why $C_T$ is a more reliable measure of starting copy number than an endpoint measurement of the amount of accumulated PCR product.

During the exponential phase, none of the reaction components is limiting; as a result, $C_T$ values are very reproducible for reactions with the same starting copy number and reaction stoichiometry. This leads to greatly improved precision in the quantitation of DNA and RNA. On the other hand, the amount of PCR product observed at the end of the reaction is very sensitive to slight variations in reaction components. This is because endpoint measurements are generally made when the reaction is beyond the exponential phase, and a slight difference in a limiting component can have a drastic effect on the final amount of product.

For example, side reactions, such as formation of primer dimers, can consume reagents to different extents from tube to tube. Thus, it is possible for a sample with a higher starting copy number to end up with less accumulated product than a sample with a lower starting copy number. The differences between endpoint and real-time detection are graphically illustrated in Figure 3, which shows amplification of 96 identical samples. The overall change in reporter signal, as measured at cycle 40, varies widely among the replicates. However, the amplification plots are remarkably similar between cycles 22 and 25, during which the $C_T$ values are determined.

### *Advantages of real time*

The development of competitive PCR was driven by a reliance on endpoint measurements. Determining $C_T$ values by following the real-time kinetics of PCR eliminates the need for a competitor to be co-amplified with the target. Quantitation can be performed by the more basic method of preparing a standard curve and determining unknown copy number by comparison to the standard curve. Compared to endpoint measurements, the use of $C_T$ values also greatly expands the dynamic range of quantitation because data are collected for every cycle of PCR. Figure 2b shows a linear relationship between $C_T$ and log starting copy number for over six orders of magnitude, compared to the one or two orders of magnitude typically observed

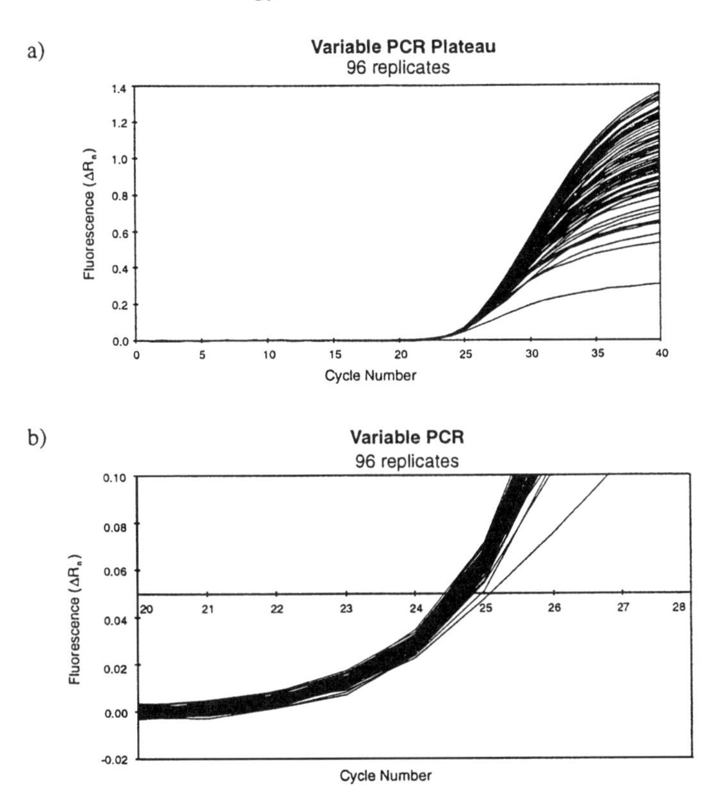


FIGURE 3. Amplification of a segment of the b-actin gene from human genomic DNA. Samples contained 10 ng human genomic DNA (corresponds to 3300 copies of a single copy gene) and were amplified using the components of TaqMan® PCR Reagent Kit (Perkin-Elmer).

(a) Amplification plots of 96 replicates.

(b) Detail of cycles 20–28. The abscissa is placed at a $\Delta R_n$ value of 0.05 to show the threshold used for calculation of $C_T$. The average final $\Delta R_n$ value at cycle 40 is 1.03 ± 0.22 (c.v. = 21.4%). The average $C_T$ value is 24.64 ± 0.11. A standard deviation of 0.11 for $C_T$ corresponds to a c.v. of 7.9% for calculated starting copy number.

with an endpoint assay. Quantitation with the 7700 detector eliminates post-PCR processing, which not only increases throughput and reduces the chances for carryover contamination, but also removes post-PCR processing as a potential source of error. Although not immune, $C_T$ values are less sensitive than endpoint values to the effects of PCR inhibitors, again, because measurements are from the exponential phase where reaction components are not limiting. In fact, the only way to identify and examine the exponential phase of PCR is to do quantitation in real time.

### *Future development: internal controls for real time*

Unlike competitive PCR, quantitation with the 7700 system does not require co-amplification of an internal reference. Other reasons remain for using an internal reference: normalizing for differences in the amount of total DNA or RNA added to a reaction, or detecting the presence of PCR inhibitors. In fact, controls for normalization or inhibition do not need to be run in the same tube as the target, but some researchers may prefer to do this. Although the availability of multiple reporter dyes means that amplification of more than one product can be detected in a single tube using the 7700 system, the inherent difficulty is keeping amplification of the control from interfering with target amplification. For the kinetic analysis of PCR, this is best done by amplifying each amplicon with a distinct pair of primers. If two amplicons share primers, as in competitive PCR, then when the more abundant species reaches plateau phase, amplification of the minority species is forced to plateau as well. This is because exhaustion of primers by the majority component limits the concentration of primers available for amplification of the minority component. Exhaustion of shared primers is a major reason that competitive PCR has a limited dynamic range. When different pairs of primers are used in the same reaction, the effect of limiting primers can, in fact, be utilized to keep the two amplifications from interfering with each other. Using kinetic analysis with an intercalator, Higuchi et al. demonstrated that reducing primer concentrations limits the yield of PCR product.

Figure 4 shows PCR amplifications with decreasing concentrations of primers run on the 7700. At 120 and 80 nM, the amplification plots are similar, indicating that the reactions are not limited by the amount of primers. The remaining plots show that the more dilute the primer concentration, the lower the plateau fluorescence level at the end of the reaction. This demonstrates that a lower primer concentration limits the reaction, forcing it to plateau at a lower level of product. In terms of kinetic analysis, however, all the reactions except 4 nM have the same $C_T$ value. Thus, the strategy for performing two independent reactions in the same tube is the following: adjust the primer concentrations such that accurate $C_T$s are obtained, but soon after that, the exhaustion of primers defines the end of the reaction. In this way, amplification of the majority species is stopped before it can limit the common reactants available for amplification of the minority species. The simplest use for a second reaction in the same tube will be a +/− positive internal control to indicate whether or not amplification is

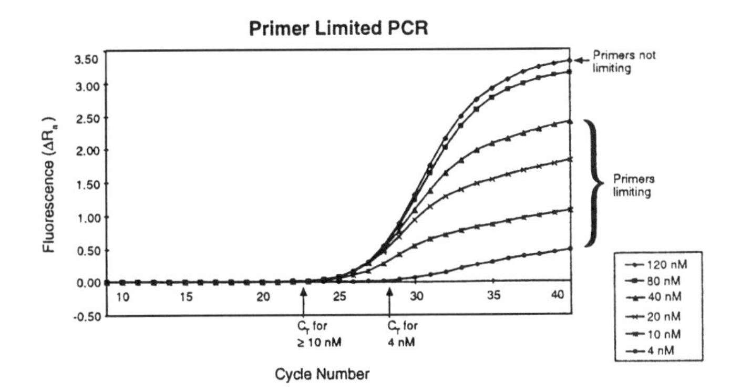


FIGURE 4. Amplification with different concentrations of primers. Samples of a plasmid template (5 pg per reaction) were amplified using primer concentrtions ranging from 4 nM to 120 nM. Both forward and reverse primers were present at the indicated concentrations. The enzyme used for amplification was AmpliTaq Gold™.

inhibited in any given sample. If there is no amplification of target in a sample, then detection of the positive internal control in that tube will show that the sample is a true negative, not a PCR failure. Work is in progress to implement this type of internal control strategy into the 7700 detection system.

## *Future development: universal assay conditions*

The experience gained in the development of the 5′ nuclease PCR assay has made it possible to identify generic conditions for the quick and reproducible development of these assays. These conditions have converted the haphazard development of 5′ nuclease assays into a straightforward process with predictable success.

At the core of this 5′ nuclease assay design strategy are a universal 5′ nuclease PCR Master Mix formulation and universal thermal cycler protocol for DNA amplification. The universal Master Mix formulation includes excess concentrations of dNTPs, AmpliTaq Gold™, and $Mg^{+2}$. It also includes the AmpErase UNG™ enzyme to prevent amplicon contamination. The Master Mix formulation is shown in Table 1.

The universal thermal cycler protocol shown in Table 2 runs the elongation step at 60°C. These universal conditions allow multiple targets to be assayed in the same 5′ nuclease PCR run.

The assay-specific parts of this strategy focus on designing primers and probes, as well as optimizing concentrations. The first task is to accurately

TABLE 1. DNA PCR Master Mix Formulation.

| Reagent | Final Concentration in Reaction |
|---|---|
| 20% Glycerol | 6% |
| 10X TaqMan Buffer A | 1X |
| 25 mM MgCl2 | 7.5 mM |
| dATP (10 mM) | 200 μM |
| dCTP (10 mM) | 200 μM |
| dGTP (10 mM) | 200 μM |
| dUTP (20 mM) | 400 μM |
| AmpliTaq Gold (5 U/μL) | 0.05 U/μL |
| AmpErase UNG (1 U/μL) | 0.01 U/μL |

TABLE 2. Universal Thermal Cycler Protocol.

| Step | Time (min) | Temp (°C) | Purpose |
|---|---|---|---|
| HOLD | 2 | 50 | AmpErase UNG reaction |
| HOLD | 10 | 95 | AmpErase UNG inactivation<br>AmpliTaq Gold activation |
| 40 CYCLES | | 95 | DNA melting |
| | 1 | 60 | Elongation |

identify the target sequence. Target amplicons, including the two primers and the probe, should be as short as possible. Amplicon lengths of less than 150 bp yield consistent results.

Successful primers and probes have been designed within a % GC range of 30–80%. Primer and probe designs have specific requirements for structure and $T_m$. A requirement common to both primers and probes is that no more than three contiguous Gs be present in their sequence. The TaqMan Probe document in the Primer Express™ primer and probe design software facilitates the design of primers and probes for 5′ nuclease assay applications.

The most efficient procedure calls for the probe to be designed first. The universal assay conditions include a probe design that incorporates the following characteristics. It must not have a G on the 5′ end because this G would quench the reporter dye signal. The probe should hybridize to the strand that has more Cs than Gs. The $T_m$ for this probe (calculated by Primer Express) must be 68–70°C.

Forward and reverse primers should be designed so that they are positioned as closely as possible to the probe. Their sequences should have less than three G + Cs within the last five nucleotides on the 3′ end. The $T_m$ for these primers (calculated by Primer Express) must be 58–60°C.

Finally, the primer and probe concentrations are optimized for normalized reporter signal ($R_n$) and $C_T$ using experimental matrices with which a range of effective $T_m$ may be tested for the primers and probes designed. The outcome of this optimization process will be a 5′ nuclease assay with a signal-to-noise ratio of 200 or higher for 500 starting copies. Assays using this strategy have been optimized for over 35 different target sequences.

## Conclusions

Compared to endpoint quantitation methods, real-time PCR offers streamlined assay development, reproducible results, and a large dynamic range. Real-time PCR eliminates the need for in-tube standards with identical primer sets as targets. Thus, the process of creating quantitative assays is simplified because the construction and characterization of an in-tube standard is no longer required. Real-time PCR now makes quantitation of DNA and RNA much more precise and reproducible because it relies on $C_T$ values rather than endpoint measurements. In addition, the use of $C_T$ values allows a larger dynamic range.

For the first time, a system is commercially available for the real-time detection of PCR amplification. The 7700 Sequence Detection System consists of four major elements: (1) uniquely designed fluorogenic oligonucleotide probes, (2) a PCR assay that exploits the 5′ nuclease activity of *Taq* DNA polymerase, with its polymerization-dependent cleavage of the target-specific probe, in order to detect the accumulation of specific PCR product, (3) instrumentation that measures the fluorescence signal from 96 samples during the thermal cycling reaction, and (4) software that processes and analyzes the fluorescence data.

The 7700 sequence detector delivers the high-throughput, increased reproducibility, precision, and dynamic range inherent in the kinetic analysis of PCR. Because of the use of target-specific fluorogenic probes, only amplification of the intended sequence is measured. With detection through closed tubes, there is much less chance for carryover contamination. As a result, the 7700 gives researchers the power to perform precise quantitative nucleic acid assays with the necessary throughput.

## REFERENCES

Gelfand DH, Holland PM, Saiki RK, and Watson RM (1993): U.S. Patent 5,210,015.

Higuchi R, Dollinger G, Walsh PS, and Griffith R (1992): Simultaneous amplification and detection of specific DNA sequences. *Biotechnology* 10:413–17.

Higuchi R, Fockler C, Dollinger G, and Watson R (1993): Kinetic PCR: Real time monitoring of DNA amplification reactions. *Biotechnology* 11:1026–30.

Holland PM, Abramson RD, Watson R, and Gelfand DH (1991): Detection of specific polymerase chain reaction product by utilizing the 5′ to 3′ exonuclease activity of Thermus aquaticus DNA polymerase. *Proc Natl Acad Sci USA* 88:7276–80.

Lee LG, Connell CR, and Bloch W (1993): Allelic discrimination by nick-translation PCR with fluorogenic probes. *Nucl Acids Res* 21:3761–6.

Livak KJ, Flood SJA, Marmaro J, Giusti W, and Deetz K (1995): Oligonucleotides with fluorescent dyes at opposite ends provide a quenched probe system useful for detecting PCR product and nucleic acid hybridization. *PCR Methods Appl* 4:357–62.

Lyamichev V, Brow MAD, and Dahlberg JE (1993): Structure-specific endonucleo-lytic cleavage of nucleic acids by eubacterial DNA polymerases. *Science* 260:778–83.

# Statistical Estimations of PCR Amplification Rates

Jean Peccoud and Christine Jacob

## Introduction

Quantitative applications of the Polymerase Chain Reaction (PCR), also known as Quantitative-PCR (Q-PCR) are intended either to determine the number of copies of a given nucleic acid sequence, or more generally, to determine the relative abundance of two sequences. Current methods to determine exact numbers of molecules overcome the difficulty of determining the amplification rate by assuming identical amplification rates for a target DNA sequence and a standard of known quantity introduced into the experiment design, so that only the ratio of amplified products need be determined. Violations of the hypothesis of identical amplification rates for two sequences will result in a systematic bias in the experimental results that underestimates or overestimates the initial copy numbers. Acquisition of kinetic PCR data was pioneered by Higuchi et al. (1992, 1993), and commercial instruments have been available since early 1996. Kinetic data provide a new way to determine the amplification rate, and we can foresee that their availability will rekindle interest in the algorithms used to compute the initial quantities of DNA sequences. Analysis of kinetic PCR patterns will soon make its way into the family of recipes that have been in use for some years in this field. This chapter provides evidence that a statistical analysis of the amplification rate is critical to ensuring a reliable estimate of the initial copy number.

## PCR Amplification Leads to Stochastic Fluctuations

PCR is an exponential amplification of a DNA target molecule population of initial quantity $N_0$. If every molecule were duplicated at each cycle, the population size at cycle $n$, $N_n$, would then be twice the size of the population at cycle $n - 1$. Following this reasoning leads to the formula $N_n = 2^n N_0$, which would be a convenient basis for the derivation of the initial copy number from the size of the population after $n$ cycles of amplification.

$$N_0 = \frac{N_n}{2_n} = 2^{-n} N_n$$

Unfortunately, things are not that simple. The yield of the amplification reaction is not 100%, and thus the amplification rate $m$ is less than 2, and in practice $1 < m < 2$. As we shall see, the surprising consequence is that the behavior of the PCR reaction is no longer deterministic. When the yield of the reaction is 100%, there is a nonambiguous relation between the initial copy number and the number of molecules after $n$ cycles of amplification. As soon as it is possible to measure the number of the amplification products, one can extrapolate to the initial number of molecules that were amplified. The only other deterministic case occurs when the amplification rate is 1, which means that no amplification occurs at all, but this case has no more practical interest than the previous one.

When the amplification rate is between 1 and 2 — the usual case — then reaction dynamics become stochastic, as the following discussion will show. The reaction yield $r = m - 1$ describes a probability that a given molecule will be duplicated during one cycle of amplification. Suppose that a single initial molecule undergoes PCR with a reaction yield $r$; then, after one cycle of amplification, the number of molecules can either be 2 with a probability $r$ or remain 1 with a probability $1 - r$. As the reaction proceeds, the number of molecules after cycle $n$ will be randomly distributed between 1 and $2^n$. Suppose now that instead of a single initial molecule, there were 2 initial molecules. Then, the number of molecules after $n$ cycles in the reaction would be randomly distributed between 2 and $2 \times 2^n$. Consider the consequence of this with respect to determining the initial copy number: for any PCR that results in a number of molecules $N_n$ greater than 2 but less than $2^n$, it is no longer possible to determine with certainty whether the initial copy number was 1 or 2 molecules. Some information has been lost during the course of the amplification.

It is worth stressing that this argument is not based on any measurement error in the amplification rate or in the number of amplification products at cycle $n$, and introducing such errors would make the determination of the initial quantity even more challenging.

It is the random or stochastic behavior of PCR itself that requires a suitable statistical analysis. Initial copy numbers cannot be determined;

they can only be statistically estimated. This means for instance that the relation below is not rigorously correct:

$$N_0 = \frac{N_n}{m^n}$$

Instead, it holds approximately:

$$N_0 \cong \frac{N_n}{m^n} \tag{1}$$

The meaning of the $\cong$ sign can be made more precise:

1. The larger $N_0$ is, the better is the approximation of $N_0$ by $N_n m^{-n}$: at any cycle $n$, $\lim_{N_0 \to \infty_0} \frac{N_n m^{-n}}{N_0} = 1$
2. The mean value $E_{N_0}(N_n m^{-n})$ of the $N_n m^{-n}$ obtained from independent replicate amplifications, each starting from $N_0$ molecules and each having the same amplification rate $m$, is equal to $N_0$ whatever the value of $N_0$:
   for any $N_0$, $E_{N_0}(N_n m^{-n}) = N_0$
3. When $n$ is large enough, $N_n m^{-n}$ is approximately equal to a random variable $W_{N_0,m}$ with expectation (mean value) $N_0$ and variance $(2-m)m^{-1}$:

$$\lim_{N \to \infty} \frac{N_n}{m_n} = W_{N_0,m} \text{ with } \mathrm{E}\left(W_{N_0,m}\right) = N_0, \sigma^2\left(W_{N_0,m}\right) = \frac{2-m}{m} N_0$$

Although these properties may seem a bit technical at first glance, they are a very good expression of how far an estimation of the initial copy number based on (1) might be from the actual value of $N_0$. The ratio of the standard deviation of the estimate over its mean value is a simple indicator of the dispersion of this estimation. In practice, kinetic PCR experiments usually involve enough cycles prior to observation so that this limit property can be used in practice. If 1000 molecules are amplified with an amplification rate of 1.80, then the dispersion of the estimation of the initial copy number based on (1) is:

$$\frac{\sqrt{\frac{2-m}{m}N_0}}{N_0} = \sqrt{\frac{2-m}{mN_0}} = \sqrt{\frac{2-1.80}{1.80 \times 1000}} = 1\%$$

This computation demonstrates that for initial copy numbers greater than 1000, Equation (1) allows them to be estimated rather precisely. However, for initial copy numbers less than 1000 the precision of the estimation may become limiting and should then be provided with a confidence interval for the actual initial copy number. For instance, the estimation of the initial copy number computed from an amplification starting with a single molecule and with a rate of amplification 1.80, has a relative dispersion equal to $\sqrt{0.2/1.80} = 33\%$. Actually, the relative uncertainty of the quantitative measurement based on a PCR amplification can be derived from the formula used to compute the confidence interval of the initial copy number estimation. These results demonstrate that for low copy numbers, the measurement uncertainty is significantly greater than the dispersion indicator computed here. Uncertainties range from 100% for a few copies to 10–20%, depending on the amplification rate, for initial copy numbers close to 100 (Peccoud and Jacob, 1996). Several authors have reported on the difficulty of obtaining reproducible amplification results when starting from low copy numbers (Lantz and Bendelac, 1994; Piatak et al., 1993; Karrer et al., 1995). Apart from the inherent sampling errors that result from the manipulation of such low numbers of molecules, the inherent stochastic fluctuations of PCR dynamics itself may explain a large part of the dispersion in their results.

## Estimations of Initial Copy Numbers Must be Based on Amplification Rate Estimations

Since most researchers need to quantitate copy numbers below 1000, this section focuses on amplifications starting with high copy numbers (>1000). It will then be considered that:

$$N_0 = \frac{N_n}{m^n} \tag{2}$$

Even in this restricted perspective, the estimation of the amplification rate could be extremely valuable. Since kinetic data have not been previously available, rigorous estimation of the amplification rate was difficult, so methods have been developed to bypass this step in the analysis of amplification data. Most of these techniques rely on standard sequences introduced in known quantities into the experimental design. Assuming that the standard and the target sequence (the one that must be quantified) have identical amplification rates, one can determine the initial copy number of the target, $N_{n,T}$, from the initial quantity of the standard, $N_{0,S}$, and the measurements of the amount of two amplified sequences, $N_{0,S}$ and $N_{n,T}$, without any direct computation of the amplification rate. The basis of this approach is the next relation, which can easily derived from Equation (2):

$$\frac{N_{n,S}}{N_{n,T}} = \frac{N_{0,S}}{N_{0,T}} \left( \frac{m_S}{m_T} \right)^n$$

Given the hypothesis that $m_s = m_T$, the ratio of the standard molecule number over the number of target molecules remains constant after any number of amplification cycles. However, violations of this hypothesis will result in a significant evolution of the ratio over time, changing as $(m_S m_T^{-1})^n$. For instance, when $m_S = 1.9$, $m_T = 1.8$, and $n = 25$, then there is a 3.86-fold difference between $N_{n,S}/N_{n,T}$ and $N_{0,S}/N_{0,T}$. Factors causing different amplification rates for the standard and the target are probably more numerous than factors causing exactly the same amplification rates. Disparity may arise from minor differences in the sequences, from tube–to–tube differences, from sample– to–sample differences, and so on. Since the methodology based on ratios arose at a time when the amplification rates were difficult to measure, it is likely that violations of the hypothesis would not have been detected and taken into account, and quantitative estimations would then have been contaminated with a systematic error.

When Q-PCR experiments are conducted for relative quantification purposes such as the comparison of the quantities of two molecules—a common situation in gene expression and mRNA quantification experiments—the same argument applies. Quantitative differences between the two sequences might be underestimated or exaggerated by small variations in their respective amplification rates.

## The Amplification Rate Estimators

The previous section emphasized the need for a method to estimate the amplification rate. When examined carefully, the requirements for such a method are very stringent.

1. The estimation of the amplification rate of a reaction must be based on the data collected from this reaction only. It cannot be based on a set of related reactions, since there are variations of the amplification rate from one reaction to another.
2. The estimation must be able to detect the end of the so-called "exponential phase," the early phase of the reaction during which the amplification sustains a steady rate of amplification.
3. The estimation must be computed from the measurement of the DNA molecule numbers, and not from the molecule numbers themselves.

Requirement 3 may appear naive but is important since real-world measurements are always contaminated with errors. The most common model of a measurement assumes that a measure, $X_n$, is proportional to the measurand, which in this case is the number of molecules $N_n$, plus a random value $\varepsilon_n$ with a constant statistical distribution that is usually assumed to be Gaussian, $N(0,\sigma^2)$.

$$X_n = a \cdot N_n + \varepsilon_n \tag{3}$$

A convenient way to meet requirement 1 is to use a set of kinetic data collected during a single amplification. Requirement 2 will be met only if the estimator can be computed on a subset of the kinetic data and if it is sensitive enough to detect the decay of the amplification rate in data collected during the late phase of the reaction.

We have previously proposed and characterized the estimator $\hat{m}_n$ (Jacob and Peccoud, 1996a; 1996b), which fits the three requirements aforementioned:

$$\hat{m}_n = \frac{X_{n-2} + X_{n-1} + X_n}{X_{n-3} + X_{n-2} + X_{n-1}} \tag{4}$$

As $n$ grows, this estimator converges exponentially towards the actual value of the amplification rate, $\lim_{n\to\infty} \hat{m}_n = m$. In principle, the speed of convergence depends on two parameters: the amplification rate and the initial copy number. However, simulations of kinetic PCR data with increasing errors demonstrate that the measurement error can significantly delay the observation of this convergence (see Figure 3). Another interesting use of $\hat{m}_n$ is the ability to monitor on a cycle–by–cycle basis, the evolution of the amplification rate during the late phase of the reaction. In this case, the estimator is very close to the actual value of $m$ since the number of cycles is high enough. This capability is particularly valuable for detecting the end of the exponential phase.

In order to avoid confusion that may arise from a previous paper (Peccoud and Jacob, 1996), it is worth noting that $\hat{m}_n$ is not the only possible estimator. There is in fact an entire family of valid estimators that have identical limit properties, although they may differ with respect to their sensitivity to measurement errors. For technical reasons beyond the scope of this paper, only the estimator Equation (4) will be considered here. Finding the most suitable estimator to use on noisy data is still an interesting field of investigation.

Before we proceed to an estimation of amplification rates on real data, another tool is yet required to measure the convergence of the estimator. Graphical representations of the estimations will provide a visual appreciation of their convergence, but an index is needed to quantify the convergence, and one natural index is based on the successive differences of the estimations:

$$\delta_n = \sqrt{\left(\hat{m}_n - \hat{m}_{n-1}\right)^2} = \left|\hat{m}_n - \hat{m}_{n-1}\right|$$

Here $\delta_n$ is the absolute value of the difference between two successive estimations. The lower its value is, the closer are two successive estimations of the amplification rate.

## Data Set

The ABI PRISM™ 7700 Sequence Detector system by PE Applied Biosystems is the first commercially available instrument to produce kinetic data of PCR amplifications. A data set representative of the performance of this instrument was kindly provided to us by Ray Lefebvre and Lincoln McBride of Perkin-Elmer.

A detailed description of the 7700 system can be found elsewhere in this book, in the chapter by Lincoln McBride et al. At installation, the Install Kit is run to validate instrument performance. The details of the protocol used to amplify the kit are provided in the instrument manual. For our purpose, only the general structure of the experiment is of interest

TABLE 1. The 7700 Installation Kit plate layout

| Wells | Initial Copy Number |
|---|---|
| 1–4 | No Template Control |
| 5–8 | 1000 copies standard |
| 9–12 | 2000 copies standard |
| 13–16 | 5000 copies standard |
| 17–20 | 10,000 copies standard |
| 21–24 | 20,000 copies standard |
| 25–60 | 10,000 copies |
| 61–96 | 5000 copies |

(Table 1). The No Template Control (NTC) is a reaction conducted in normal conditions, except that the DNA template solution is replaced by TE buffer. The last two series of replicates (10,000 and 5000) are treated as unknowns. Their initial copy numbers are derived from the standard curve generated from the standards, and compared to their known copy numbers to validate the instrument performance.

The data that will be analyzed in this paper are called the "clipped data" by Perkin Elmer, which is the normalized fluorescence from the reporter dye at the end of each extension phase (FAM dye was the reporter in this case).

## Analysis of the No Template Control and Correction for the Background Trend

Before proceeding to an estimation of the amplification rate, it is necessary to ascertain as much as possible that the data fit the model, Equation (3). One way to do this is to carefully analyze the data of the NTC replicates. Since no reaction occurs in these wells, their fluorescence should not increase during the 40 cycles of amplification; they should only be subject to random fluctuations that result from measurement errors of the instrument.

When the NTC data are plotted together (the four replicates are pooled into a single data set), it appears that their fluorescence has a very significant growth trend which can be observed in Figure 1A. The origin of the growth of the background fluorescence is difficult to figure out. A nonenzymatic degradation of the TaqMan probe may occur during the thermocycling as a result of the incubation at high temperature, or the laser illumination used to excite the fluorescent dyes may gradually break some

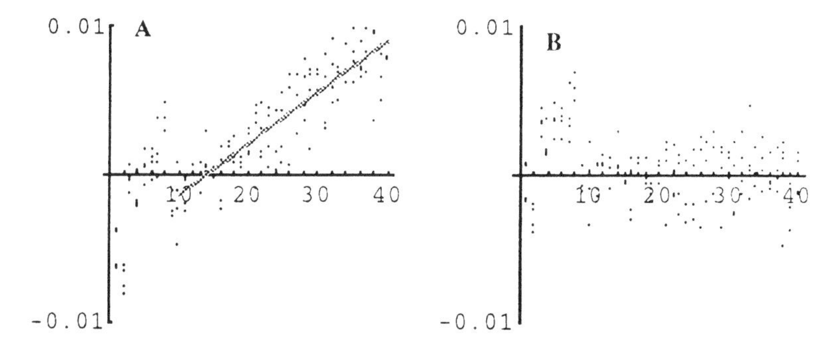


FIGURE 1. Correction of the NTC trend: The No Template Control data are pooled and plotted on A. There is a significant trend that is materialized by the gray line. On B, the data were corrected for the trend. All other data were corrected similarly.

of the bonds of the probe over time. This phenomenon does not seem to be documented so far. The growth of the background fluorescence is not completely linear; there is a peak between cycles 1 and 10, and afterwards the growth is more regular. Since the data collected before cycle 10 are extremely noisy, and thus useless in the computation of the initial copy number, it is not necessary to estimate this peak precisely, and these points can be set aside. The background fluorescence measured between cycle 10 and cycle 40 is fitted to a linear model that can be regarded as the average growth of background fluorescence, shown in gray in Figure 1A. The linear model can be used to refine the data, and corrected background fluorescence is plotted in Figure 1B. The correction is effective since a trend is no longer visible. This can be confirmed statistically by computing the mean value of the refined data collected after cycle 10, which is equal to $-9.7\times10^{-18}$. This is not significantly different from 0. The standard deviation of this subset of the refined background fluorescence is $1.66\times10^{-3}$, a value that can be regarded as the standard deviation of the instrument measurement error, a point that will be addressed in more detail below. All the data are corrected in a similar way, and only this refined data set will be discussed in the following sections of this chapter.

## Analysis of One Amplification Reaction

Let us now apply the tools that were introduced in the previous sections. One reaction, well 7, was chosen to construct Figure 2. The series of $\hat{m}_n$ is

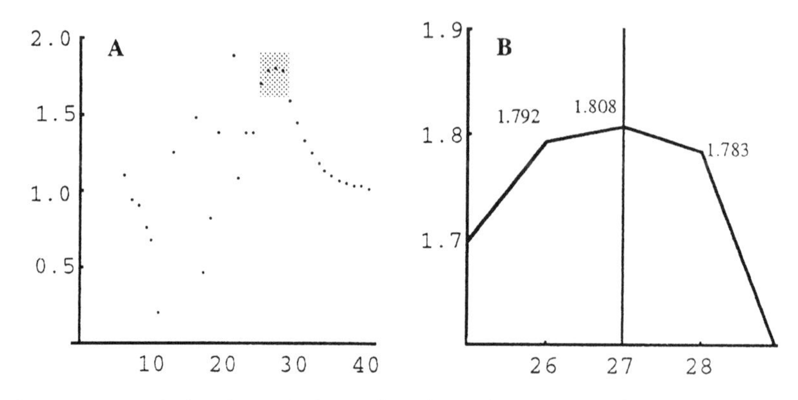


FIGURE 2. Analysis of a typical amplification: The amplification rate is estimated from well 7. Part B is a zoom on the gray area of part A.

first computed. Note that this computation is very simple to implement and could readily be programmed in a spreadsheet application.

The general behavior of the series can be understood from its graphical representation in Figure 2A. Three phases can be distinguished. The first phase extends from cycle 1 to cycle 22; in this range the estimator behavior is extremely erratic, jumping from very low negative values (–4.74 at cycle 4) to very high positive values (2.7 at cycle 20). During this phase, the signal does not rise significantly above the background, and the large fluctuations of the estimator result from the ratios of relatively small numbers. Since the noise fluctuates around 0, the sum of three successive measurements can be either negative or positive; thus, the sign of $\hat{m}_n$ changes frequently. The mean value of the estimator in this phase is 0.95, indicating that despite the large fluctuations, the average amplification rate estimation is close to 1. At this stage, then, no amplification has yet been detected.

The second phase extends from cycle 23 to cycle 27, corresponding to the rising slope of the estimator curve that reaches its maximum at cycle 27. The signal itself starts to rise above the background noise, along with the amplification rate estimator. But since noise is still a large fraction of the signal, the rate estimator grows toward its limit value. Because of the noisy component of the measure, the growth is not always very smooth.

The third and last phase of the reaction appears as an exponential decay of the estimator values, resulting from the end of the exponential phase of the reaction. Since the real amplification rate has started to decrease, its estimator shows a similar evolution, as we would expect. When the ascending and descending slopes of the peak are compared, the latter is clearly much smoother than the former, since, at this time, the estimator has reached its asymptotic behavior, and the noise is no longer large enough to seriously perturb the estimator behavior, at least on the scale of this plot.

Why was the end of the exponential phase set to cycle 27? This is not because the peak reaches its maximum at cycle 27, but rather because it seems that the estimator most closely approaches its limit at cycle 27. If the amplification rate were constant during 40 cycles, then estimator fluctuations would decrease so much that they would allow a very precise determination of the amplification rate value (see Figure 4 and the next section). On this particular data set, the situation is more complex. There are very few cycles in the exponential phase, where the noise is already negligible. Prior to this phase the estimator is erratic, and afterwards it simply follows the decay of the amplification rate. Plotting the evolution of $\delta_n$ helps one to determine the end of the exponential phase. The convergence of $\hat{m}_n$ results in a local minimum for $\delta_{n,}$ and there is actually such a minimum at cycle 27 (see Figure 3). The problem is that there is also a minimum at cycle 24 that is even lower than the one at cycle 27. Why would we not consider cycle 24 as the end of the exponential phase? From Figure 2A, one can see that the small plateau at cycle 24 is more likely due to a random fluctuation than to the convergence of the estimator, because it is too far from the peak maximum. Instead of using the $\delta_n$ plot, the highly magnified zoom on the top of the peak (gray rectangle on Figure 2A) shown in Figure 2B can be used to reach a similar conclusion.

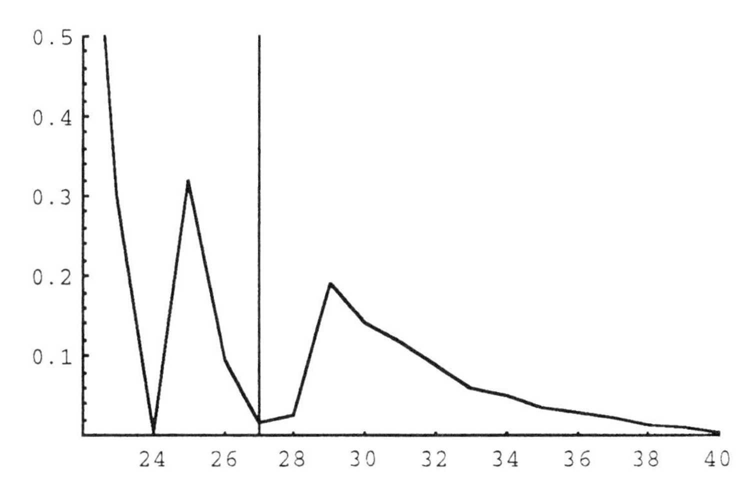


FIGURE 3. Analysis of the $\hat{m}_n$ convergence: The convergence index $\delta_n$ minimum at cycle 27, which is the end of the exponential phase. The minimum at cycle 24 is ascribed to random fluctuations.

TABLE 2. Amplification rate estimations for wells 5 to 24

| Well | Cycle | $\hat{m}_n$ | $\delta_n$ |
|---|---|---|---|
| 5 | 28 | 1.97 | 0.028 |
| 6 | ND | | |
| 7 | 27 | 1.81 | 0.016 |
| 8 | 27 | 1.79 | 0.021 |
| 9 | ND | | |
| 10 | 27 | 1.92 | 0.30 |
| 11 | 28 | 1.78 | 0.31 |
| 12 | ND | | |
| 13 | ND | | |
| 14 | 26 | 1.88 | 0.010 |
| 15 | 26 | 1.85 | 0.032 |
| 16 | 26 | 1.83 | 0.028 |
| 17 | 25 | 1.86 | 0.042 |
| 18 | 24 | 1.89 | 0.023 |
| 19 | ND | | |
| 20 | 23 | 1.88 | 0.009 |
| 21 | 23 | 1.85 | 0.030 |
| 22 | 24 | 1.83 | 0.001 |
| 23 | 24 | 1.79 | 0.028 |
| 24 | 23 | 1.87 | 0.003 |

It is possible to apply this method to analyze all the data collected for the standards in wells 5–24. The results are presented in Table 2. When it was not possible to find a cycle $n$ such that $\delta_n < 0.05$, the result is reported as being Not Determined (ND), meaning that the convergence of the estimator was too perturbed by the measurement error to be reliably observed. When it is possible to determine, the estimation of the end of the exponential phase is reproducible and consistent with the initial copy number. There is a difference of about 4 or 5 cycles between the 20,000 copy reactions and the reactions starting from 1000 copies. The well-to-well variation of the amplification rate estimation is less than $\pm$ 0.04 which compares well with the $\delta_n$ values, so that at this stage of the analysis one cannot evaluate a possible well-to-well difference in the amplification rate. Instead, since the mean value of the amplification rate estimation is 1.855, the differences in the estimations of the end of exponential phase are explained reasonably well by the 20-fold dilution factor, since $1.855^4 = 11.84$, and $1.855^5 = 21.96$.

## Validation of the Model and Fluorescence Calibration

In order to confirm the validity of the model used to build the statistical estimators, it is worthwhile trying to compare the analysis of experimental data with a corresponding analysis of simulated data. Data from well 22 are used for the comparison, since they converge well and thus allow a precise estimate of the amplification rate. In this section, data will be simulated with the addition of an increasing level of noise until the analysis of simulated data matches well with the pattern observed in experimental data.

The computation of the $\hat{m}_n$ allows one to determine that the end of the exponential phase is cycle 24, where the amplification rate estimation is 1.83. For this trajectory, the initial copy number is 20,000. The only unknown parameter is the standard deviation of the measurement error. Noise of increasing levels over a range of several orders of magnitude was successively added to simulated data. The effect of the noise on the convergence of the estimator can be observed in Figure 4. As anticipated, when the noise level becomes too large, it interferes with the ability of the estimator to converge before the end of the exponential phase. For noise with standard deviation of $4 \times 10^8$ molecules, the analysis of the simulated data and the analysis of the experimental data look rather similar. Of course, this appreciation is mainly visual (Figure 5). Since the calibration function of the 7700 is not yet available, data produced by the simulation algorithm were expressed in molecule numbers. The standard deviation of the measurement error used to construct Figure 5 is also expressed in molecule numbers.

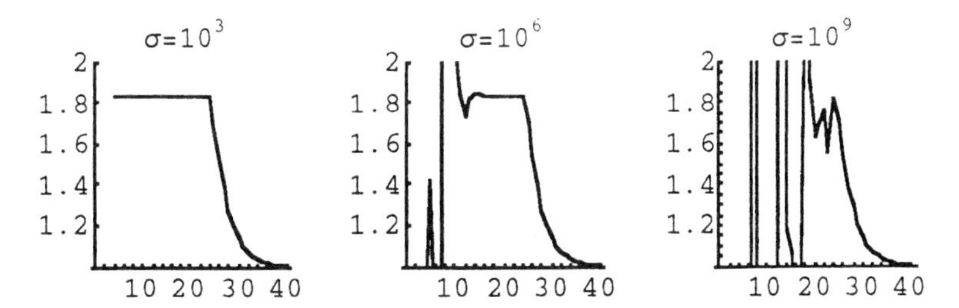


FIGURE 4. Effect of the noise level on the convergence of the amplification rate estimator. Analysis of three sets of simulated data. The parameters of the simulation.

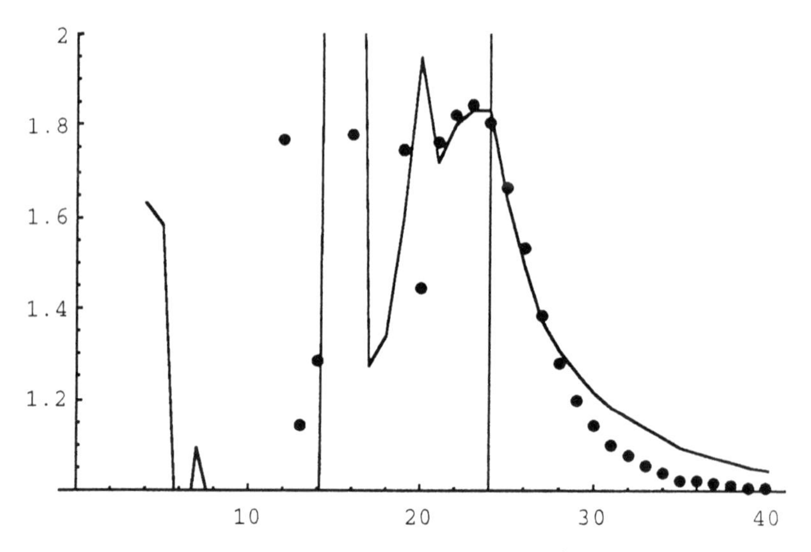


FIGURE 5. Comparison of simulated data with a $4\times10^8$ molecule noise standard deviation with the analysis of the data collected from well 22. For this level of noise, it is possible to obtain analysis patterns that look similar, except in the late phase of the reaction. This comparison is a validation of the model used to compute the amplification rate estimations.

How can these simulated data be converted into the corresponding fluorescence units? Since the initial copy number is high, the fluctuations due to noise are limited. Thus, faithful simulation parameters should ensure that at the end of the exponential phase, i.e. at cycle 24, the data of the simulated amplification, $3.98\times10^{10}$ molecules, and the data measured on well 22, 0.168 fluorescence units, should be approximately equal but expressed on different scales. The ratio of these two values suggests a scaling factor equal to $4.22\times10^{-12}$ fluorescence units/molecule. We can check that this ratio is consistent with other features of our model that have been estimated on different data since it is also possible to compare the two expressions of the standard deviation of the measurement error.

In order to adjust the analysis of simulated data to match the analysis pattern of experimental data, it is necessary to introduce a measurement error of $4\times10^8$ molecules. But the analysis of the NTC data provides another estimation of the error standard deviation: $1.66\times10^{-3}$ fluorescence units. The product of the scaling factor and the error standard deviation, expressed in molecules, should be approximately equal to $1.66\times10^{-3}$:

$$4.22\times10^{-12} \times 4\times10^{8} = 1.68\times10^{-3}.$$

Although the two numbers do not match exactly, they do look very consistent with one another. In summary then, two independent lines of reasoning and computation give consistent results and tend to demonstrate the validity of the model upon which the analysis is based.

## Analysis of the Amplification Rate Decay

In the simulation algorithm used in the previous section, the decay of the amplification rate is assumed to be exponential. The hypothesis is that the amplification rate loses a few percent every cycle after the end of the exponential phase. If $ne$ is the last cycle in the exponential phase, and if the rate decreases by $\tau$% per cycle, then for any $n > ne$, the amplification rate used in the simulation is:

$$m_n = m \cdot (1-\tau)^{n-ne}.$$

It was puzzling that this type of decay did not permit us to match the late phase of the simulations with the experimental data. This is visible in Figure 5, where it is apparent that there is a discrepancy between the two plots in the late phase of the reaction. This was confirmed when $\log(\hat{m}_n)$ was plotted as a function of $n$. It became obvious that the relationship between these two quantities was not linear. Since it still seemed to be decreasing very rapidly, we tried plotting $\log(\log(\hat{m}_n))$ against $n$, and in this case a linear relationship finally appeared. The slope of this line is on the order of –0.18. Analysis of several trajectories gives similar results. This fit means that the time evolution of the amplification rate in the late phase of the reaction can be represented by:

$$m_n = m \cdot \exp\left(\exp\left(-\tau\left(n-ne\right)\right)\right).$$

This is a spectacular decay. The end of the exponential phase is not marked by a slightly decreasing amplification rate, but rather by a total collapse of the reaction yield. It will be interesting to see what a possible biochemical or biophysical explanation of this observation might be.

## Perspective

The analysis of the data collected on only a few wells is presented in this paper. Some of the wells (5 out of 16) did not show a convergence strong enough to allow one to reliably determine the end of the exponential phase or the amplification rate. Moreover, at this stage of our research there is still a need for an objective criterion that could be used to completely automate the analysis of kinetic PCR data. The measurement error strongly affects the convergence behavior of the amplification rate estimator. In many cases it seems that the amplification rate starts to decrease before the signal-to-noise ratio increases enough to allow the amplification rate to be reliably determined.

The relative measurement uncertainty of the 7700 can be computed and is approximately equal to 1% at the end of the exponential phase. For a DNA quantification protocol, this precision is extremely good. However, it still limits the ability to determine the amplification rate with the greater accuracy needed for a precise estimation of the initial copy number. Another possibility currently being investigated is the use of estimators of the amplification rate that are less sensitive to measurement errors. In the near future, we expect that a combination of more effective estimators applied on data of higher quality will allow one to determine the amplification rate with higher precision.

Though it is still in its infancy, the statistical estimation of the amplification rate provides valuable results. The analysis of the reactions can be performed without any standard. Analysis of several replicates of identical reactions (wells 25–60 and 61–96) tends to indicate that there may be significant well-to-well variability in the amplification rate (data not shown), even though experimental conditions are identical. If confirmed, this result would indicate that the reaction is sensitive to parameters that are not controlled by current experimental setup. This would be another strong argument in favor of individual amplification rate estimations. In the most favorable reactions, it is already possible to determine the amplification rate with a $10^{-2}$ accuracy. As soon as the data measured by the kinetic PCR instrument can be calibrated, these amplification rate estimations could be used for an absolute quantification of the initial copy number. Comparison of the analysis of NTC data with simulated data might produce an original and powerful way to calibrate the instrument.

Finally, whatever the current limits of the analysis of kinetic PCR data as presented in this paper, it has a very nice quality. It is self-validating and does not rely upon assumptions or hypotheses that cannot be verified. The only hypotheses used are qualitative and related to the dynamics of the Polymerase Chain Reaction and to the measurement of DNA molecule numbers. The comparison of simulated data with experimental data demonstrates their validity. Measurement errors are still limiting the accuracy of the amplification rate estimation. It is possible to compute a convergence index that reflects the quality of the estimates. High values of this index mean that the data are too noisy to be analyzed accurately. This convergence index should provide a solid basis by which to compare the quality of the data collected on different kinetic PCR machines.

## REFERENCES

Higuchi R, Dollinger G, Walsh PS, and Griffith R (1992): Simultaneous amplification and detection of specific DNA sequences. *Bio/Technology* 10:413–17.

Higuchi R, Fockler C, Dollinger G, and Watson R (1993): Kinetic PCR analysis: real-time monitoring of DNA amplification reactions. *Bio/Technology* 11:1026–30.

Jacob C and Peccoud J (1996a): Estimation of the offspring mean for a supercritical branching process from partial and migrating observations. *C R Acad Sci Paris Série I* 322:736–68.

Jacob C and Peccoud J (1996b): Inference on the initial size of a supercritical branching process from partial and migrating observations. *C R Acad Sci Paris Série I* 322:875–80.

Karrer EE, Lincoln JE, Hogenhout S, Bennett AB, Bostock RM, Martineau B, Lucas WJ, Gilchrist DG, and Alexander D (1995): *In situ* isolation of mRNA from individual plant cells: Creation of cell-specific cDNA libraries. *Proc Natl Acad Sci USA* 92:3814–18.

Lantz O and Bendelac A (1994): An invariant T cell receptor alpha chain is used by a unique subset of major histocompatibility complex class I-specific CD4+ and CD4– CD8– T cells in mice and humans. *J Exp Med* 180:1097–106.

Peccoud J and Jacob C (1996): Theoretical uncertainty of measurements using quantitative polymerase chain reaction. *Biophys J* 71:101–8.

Piatak M, Saag MS, Yang LC, Clark SJ, Kappes JC, Luk KC, Hahn BH, Shaw GM, and Lifson JD (1993): High Levels of HIV-1 in plasma during all stages of infection determined by competitive PCR. *Science* 259:1749–54.

# Fluorescence Monitoring of Rapid Cycle PCR for Quantification

Carl Wittwer, Kirk Ririe, and Randy Rasmussen

## Introduction

The polymerase chain reaction (PCR) benefits from rapid temperature cycling (Wittwer et al., 1994). In particular, rapid cycling appears to improve the quantitative PCR of rare transcripts (Tan and Weis, 1992). The glass capillaries used as sample containers for rapid cycling are natural cuvettes for fluorescence analysis. Fluorometric monitoring of PCR has been reported with double-stranded DNA (dsDNA) dyes (Higuchi et al., 1992; Higuchi et al., 1993; Ishiguro et al., 1995; Wittwer et al., 1997a) and sequence-specific probes (Lee et al., 1993; Livak et al., 1995; Wittwer et al., 1997a). We have integrated a fluorimeter with a rapid temperature cycler for fluorescence monitoring during amplification (Wittwer et al., 1997b). Both cycle-by-cycle fluorescence monitoring and continuous (within cycle) monitoring offer unique quantitative information.

## Instrumentation

The design of the fluorescence temperature cycler is based on flow cytometry optics and earlier rapid temperature cycling instruments (Wittwer et al., 1991; Wittwer et al., 1994). Instead of a high-intensity halogen bulb for heating, hot and/or cool air are continuously pumped into the sample chamber for temperature control (Figure 1). The samples are placed in glass capillary tubes arranged vertically around a carousel that positions the tubes for fluorescence acquisition. A xenon arc is filtered to 450–490 nm for illumination of a 5–7 mm length of capillary. Collection optics include filters for SYBR® Green I (520–580 nm), fluorescein (520–550 nm), rhodamine (580–620 nm), and Cy5 (660–680 nm).

Instrument control and data display are handled by a custom LabView graphical interface (National Instruments, Austin, TX). For multiple tubes, the carousel rapidly positions each tube sequentially at the optical focus for 100 msec acquisitions. For continuous monitoring of a single tube, data are averaged and acquired every 200 msec. Time, temperature, and two channels of fluorescence are continuously displayed as fluorescence vs. cycle

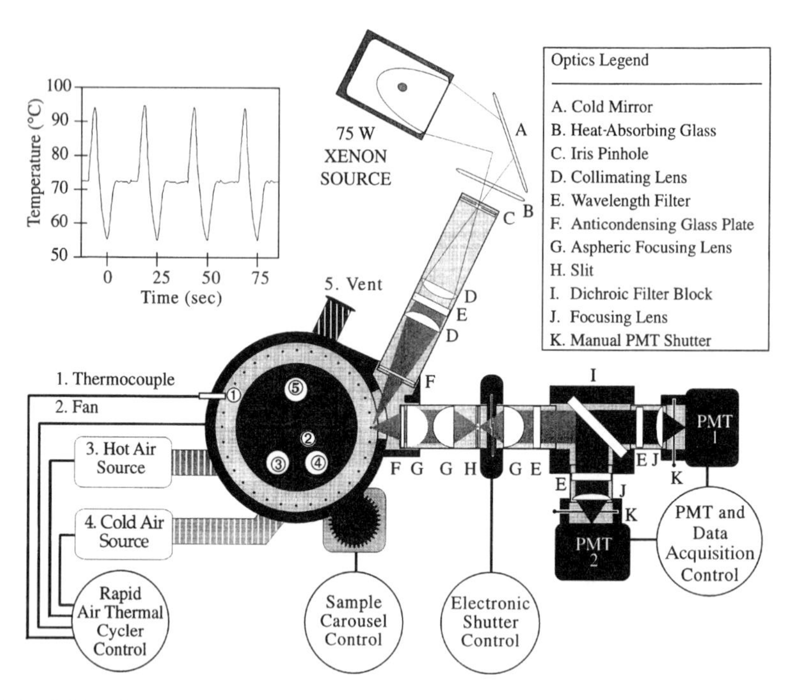


FIGURE 1. Diagram of the rapid temperature cycler with fluorescence capability. The temperature cycler is on the left, excitation optics on the top, and collection optics on the right. The sample temperature is monitored by a tubular thermocouple (1) that is matched in temperature response to samples placed vertically around a circular sample chamber. Temperature homogeneity is achieved by a central high–velocity fan (2). The temperature is controlled by hot air (3) and cold air (4) blown into the chamber. A vent (5) provides an air exit. All air paths are through serpentine tubes that exclude outside light. Typical sample temperature vs. time traces are shown at upper left (programmed for 94°C maximum, 55°C minimum, and 10 sec at 72°C). A 75–watt xenon light source (top) is focused to an entrance iris with an elliptical reflector. Most infrared radiation is removed through a cold mirror and heat–absorbing glass. After collimation and spectral filtering as desired, light is focused on a 5–10 $\mu$l amplification sample, positioned by stepper motor control of the carousel. Emitted light is imaged through a slit, collimated, spectrally separated by dichroic and secondary filters, and finally focused onto photomultiplier tubes. Reprinted with permission of Eaton Publishing from *BioTechniques*, Wittwer et al., 1997b.

number and fluorescence vs. temperature plots. Details of the instrumentation are described elsewhere (Wittwer et al., 1997b). Melting and reannealing curves were obtained on a commercial instrument (Light Cycler LC24, Idaho Technology, Idaho Falls, ID).

## Cycle-by-Cycle Monitoring

The analysis of PCR products is usually performed after amplification is complete. When used for quantitative PCR, these endpoint methods require good estimates of initial template concentration for accurate results. The potential utility of fluorescence monitoring every cycle for quantitative PCR was first demonstrated with ethidium bromide (Higuchi et al., 1993). Fluorescence is usually acquired once per cycle during a combined annealing/extension phase. Ethidium bromide fluorescence is enhanced when intercalated into dsDNA and is a measure of product concentration. Monitoring the amount of product once per cycle with dsDNA dyes is an important advance that allows a wide dynamic range of initial template concentrations to be analyzed.

*Double–Stranded DNA Dyes.* We have used SYBR® Green I (Molecular Probes, Eugene, OR) as a sensitive dye for following PCR product accumulation (Wittwer et al., 1997a). As shown in Figure 2, quantification of initial template copy number is possible over a $10^8$-fold range. The fluorescence curves are displaced horizontally according to the initial template concentration. The sensitivity is limited at low initial template concentrations because amplification specificity is not perfect. When no template is present, undesired products such as primer dimers are eventually amplified. There is little difference between 0, 1, and 10 average initial template copies.

Although quantification of low initial copy numbers is difficult with dsDNA dyes (Higuchi et al., 1993; Wittwer et al., 1997a), the simplicity of using these dyes is attractive. The dyes can be used for any amplification and custom fluorescently-labeled oligonucleotides are not necessary. Quantification of very low copy numbers with dsDNA dyes will require either improved amplification specificity or a means to differentiate the desired product from nonspecific amplification.

*Sequence–Specific Fluorescence Monitoring.* Sequence–specific detection of PCR products can be obtained by including two different fluorophores on oligonucleotide probes. Two general schemes are possible (Figure 3). Both schemes require a donor with an emission spectrum that overlaps the absorbance spectrum of an acceptor.

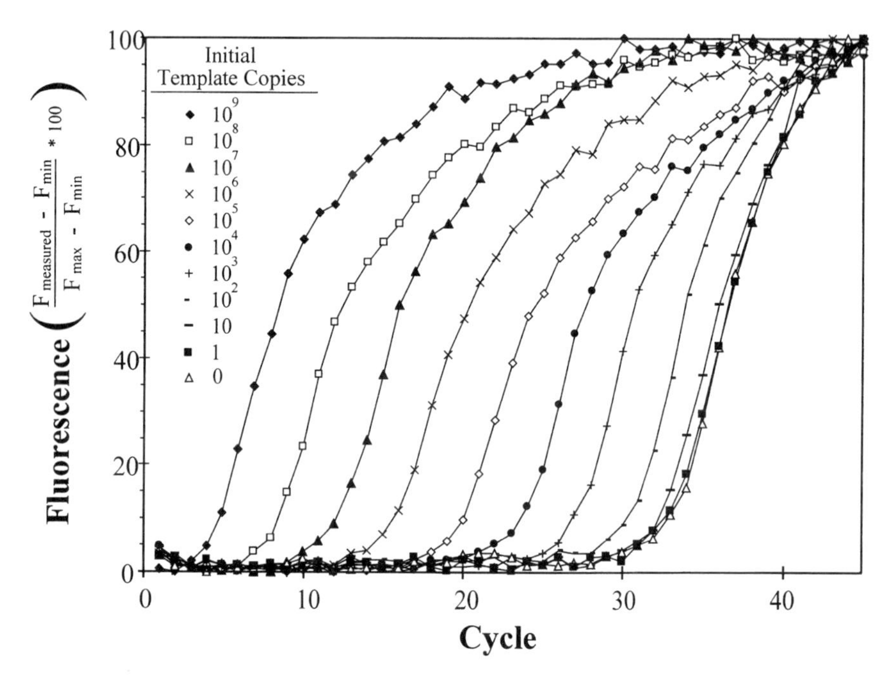


FIGURE 2. Fluorescence vs. cycle number plot of DNA amplification monitored with the dsDNA dye SYBR® Green I. A 536 base pair fragment of the human *β*-globin gene (Wittwer et al., 1991) was amplified from 0 to $10^9$ average template copies in the presence of a 1:10,000 dilution of SYBR® Green I. Purified template DNA was obtained by PCR amplification, phenol/chloroform extraction, ethanol precipitation, and ultrafiltration (Centricon 30, Amicon, Danvers, MA) and quantified by absorbance at 260 nm. Each temperature cycle was 28 sec long (95°C maximum, 61°C minimum, 15 sec at 72°C, average rate between temperatures 5.2°C/sec). All samples were amplified simultaneously and monitored for 100 msec between seconds 5 and 10 of the extension phase. The display was updated each cycle for all tubes, and 45 cycles were completed in 21 min. The data for each tube were normalized to a percentage of the difference between minimum and maximum values for each tube (y-axis). Reprinted with permission of Eaton Publishing from *BioTechniques*, Witter et al., 1997a.

If the donor and acceptor are synthesized on the same oligonucleotide, the acceptor dye quenches the fluorescence of the donor because of their proximity (Figure 3A). During temperature cycling, some of the probe hybridizes to single–stranded PCR product and may be hydrolyzed by polymerase 5′-exonuclease activity (Lee et al., 1993; Livak et al., 1995). Because the donor and acceptor are no longer linked, the donor is released from acceptor

A. Release from quenching by hydrolysis.

• Quenching of donor by acceptor

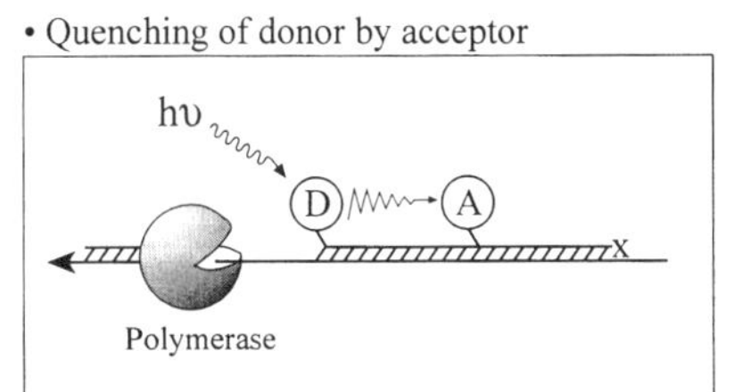


• Increased donor fluorescence after hydrolysis

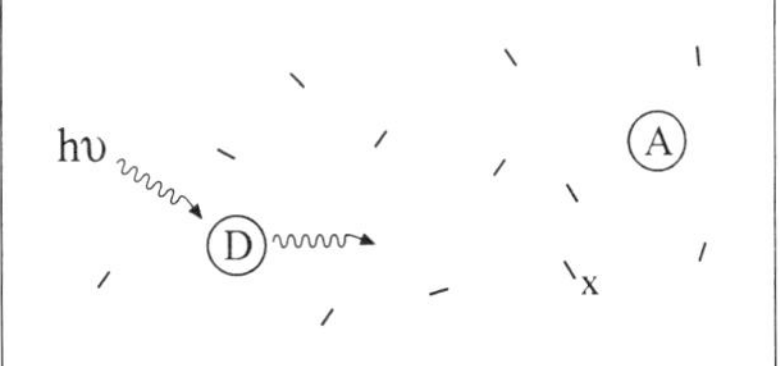


B. Increased resonance energy transfer by hybridization.

•Baseline donor fluorescence

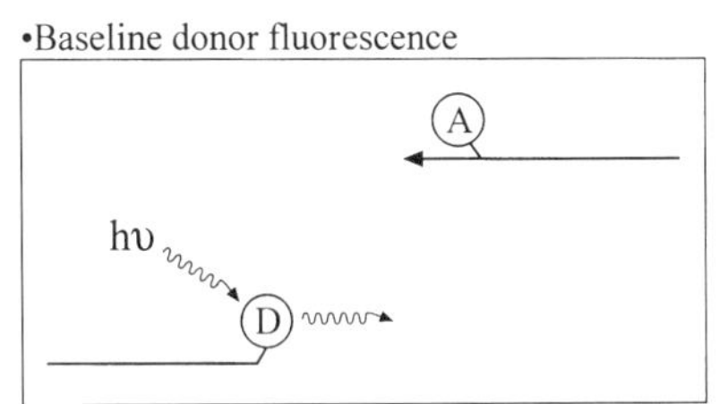


•Increased donor transfer to acceptor

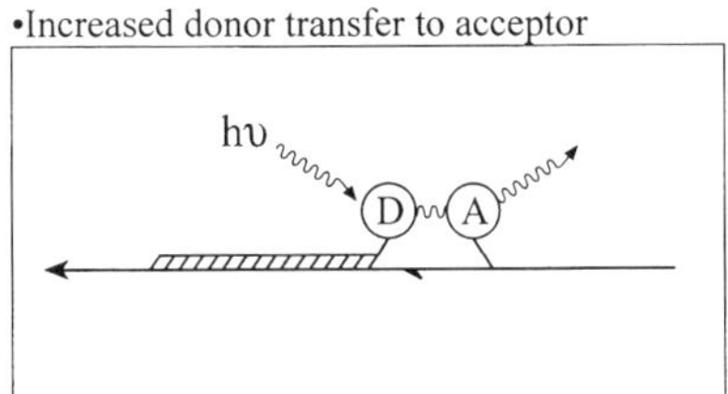


FIGURE 3. Two different schemes for sequence-specific fluorescence monitoring during PCR. In scheme A, a donor (D) is initially quenched by an acceptor (A), because both are together on a single oligonucleotide probe. After hydrolysis by polymerase 5′-exonuclease activity, the donor and acceptor are separated and the fluorescence of the donor increases. In scheme B, the donor is on a probe, and the acceptor is on one of the primers. The dyes on the probe and primer are approximated by hybridization. This results in resonance energy transfer to the acceptor, increasing the acceptor fluorescence. Portions of this figure were reprinted with permission of Eaton Publishing from *BioTechniques*, Wittwer et al., 1997a.

quenching, and the donor fluorescence increases. Since the fluorescence signal is dependent on probe hydrolysis instead of hybridization, we refer to this class of probes as "hydrolysis" probes.

A different fluorescence scheme is possible if the donor and acceptor are placed on different oligonucleotides (Wittwer et al., 1997a). In Figure 3B, the donor is placed at the 3′ end of a probe, and one primer is labeled with the acceptor. There is minimal interaction between the dyes because they are on separate oligonucleotides. During temperature cycling, the probe hybridizes near the labeled primer, and resonance energy transfer occurs. Because the fluorescence signal is dependent on probe hybridization, we refer to this class of probes as "hybridization" probes.

Whether the donor fluorescence is merely quenched or actually increases acceptor fluorescence depends on the specific fluorophores and solvent conditions. An increase in donor fluorescence is usually observed during PCR with hydrolysis probes, whereas an increase in acceptor fluorescence is usually monitored with hybridization probes.

*Hydrolysis Probes.* In contrast to dsDNA dyes, sequence–specific probes can easily quantify very low initial template numbers (Figure 4). A single template copy can apparently be distinguished from the absence of template. The efficiency of amplification decreases as the initial copy number drops below 1500. Although the reason for this drop in efficiency is unknown, one possibility is that amplification of alternate, undesired tem-

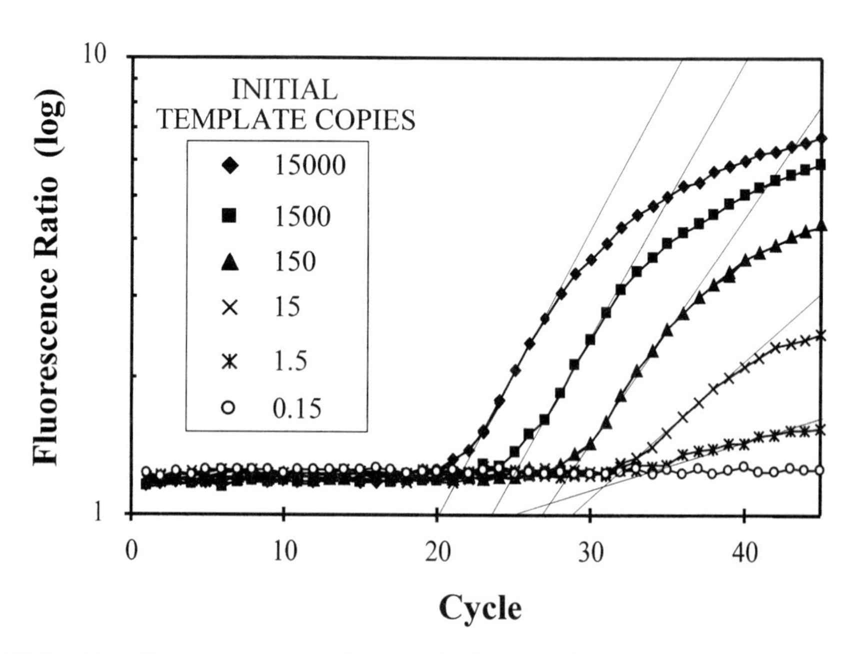


FIGURE 4. Fluorescence ratio (fluorescein/rhodamine) vs. cycle number plot of DNA amplification monitored with a dual-labeled hydrolysis probe. The thin lines indicate the log-linear portion of each curve. A 280 base pair fragment of the human $\beta$-actin gene was amplified from 0.15 to 15,000 (average) copies of the human genomic DNA per tube. The hydrolysis probe (included at 0.2 $\mu$M) was from Perkin Elmer (Foster City, CA). Each temperature cycle was 26 sec long (94°C maximum, 60°C for 15 sec, average rate between temperatures 6.2°C/sec). All samples were amplified simultaneously and monitored for 100 msec between seconds 5 and 10 of the annealing/extension phase. The display was updated each cycle for all tubes, and 45 cycles were completed in under 20 min.

plates consumes limiting reagents. Note that the fluorescence signal continues to increase even after many cycles, because hydrolysis is cumulative and not strictly related to product concentration.

*Hybridization Probes.* In contrast to hydrolysis probes, the fluorescence signal from hybridization probes is not cumulative and develops anew during each annealing phase. The fluorescence is a direct measure of product concentration because the hybridization is a pseudo-first-order reaction (Young and Anderson, 1985). Because the concentration of probe is much greater than the product, the fraction of product hybridized to probe is independent of product concentration.

The fluorescence signals from SYBR® Green I, hydrolysis probes, and hybridization probes are directly compared in Figure 5. All probes have nearly the same sensitivity with detectable fluorescence occurring around cycle 20. With extended amplification, the signal continues to increase with the hydrolysis probe, is level with SYBR® Green I, and slightly decreases with the hybridization probe. The decreasing signal after 35 cycles with the hybridization probes may be caused by probe hydrolysis from polymerase exonuclease activity. Although the change in fluorescence ratio from the hydrolysis probe is greater than that from the hybridization probe (Figures 5B and 5C), the coefficient of variation of fluorescence from the hydrolysis probes is greater (Figure 5D). That is, fluorescence measurements resulting from hybridization probes are more precise than ones obtained with hydrolysis probes, even though the absolute signal levels are lower.

*Algorithms for Quantification Using Cycle-by-Cycle Monitoring.* Although it is obvious that quantitative information about initial template concentrations is contained in fluorescence curve sets such as Figures 2 and 4, it is not obvious how to best extract this quantitative information. Consider Figure 6, where curves for the amplification of $10^4$ and $10^5$ copies of purified template (taken from Figure 2) are compared to the amplification of 50 ng of genomic DNA. The genomic DNA curve is just to the left of the $10^4$ curve, implying slightly more than $10^4$ copies of the target. But how much more? What is the best estimate from the fluorescence curves?

There are two general approaches to this interpolation problem. We can interpolate entire curves, attempting to use all relevant data points. This curve-fitting approach should give the most accurate answer. Alternately, we can interpolate the unknown along single vertical or horizontal lines. Single vertical interpolation is similar to conventional endpoint analysis. At a certain cycle number, the unknown fluorescence is interpolated between known values. Table 1 lists results from vertical interpolation of cycles 20–24. Interpolation along horizontal lines requires establishing a fluorescence threshold. Interpolation from fluorescence thresholds of 10, 20, 30, and 40% are also listed in Table 1.

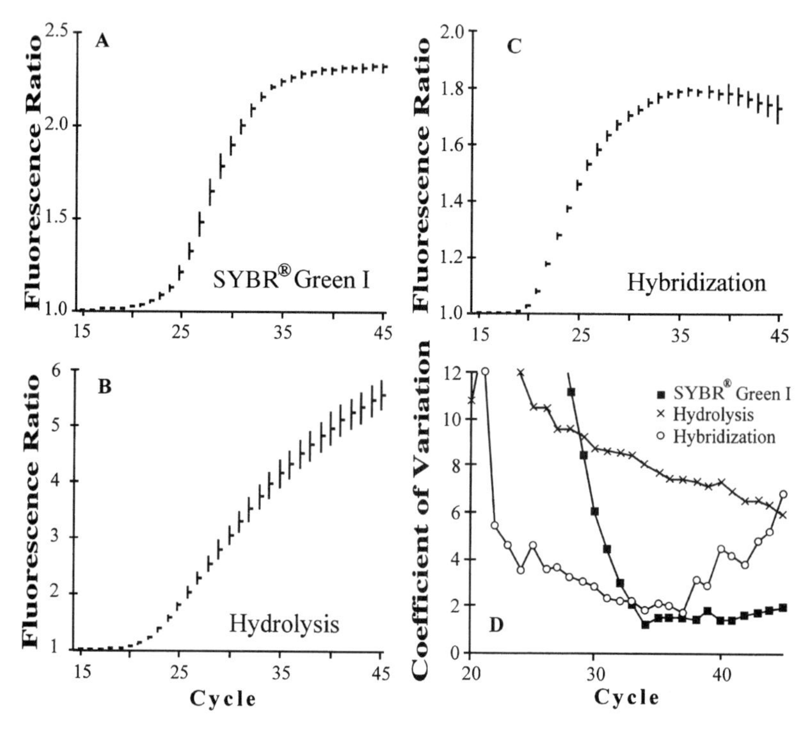


FIGURE 5. Comparison of three fluorescence monitoring techniques for PCR. The fluorescence probes are the dsDNA dye SYBR® Green I (A), a dual-labeled fluorescein/rhodamine hydrolysis probe (B), and a fluorescein-labeled hybridization probe with a Cy5-labeled primer (C). All amplifications were performed in ten replicates with 15,000 template copies (50 ng of human genomic DNA/10$\mu$l). The temperature cycles were 31 sec long (94°C maximum, 60°C for 20 sec, average rate between temperatures 6.2°C/sec). Fluorescence was acquired for each sample between seconds 15 and 20 of the annealing/extension phase. The mean +/– standard deviations are plotted for each point. SYBR® Green I was used at a 1:10,000 dilution in the amplification of a 205 base pair human $\beta$-globin fragment from primers KM29 and PC04 (Wittwer et al., 1989). The hydrolysis probe and conditions are described in Figure 4. The hybridization probe, TCTGCCGTTACTGCCTGTGGGGCAAG-fluorescein (from fluorescein-CPG, Glen Research, Sterling, VA) was used with KM29 and the Cy5-labeled primer CAACTTCATCCACGTXCACC, where X was an amino-modifier C6dT (Glen Research) labeled with Cy5 (MonoReactive, Amersham, Arlington Heights, IL). The precision of the three fluorescence monitoring techniques are compared in (D). The data are plotted as the coefficient of variation (standard deviation/mean) of the fluorescence ratio above baseline (taken as the average of cycles 11–15).

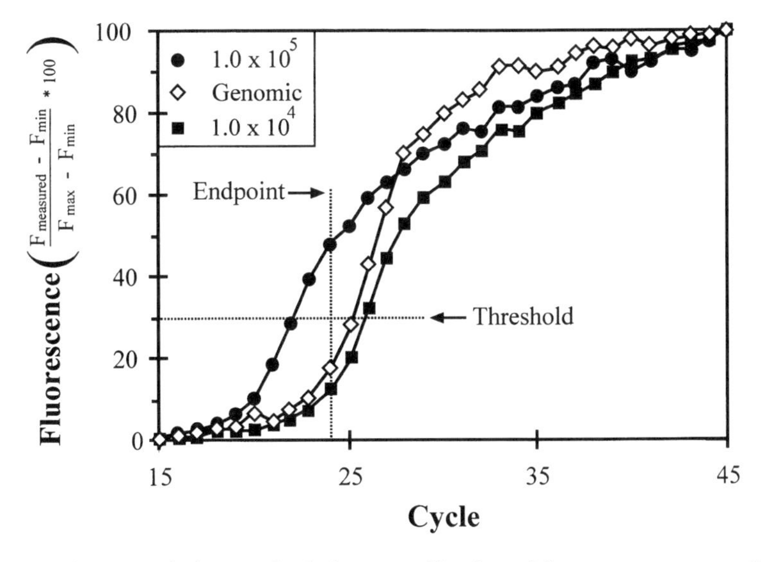


FIGURE 6. Interpolation methods for quantification of fluorescence curves. Endpoint interpolation compares the fluorescence of unknowns and standards at a given cycle number. Threshold interpolation determines the fractional cycle number at which each curve exceeds a given fluorescence.

TABLE 1. Single–line interpolation methods for fluorescence quantification (see Figure 6).

| EndPoint | | Threshold | |
|---|---|---|---|
| Cycle | Copies | Fluorescence (%) | Copies |
| 20 | 41,000 | | |
| 21 | 12,000 | 10 | 17,000 |
| 22 | 12,000 | 20 | 14,000 |
| 23 | 13,000 | 30 | 15,000 |
| 24 | 14,000 | 40 | 18,000 |
| Avg | 18,000 | Avg | 16,000 |

Endpoint or threshold interpolations do not use much of the available data and can be highly affected by single aberrant data points, e.g., endpoint interpolation at cycle 20. A better approach is to interpolate curves that fit the data well. For example, an exponential fit for data within the log-linear portion of each curve should follow: $F = AN(1 + E)^n$, where $F$ is the fluorescence signal, $A$ is a fluorescence scaling factor, $N$ is the starting copy number, $E$ is the amplification efficiency, and $n$ is the number of cycles. $A$ and $E$ could first be determined for values of $N$ that bracket the unknown. Then, the best value of $N$ for the unknown could be determined. Using this approach, the gene copy number in 50 ng of human genomic DNA is $1.7 \times 10^4$, near the middle of values in Table 1 and close to the accepted value of $1.5 \times 10^4$ (assuming 6.6 pg/cell). Other curve fits may be better than the exponential example given above. The exponential fit necessarily relies only on the log-linear cycles (7 out of 45), and these fall in a region where the fluorescence precision is poor (Figure 5D). Ideally, a sigmoidal curve could be used, which would include parameters that described the lag phase, the log-linear phase, and the plateau phase, thus using all the available data. The fit could be weighted to reflect the expected precision of the data points. Different parameters would be necessary for different fluorescence probes because of variation in the curve shapes (Figure 5).

## Continuous (Within Cycle) Monitoring

Fluorescence monitoring during PCR is usually done once each cycle at a constant temperature. A stable temperature is important because fluorescence changes as a function of temperature. However, monitoring fluorescence during temperature changes can be very informative. For example, if fluorescence is acquired continuously throughout temperature cycling, hydrolysis probes show a linear change in fluorescence ratio with temperature and a parallel increase in fluorescence as more probe is hydrolyzed (Figure 7A). In contrast, the fluorescence ratio from hybridization probes varies radically with temperature (Figure 7B). During probe hybridization at low temperatures, the ratio increases, followed by a decrease to baseline when the hybridization probe melts off the template.

The temperature dependence of product strand status during PCR is revealed by fluorescence vs. temperature plots using SYBR® Green I (Figure 8). As amplification proceeds, temperature cycles appear as rising loops between annealing and denaturation temperatures. As the sample is heated, fluorescence is high until denaturation occurs. As the sample cools, fluorescence increases, reflecting product reannealing. When the temperature is constant during extension at 72°C, increasing fluorescence correlates with additional DNA synthesis.

The ability to monitor product denaturation and product reannealing suggests additional methods for DNA quantification. These methods require fluorescence monitoring within individual temperature cycles and cannot be used when fluorescence is only acquired once per cycle. For

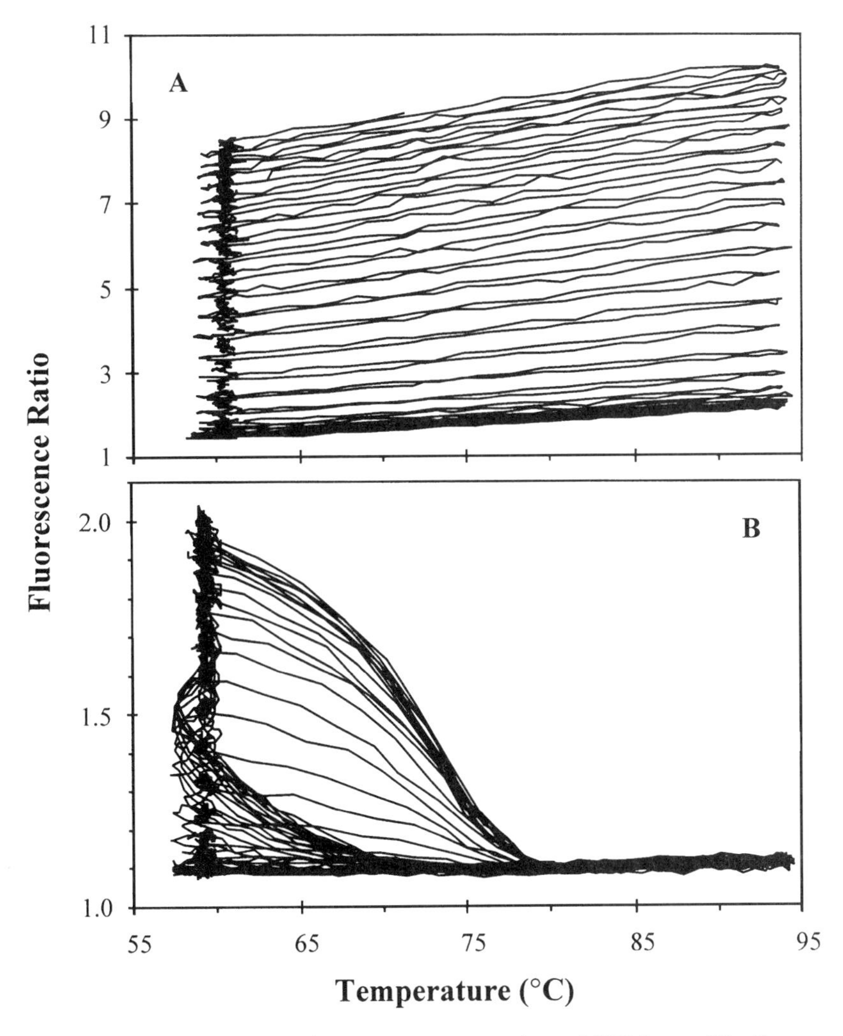


FIGURE 7. Fluorescence ratio vs. temperature plots of DNA amplification continuously monitored with hydrolysis (A) and hybridization (B) probes. Experimental conditions for the hydrolysis probe are described in Figure 4 and those for the hybridization probe in Figure 5. Part A is reprinted with permission of Eaton Publishing from *BioTechniques,* Wittwer et al., 1997a.

example, PCR products have characteristic melting curves dependent on product GC/AT ratio and length (Figure 9). Indeed, empirical algorithms used to predict melting temperatures (Wetmur, 1995) indicate that various PCR products should melt over a 50°C range. Since nonspecific products

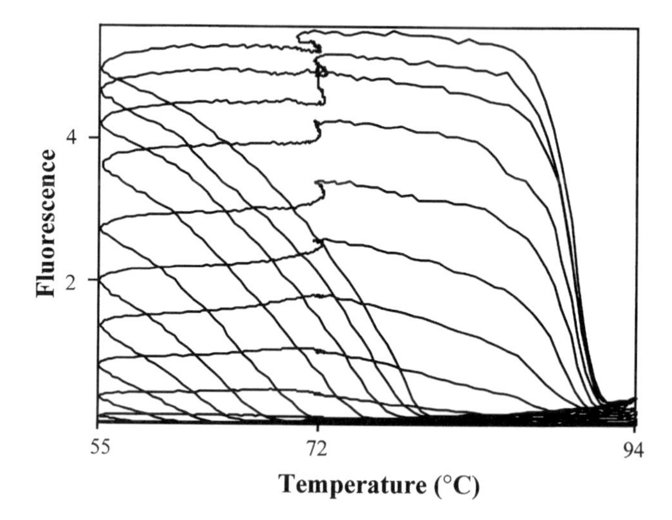


FIGURE 8. Fluorescence vs. temperature plot of DNA amplification with the dsDNA dye SYBR® Green I. A 536 base pair fragment of the human *β*-globin gene was amplified from $10^6$ copies of purified template and a 1 : 30,000 dilution of SYBR® Green I. Other conditions are described in Figure 2. Cycles 15–25 are displayed. The fluorescence vs. temperature data have been transformed to remove the effect of temperature on fluorescence.

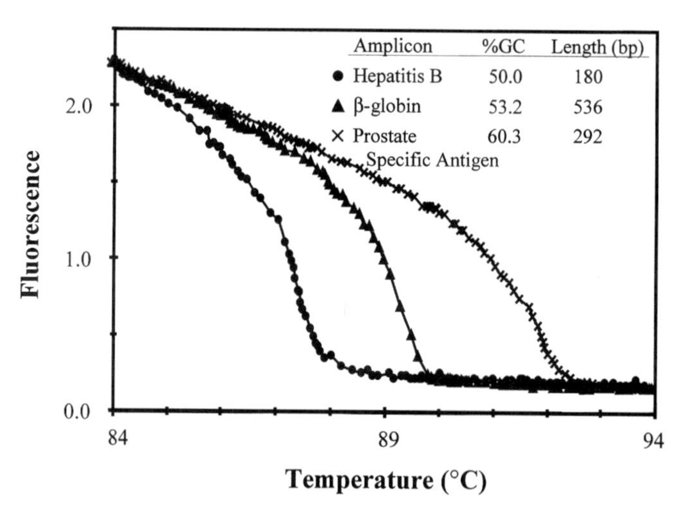


FIGURE 9. Melting curves of three purified PCR products. PCR products were purified and quantified as in Figure 2. A 1 : 10,000 dilution of SYBR® Green I was added to 50 ng of DNA in 10 $\mu$l of 50 mM Tris, pH 8.3, and 3 mM $MgCl_2$. The temperature was increased at 0.2°C/sec and the fluorescence acquired and plotted against temperature.

will melt differently from the desired PCR product, dsDNA fluorescence can be separated into specific and nonspecific components (Ririe et al., 1997). This should allow quantification of very low copy numbers using generic dsDNA dyes such as SYBR® Green I or ethidium bromide.

*Melting Curves for Competitive Quantification.* An additional use for melting curves can be envisioned for competitive quantitative PCR. This technique requires a competitive template for co–amplification with the native template. Control templates could be designed that differ in melting temperature from the natural amplicon. A melting temperature shift of 3°C can be obtained by changing the product GC percentage by 7–8% (Wetmur, 1995). A 3°C shift in the melting temperature should be enough to separate the native and competitive components of the product melting curve (Figure 9). These melting curves, acquired during amplification, would be differentiated according to melting peaks (Hillen et al., 1981) and used to determine the relative amounts of competitor and template. This should eliminate any need to analyze the sample after temperature cycling.

*Hybridization Kinetics for Absolute Quantification of DNA Generated During PCR.* Monitoring of product-product reannealing after denaturation provides a potential method for direct, absolute quantification of amplified product. If the sample temperature is quickly dropped from the denaturation temperature and held at a lower temperature, the rate of product reannealing should follow second-order kinetics (Young and Anderson, 1985). When different concentrations of DNA are tested, the shape of the reannealing curve is characteristic of the DNA concentration (Figure 10). For any given PCR product and temperature, a second-order rate constant can be measured. Once the rate constant is known, an unknown DNA concentration can be determined from experimental reannealing data. Initial determination of a rate constant is demonstrated in Figure 11. Cooling is not instantaneous, and some reannealing occurs before a constant temperature is reached. Rapid cooling maximizes the amount of data available for rate constant or DNA concentration determination. The technique requires pure PCR product, but this can be ensured by melting curves also obtained during temperature cycling. This method of product quantification by reannealing kinetics is independent of any signal variation between tubes. With appropriate controls, absolute product quantification can be used to estimate the initial target copy number.

## Summary

Fluorescence monitoring of PCR is a powerful tool for DNA quantification. Cycle-by-cycle monitoring using dsDNA dyes does not require unique probes, and allows quantification over a large dynamic range. Sequence-specific

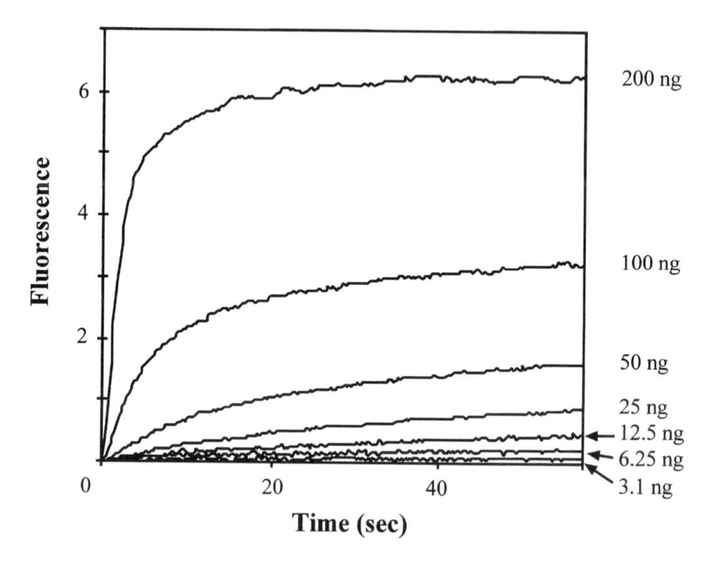


FIGURE 10. Reannealing curves for different concentrations of PCR product. Different amounts of a purified 536 base pair fragment of DNA (Figure 2) were mixed with a 1:30,000 dilution of SYBR® Green I in 5 $\mu$l of 50 mM Tris, pH 8.3 and 3 mM $MgCl_2$. The samples were denatured at 94°C and then rapidly cooled to 85°C.

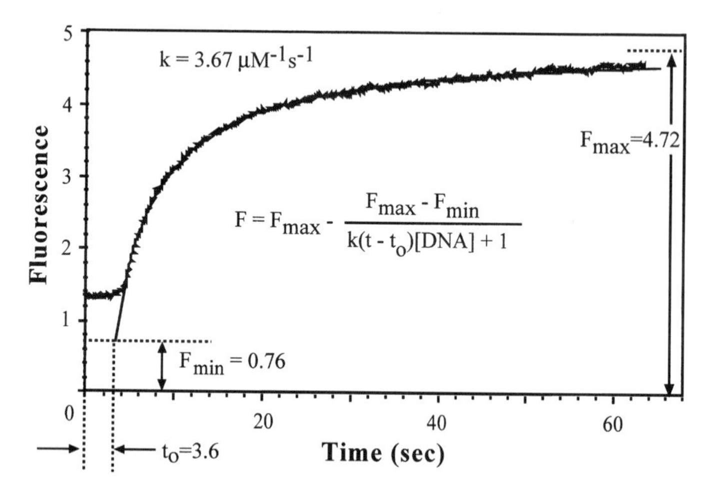


FIGURE 11. Determination of a second–order reannealing rate constant. The curve was fit by nonlinear least squares regression with $F_{max}$, $F_{min}$, $t_o$, and $k$ as the floating parameters. With the rate constant ($k$) determined, DNA concentrations can subsequently be determined on unknown samples.

fluorescent probes offer enhanced specificity and can be used for the quantification of only a few copies of template per reaction. There are at least two classes of sequence-specific, fluorescent oligonucleotide probes that are useful in PCR. The mechanism of signal generation naturally separates them into either hydrolysis or hybridization probes. By monitoring the fluorescence once each cycle, either probe system can be used for quantification. Continuous monitoring of fluorescence allows acquisition of melting curves and product annealing curves during temperature cycling. These techniques can be integrated with rapid cycle PCR for combined amplification and complete quantitative analysis in under 15 minutes.

## REFERENCES

Higuchi R, Dollinger G, Walsh P, and Griffith R (1992): Simultaneous amplification and detection of specific DNA sequences. *Bio/Technology* 10: 413–17.

Higuchi R, Fockler C, Dollinger G, and Watson R (1993): Kinetic PCR analysis: real time monitoring of DNA amplification reactions. *Bio/Technology* 11:1026–30.

Hillen W, Goodman T, Benight A, Wartell R, and Wells R (1981): High resolution experimental and theoretical thermal denaturation studies on small overlapping restriction fragments containing the Escherichia coli lactose genetic control region. *J Biol Chem* 256:2761–6.

Ishiguro T, Saitoh J, Yawata H, Yamagishi H, Iwasaki S, and Mitoma Y (1995): Homogeneous quantitative assay of hepatitis C virus RNA by polymerase chain reaction in the presence of a fluorescent intercalator. *Anal Biochem* 229:207–13.

Lee L, Connell C, and Bloch W (1993): Allelic discrimination by nick-translation PCR with fluorogenic probes. *Nucl Acids Res* 21:3761–6.

Livak K, Flood S, Marmaro J, Giusti W, and Deetz K (1995): Oligonucleotides with fluorescent dyes at opposite ends provide a quenched probe system useful for detecting PCR product and nucleic acid hybridization. *PCR Meth Appl* 4:357–62.

Ririe K, Rasmussen R, and Wittwer C (1997): Product differentiation by analysis of DNA melting curves during the polymerase chain reaction. *Anal Biochem*, in press.

Tan S and Weis J (1992): Development of a sensitive reverse transcriptase PCR assay, RT-RPCR, utilizing rapid cycle times. *PCR Meth Appl* 2:137–43.

Wetmur J (1995): Nucleic acid hybrids, formation and structure of. In: *Molecular Biology and Biotechnology: A Comprehensive Desk Reference.* Meyers R, ed., pp. 605–8, New York: VCH.

Wittwer C (1989): Automated polymerase chain reaction in capillary tubes with hot air. *Nucl Acids Res* 17:4353–7.

Wittwer C, Reed G, and Ririe K (1994): Rapid cycle DNA amplification. In: *The Polymerase Chain Reaction* Mullis K, Ferré F, and Gibbs R, eds., pp. 174–81, Deerfield Beach: Springer-Verlag.

Wittwer C, Herrmann M, Moss A, and Rasmussen R (1997a): Continuous fluorescence monitoring of rapid cycle DNA amplification. *BioTechniques*, in press.

Wittwer C, Ririe K, Andrew R, David D, Gundry R, and Balis U (1997b): The LightCycler™: A microvolume multisample fluorimeter with rapid temperature control. *BioTechniques* in press.

Young B and Anderson M (1985): Quantitative analysis of solution hybridisation. In: *Nucleic Acid Hybridisation: A Practical Approach.* Hames B, Higgins S, eds., pp. 47–71, Washington DC: IRL Press.

# Kinetic ELISA-PCR: A Versatile Quantitative PCR Method

Olivier Lantz, Elizabeth Bonney, Scott Umlauf,
and Yassine Taoufik

## Introduction

Among the numerous assays proposed for quantifying specific nucleic acid sequences in biological samples, PCR offers the most sensitivity and versatility. The assay for quantifying the amount of PCR products is a crucial step in any quantitative PCR method. It should be sensitive and specific, able to display a wide dynamic range, nonradioactive, easy to perform, and inexpensive. The results of the assay should also be easily digitalized. Quantification of amplicons with ELISA fulfills these criteria. It can be automated, and readers are already available in most research and clinical laboratories. This assay can be accomplished by using colorimetry, fluorometry, or luminometry, depending on the substrate used. Luminometry displays the best sensitivity and has the widest dynamic range of these three methods (Martin et al., 1995).

Quantitative PCR can be done either by measuring the amount of PCR products at a given number of cycles (end-point quantitative PCR) (Becker-Andre and Hahlbrock, 1989; Bouaboula et al., 1992; Gilliland et al., 1990; Wang et al., 1989) or by following the amount of products during the PCR at several cycles (kinetic quantitative PCR) (Dallman et al., 1991). ELISA can be used in both formats. Indeed, kinetic PCR would not be feasible without the ability given by ELISA to process many samples in parallel (Alard et al., 1993).

In this chapter, we will first describe some of the ELISA formats available to quantify amplicons. Then, we will compare kinetic vs end-point quantitative PCR. We will describe the ELISA method we have been using these past few years and give a few examples of application. In the appendices, the reader will find a protocol of our ELISA assay, a method to process the data, a troubleshooting guide, and a method to construct homologous internal standard suitable for ELISA.

## Quantifying PCR Products Using ELISA

### *The different formats*

Several ELISA formats have been described in the literature: Figure 1 outlines some of them. Amplicons can be captured onto microplates in two ways. The first is to use biotinylated primers, which makes the resulting amplicons biotinylated and allows them to be captured on avidin-coated microtiter plates (Alard et al., 1993). The second way is to hybridize the amplicons to a capture oligonucleotide (Jenizek et al., 1995). The capture oligonucleotide can be either directly coupled to plastic or bound to avidin coated plates through a biotin moiety.

When using biotinylated primers to capture the amplicons, a hybridization step with a tagged oligo-probe is required, since the PCR products

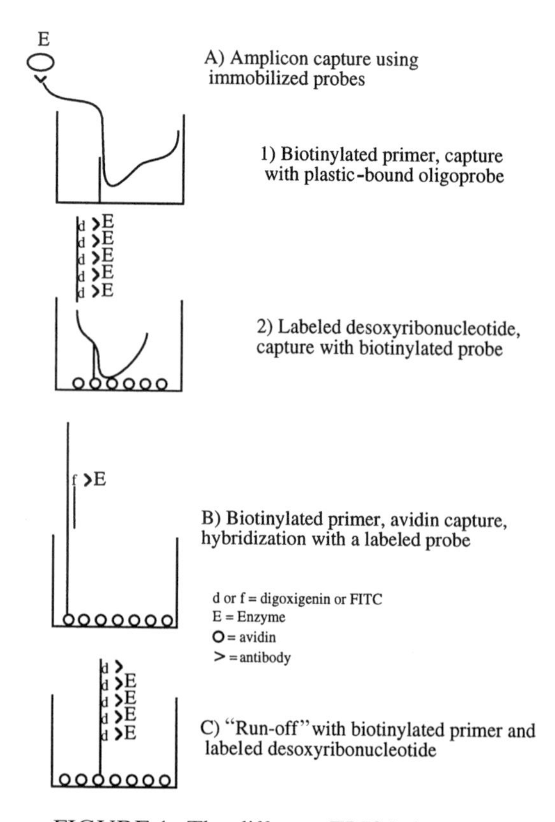


FIGURE 1. The different ELISA formats.

cannot be labeled during the PCR or else the nonspecific products, which would also be labeled, would be captured and detected. After the amplicons are captured on plates, the next step is alkaline denaturation, which will leave a single strand of DNA bound to the plate.

The single-stranded DNA is then hybridized to a digoxigenin- or FITC-labeled oligo-probe. Usually, the probe is tagged with only one label since tailing with terminal transferase greatly increases the background (Lantz, unpublished results). The main problem in this format is the biotin-binding capacity of the wells: if not sufficient, there will be competition between the nonincorporated biotinylated primers and the amplicons. Since the amplicons are much bigger, they may not be captured as efficiently as the primers. Some authors have suggested using avidin-coated microbeads (Holodny et al., 1991) or avidin bound to specially treated plastic in order to increase the binding capacity. In our experience, however, the usual high–quality plastic from Nunc or Dynatech allows enough avidin binding to capture 1 pmole of biotin per well (Alard et al., 1993). In this format, a potential problem may also come from secondary structures or breaks of single-stranded DNA, which may prevent probe hybridization.

When using the method whereby amplicons are captured with an oligo-probe, the amplicons can be tagged with many residues of the chosen label during the amplification (using digoxigenin- or FITC-dUTP for instance). Biotinylated primers can also be used to label the amplicon. In this case, one has to use capture oligo-probes that are directly coupled to plastic. These are available in some proprietary kits. The amount of labeled deoxynucleotide used during the PCR is limited by the somewhat lower PCR efficiency that most of them induce. Thus, only a few tagged nucleotides can be incorporated per amplicon. The main problem of the oligo-probe capture format is the competition between the probe and the nonhybridized strand of the amplicons, which may decrease the sensitivity.

To achieve specificity when detecting PCR products with ELISA, a hybridization step is usually performed. Compared to membrane hybridization, liquid hybridization is much faster and goes to completion. A high sensitivity can then be achieved. As shown in Figure 2, the hybridization step is very efficient since the curve obtained with a digoxigenin-labeled biotinylated oligonucleotide is almost identical to that obtained by hybridizing biotinylated PCR products with a digoxigenin-labeled oligo-probe. The plateau observed at high PCR product concentrations is related to the quite low amount (0.2 pmole) of probe in the wells.

Another format that has been recently described (Zou et al., 1996) is the so-called "ELISA run-off" with internal standard. An internal standard is co-amplified during PCR and, at the end of the amplification, the PCR tube is divided into two aliquots in which a specific biotinylated primer

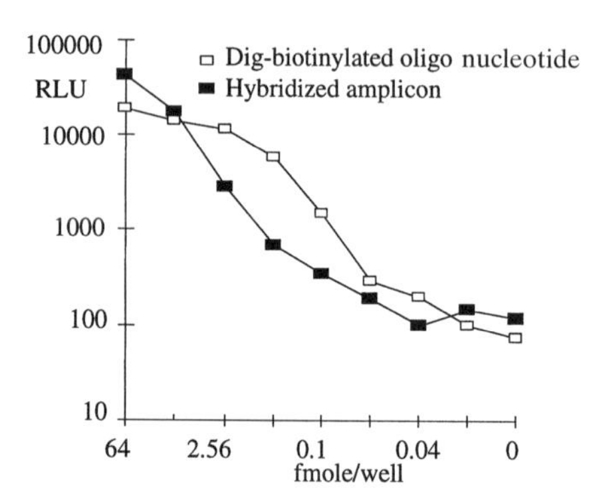


FIGURE 2. Liquid hybridization is very efficient. Serial dilutions of either digoxigeninated-biotinylated oligonucleotide or biotinylated PCR product were captured in avidin-coated microplates. The amount of captured product was revealed either directly (for the dig-biotinylated product) or after a hybridization step (for the PCR product) with an ELISA assay, as described in Figure 5.

for either the target or the standard is added, together with digoxigenin-dUTP and amplified again for one cycle. After being plotted along with an external scale of known amount of purified biotinylated products either for the target or the standard, a ratio of unknown over standard can be computed, despite potential differences in the detection efficiency between standard and target. This assay is able to detect around 0.1 fmole of amplicon per well. Although the run-off step after the PCR is rather cumbersome, this approach eliminates the hybridization step during the ELISA.

## *Critical assessment of the different ELISA formats*

As shown below, the sensitivity and the dynamic range of the assay used to quantify the amount of amplicons will determine the ease and the throughput of the quantitative PCR assay, especially when using internal standard.

There is no direct comparison of the sensitivity of the different ELISA formats in the literature, especially between the two main formats "in cycle labeling and capture by a bound probe" (Figure 1A2) and "biotinylated primer, capture to bound avidin and hybridization with a probe" (Figure 1B). In format 1B, the sensitivity is below 0.1 fmole (19 pg of 300 bp amplicons) of amplicons per well, whereas it is about 0.5 fmole in format 1A2 (Janezic et al., 1995). As stressed above, luminometry gives the best sensitivity and the widest dynamic range. Using probes directly coupled to alkaline phosphatase increases the background considerably (unpublished results).

ELISA for amplicon quantitation compares favorably with any other method of PCR product quantitation, such as ethidium bromide staining or Southern blot of agarose or acrylamide gels, hot PCR followed by agarose gel and autoradiography, or dot blot. In this discussion, we will not give a detailed comparison to other methods such as electroluminescence or proximity fluorescence with acridium ester or rare earth metals. Although their sensitivity is of the same magnitude and they are homogenous phase methods, they require either special equipment or reagents.

## Kinetic Quantitative PCR vs. End-Point Quantitative PCR with Internal Standard

### *End-point quantitative PCR with internal standard*

When performing end-point quantitative PCR, it is generally agreed upon that an internal standard must be spiked in every reaction to check the efficiency of the PCR (Ferré et al., 1994). In our opinion, the term "internal standard" should be reserved for synthetic construct amplifiable with the same primers as the ones used for the target. This internal standard can be either homologous to the target (differing by only a few nucleotides) or nonhomologous (only the two extremities match the target). Some authors have proposed the use of another gene in the same reaction tube amplified with another pair of primers, but this more closely resembles multiplex PCR than PCR with an internal standard. The multiplex format is adequate to control for RNA recovery and cDNA yield, but is not optimal to control for the potential difference in amplification efficiency between the target and the standard. It is also very difficult to control for primer arti facts and for competition phenomena, especially if the cDNA used for the control amplification is a housekeeping gene that has a much higher copy number than the experimental gene.

There are two formats when using an internal standard: In the first format, a small number of dilutions (1–4) of internal standard may be used, and the results are computed according to the ratio of unknown vs. standard amplicons (noncompetitive PCR). In the other, several (6–12) dilutions of internal standard are used, and one looks for the equivalence between the standard and the unknowns (competitive PCR). In the noncompetitive format, the amount of both amplicons is measured either during the exponential phase or at saturation. In the competitive format, the PCR is run to saturation. In the past, there have been some controversies about the characteristics of the internal standard: Should it be homologous or not to the target sequence? It is now clear that if one uses a format whereby the amount of amplicons is measured during the exponential phase, nonhomologous standard may be used. On the other hand, at saturation the amount of primers becomes limiting, and reannealing without new DNA synthesis becomes frequent. If the standard is nonhomologous to the target, heteroduplexes cannot be formed between the standard and the target, and therefore the least abundant molecular species will be amplified preferentially, as stressed by Pannetier (Pannetier et al., 1993) and Grandchamp (Grandchamp, 1995). Thus, the nonhomologous standard may still be used at saturation if the ratio of unknown over standard differs by less than 5–10-fold. However, generally, at saturation homologous standard must be used.

How different from the target may a homologous standard be? A few nucleotide differences allowing for a new restriction site or differentiation with an automatic sequencer are not harmful. The question of how many nucleotide differences can be tolerated has not been studied and will probably depend on many parameters such as the nucleotide sequences, PCR conditions, and ratio of standard vs. target sequences.

### *Kinetic quantitative PCR*

In the kinetic method, first described by Dalmann et al. (1991), an internal standard is not mandatory and an external scale is sufficient if the reproducibility of the PCR is good. The PCR efficiency of every reaction is checked by looking at the slopes of the increase in PCR products for the experimental samples and for the external scale.

One of the advantages of a kinetic PCR method is the speed with which it can quantitate a new gene, since it is not necessary to construct an internal standard. It is then possible to estimate the expression of any known gene in different samples in a matter of days. If one wants to do true quantitation, an external scale should be devised: it is most easily done by using serial dilutions of purified PCR products of the gene studied or by amplifying dilutions of the sample containing the highest quantity of the cDNA studied in the same experiment. As shown in Figure 3, the kinetic method displays also the widest dynamic range (at least 5 log10), and it is then possible to obtain meaningful results without previous knowledge of the amount of target molecules in the samples studied.

This method requires sampling every reaction at regular intervals during the PCR. At first glance this may seem cumbersome, but with multichannel pipettors and 96-well format thermocyclers, this can be easily accomplished. Moreover, the whole process can be automated with a ro-

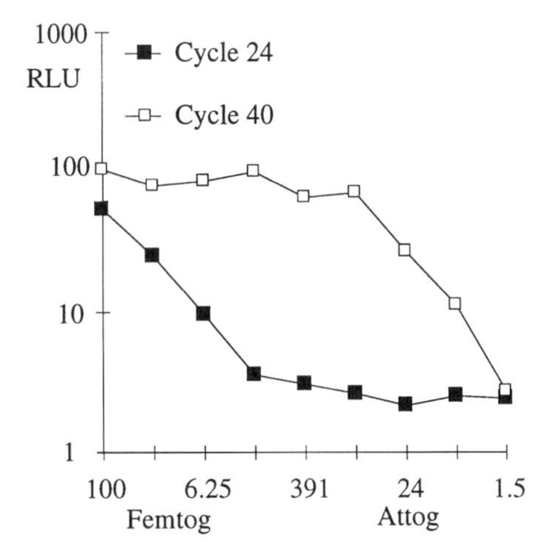


FIGURE 3. Dynamic range of kinetic quantitative ELISA-PCR. According to the cycle (24 or 40) used for the quantitation, one can study high (1.62 fg to 100 fg) or low (1.5 ag to 100 ag) copy number samples. 1 ag corresponds approximately to four copies of IL2. Serial dilutions of gel-purified PCR products were amplified and sampled at regular intervals during the PCR. The amount of amplicon was measured by ELISA and luminometry.

botic work station.

## *Kinetic vs. end-point quantitative PCR*

The advantage of using internal standards is that the PCR can be run to saturation without attendance. One of the drawbacks is that an internal standard needs to be constructed for each gene studied. The quantitation and storage of the standards are not always trivial procedures. Moreover, the sensitivity and the dynamic range of the method used for quantifying the PCR products, as well as the relative abundance of standard and experimental targets, will determine the number of PCRs to be done for each sample. Figure 4 shows that if the assay is not sensitive enough (method A in Figure 4), because of the competition phenomenon between the standard and the target, the product present in lowest amount will not reach the threshold of detection and will not be detected, which will preclude any quantitation. Thus, if samples containing very different amounts of target are studied, to detect both the target and the standard amplicons in every case, a very sensitive assay (method B) and/or several dilutions of standard should be used when performing end-point quantitative PCR with internal standard.

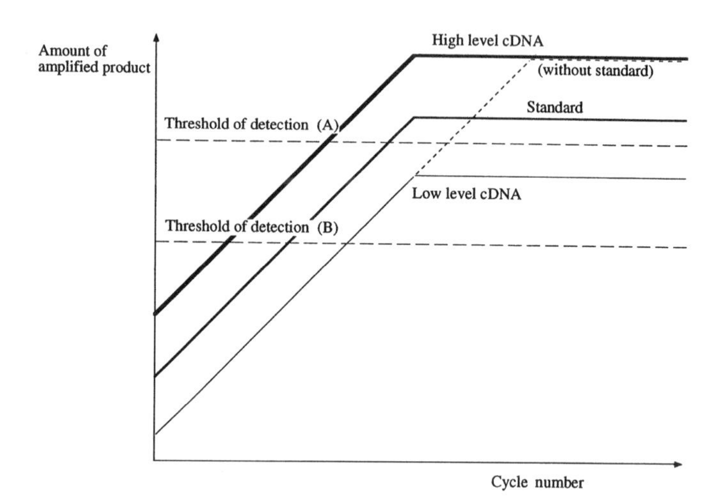


FIGURE 4. End-point quantitative PCR with internal standard requires a very sensitive method with a wide dynamic range to quantify the amount of amplicons (see text).

If the dynamic range of the assay is not broad enough, the most abundant sample will be at saturation and will be underestimated. Thus, most methods of quantitative PCR using internal standards require assaying several dilutions of the samples to be tested. This increases the cost of testing per sample and decreases the rate at which samples can be tested.

In kinetic PCR, the sensitivity of detection is not as important as in end-point PCR: for instance, with method A (Figure 4), the samples will become positive later in the PCR than with method B. However, the more sensitive the assay, the less likely that artifacts will hamper the reaction, since samples are quantified earlier in the PCR. The main problem in kinetic PCR is the necessity to sample every 3–4 cycles for each reaction: 2 min hands-on testing every 12–15 min for 1 hr, 15 min. With ELISA, the number of samples to quantify is not a problem.

For the time being, kinetic quantitative PCR with ELISA and luminometry readings is probably the most versatile and inexpensive method available: there is only one PCR per sample and the cost of the ELISA assay is small compared to the cost of PCR itself. This method is well adapted to research settings where one wants to quantitate the level of several mRNAs in a limited number of samples and to be able to rapidly change the targets. As for hospital laboratories where the number of samples is much higher and the number of genes studied less numerous and less varied, the use of internal standards may be more convenient despite its higher cost and lower throughput. Furthermore, when very small differences are studied, as for instance in the case of the loss of alleles in oncology (twofold at most), an endogenous internal standard is necessary.

## Description of Our ELISA-PCR Assay

### *The ELISA*

We will describe in this section the ELISA-PCR assay we have developed (Figure 5). In this assay, the amplification step is carried out with a pair of primers of which one is biotinylated. The PCR products are captured on avidin-coated microplates and denaturated by alkaline treatment. A digoxigenin– or FITC-labeled oligo-probe is hybridized to the captured DNA strand. The amount of probe is then measured by an alkaline-phosphatase–coupled anti-digoxigenin or -FITC antibody. Finally, the assay is revealed with paranitrophenyl phosphate (PNP) or CSPD (a luminescent dioxietan derivative). Optical density or luminometry according to the substrate used is measured. In this assay, there is no competition between the probe and the second strand of the DNA product, and there is only one digoxigenin or FITC molecule for each PCR product.

This ELISA assay can be used either in kinetic quantitative PCR or in quantitative PCR methods using internal standards. Indeed, the PCR products can be divided into two aliquots and hybridized with either the internal standard or the cDNA probe. The sensitivity of the ELISA step is better than 0.1 fmole of PCR product. Since we put in 0.2 pmole of probe per well, the dynamic range is about 3 log10 when using colorimetry. When using

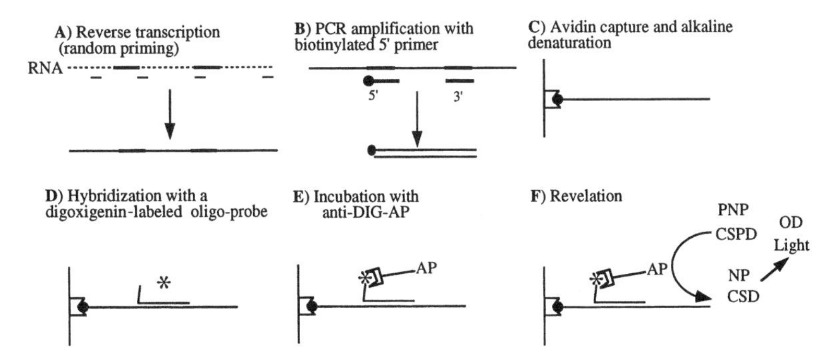


FIGURE 5. Outline of our ELISA-PCR assay (reprinted with permission of *Bio Techniques* from Alard et al., 1993).

luminometry, the sensitivity is better than 0.01 fmole and the dynamic range increases to 4 log 10.

## *mRNA quantitation with kinetic ELISA-PCR*

In our experience, if a true hot start is used (most easily done with the anti-Taq antibody from Clontech and now with the thermo-activatable Taq from Perkin-Elmer), the main source of variability in RNA quantitation using PCR is not the reverse-transcriptase (RT) step or the PCR step, but the amount of starting material and the RNA extraction. Therefore, starting with the RNA extraction step, all the samples are tested in duplicate. For cellular genes, the amount of starting material and the efficiency of RNA extraction can be checked by quantifying a housekeeping gene such as GAPDH, actin, or HPRT. However, for extracellular targets, such as viruses, one has to use an internal RNA standard spiked in the samples at the time of the RNA extraction to verify the extraction efficiency and RT yield.

Since the kinetic ELISA-PCR is very reproducible (Alard et al., 1993), quantitation can be done by comparing the signal of the unknown samples obtained at successive samplings to that of an external scale containing known amounts of the gene studied. Quantitative results can be obtained in a matter of days by using serial dilutions of the cDNA of the most concentrated sample as an external scale. If one wants absolute quantitation, serial dilutions of purified PCR products can be used. This will not assess the RT yield. Since in our experience the greatest variability comes from the number of cells and the efficiency of RNA extraction, all results are normalized to the amount of GAPDH or of some other housekeeping gene. It should be stressed that the storage and particularly the quantitation of RNA or DNA standards are not trivial tasks and require special precautions. Absolute values can only be obtained with RNA standards that are difficult to produce, to quantitate and to store. One should weigh the advantages and the difficulties of obtaining absolute vs. relative values according to one's particular scientific question.

Most of the time, only comparative results are needed. Therefore, equivalent numbers of cells are extracted, and one only has to check that the amount of cDNA is the same in the different preparations by verifying that the curves obtained after amplifying a housekeeping gene are identical. Thereafter, any shift of the curves obtained after amplifying the gene of interest will be meaningful. In other cases, normalization (because of unequal amounts of starting material for instance) is required, and the use of an external standard is required. An external scale made with serial dilutions of purified PCR products is amplified in parallel to the experimental samples.

The comparison of the signal obtained from the different samples to the external scale is not a straightforward procedure. Two methods for numerically processing the data are included in the appendices. The first method is a simple linear regression in which for each cycle the unknown values are compared to the external scale. If signal values are in the linear range of the assay (colorimetry or luminometry) for two or more successive samplings, the values are individually computed and averaged. It can easily be done using a spreadsheet. The second method is a logistic regression analysis where a four-parameter curve is fitted to the experimental curves for every sample, allowing one to compute a "corrected cycle at half–maximum." The value of the unknowns is then compared to that of the external scale by linear regression analysis. This second method is well suited for data obtained with luminometry readings and takes into account all the available information. However, a statistical package able to do curve fitting with iterative procedures is required.

## Examples of Applications

We have been using this ELISA assay in the setting of a kinetic quantitative PCR method in the last 3 years to measure the expression of a large variety of cellular genes (interleukins, growth factors, T cell receptors, for instance). One technician is able to quantify 200 duplicate samples (400 PCR total) per week. Below are a few examples of applications.

### *Interleukins*

The wide dynamic range of the kinetic ELISA-PCR assay has allowed us to study the expression of IL-2 gene expression with or without anti-CD28 costimulation, as shown in Figure 6.

### *T cell repertoire*

We have been studying the expression of a particular invariant alpha chain of the T cell receptor (TCR) in a particular murine T cell subpopulation (the $NK1^+$ T cells). This chain uses particular V$\alpha$ and J$\alpha$ segments (V$\alpha$14-J$\alpha$281) with a conserved VJ junction. In this work, we have extensively used our kinetic ELISA-PCR assay: cells from different mouse strains are sorted using FACS according to certain membrane markers and the amounts of C$\alpha$ (reflecting the number of T cells), V$\alpha$14-C$\alpha$, and

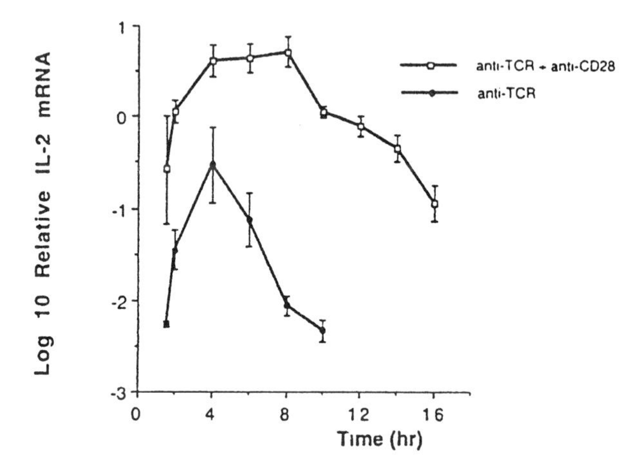


FIGURE 6. A.E7 T cell clone cells were stimulated by anti-TCR antibody with or without costimulation by anti-CD28 antibody. Cells were harvested at the indicated time and RNA extracted. IL–2 and GAPDH mRNA were quantified by ELISA-PCR. All results are normalized to GAPDH content and are expressed relative to the sample containing the highest amount of IL–2 mRNA in order to average three independent experiments (reprinted with permission of *Mol. Cel. Biol.* from Umlauf et al., 1995).

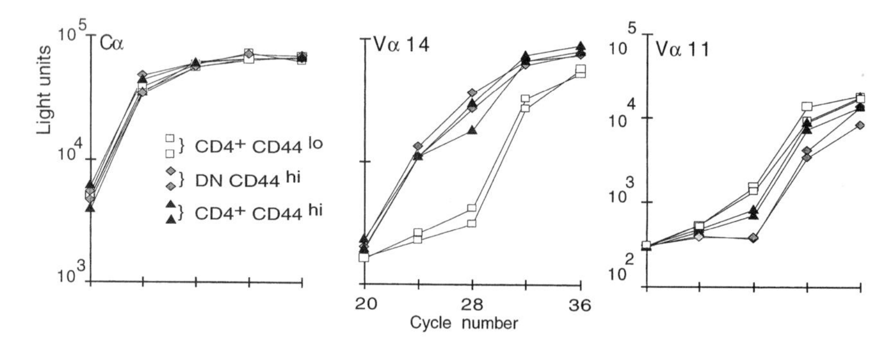


FIGURE 7. T cell repertoire analysis of mature thymocytes sorted according to CD44 and CD4/CD8 expression. Mature thymocytes were FACS sorted according to the indicated markers, and Vα14, Vα11, and Cα mRNAs were quantified (reprinted with permission of *J. Exp. Med.* from Lantz and Bendelac, 1994).

V$\alpha$14-J$\alpha$281 (hybridized either with a probe on the V$\alpha$14 or on the VJ junction) are compared in the different cell fractions. An example is displayed in Figure 7.

*Micro-chimerism*

In the course of studying the role of the persistence of antigen in different models of tolerance, we have devised a method to detect a few male cells in a large number of female cells. To do so, we amplify a Y-chromosome-specific locus in genomic DNA. One should note that the complexity of DNA preparations is about tenfold higher than that of RNA preparations, and to get few-copy sensitivity in $10^6$ cells required special procedures during the processing of the samples and high stringency PCR (high annealing temperature and true hot start). An example of the results is displayed in Figure 8.

## Conclusion

There are several formats for assaying the amount of PCR products by ELISA. Luminometry can be used in all of them and would increase the sensitivity and the dynamic range, thereby potentially also decreasing the number of dilutions to be assayed in the quantitative PCR methods using internal standards. ELISA is a

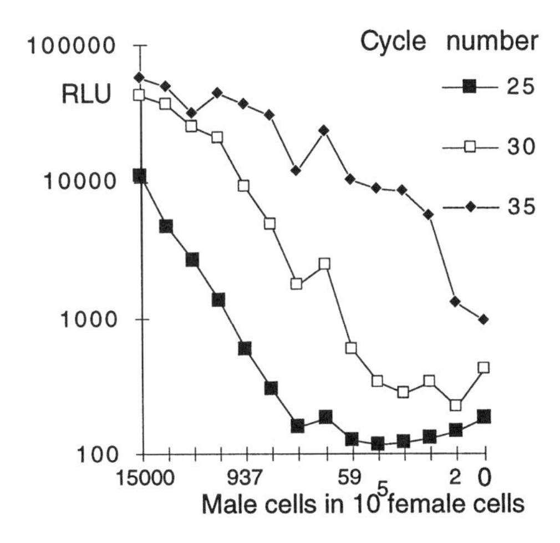


FIGURE 8. Detection of male cells diluted in female cells. Serial dilutions of male cells in $10^5$ female cells were lysed, and the amount of a Y-specific sequence (sry) was quantified. The results obtained at three different samplings are shown.

nonhomogeneous phase assay and is still somewhat labor-intensive compared with the homogeneous phase assays. The homogeneous phase Taq-Man assay on cycler (Perkin-Elmer 7700 apparatus) is now available and allows one to follow the amount of PCR products during the PCR on the thermocycler itself. Thus, kinetic PCR with or without internal standard can be done easily. In research settings, kinetic ELISA-PCR remains the most versatile and inexpensive method available for quantifying specific sequences in small samples. For clinical applications, even though the quantitative PCR step can be carried out reliably as seen above, in our opinion there is still a very strong obstacle to wider use of PCR for studying gene expression in tissue biopsies or even in cell suspensions: there is no RNA extraction method allowing the processing of many samples easily and reliably. Indeed, the most commonly used "-zol" kits are all very operator–dependent and quite time-consuming. Some other kits based on beads and spin columns generate RNA highly contaminated with genomic DNA. Thus, for wider clinical applications a more reproducible (less dependent on operator skills) and easier RNA extraction method needs to be devised.

# Appendices

## *Protocol for quantifying amplicons with ELISA*

*Reagents*

– High-binding-capacity microtiter plate (Maxisorp, Nunc). For luminometry: luminite 2 (Dynatech) or FluoroNunc Maxisorp (Nunc).

– Avidin (Sigma, ref. A9275).

– BSA containing neither biotin nor alkaline phosphatase (Sigma, ref. A6793).

– Carbonate buffer 0.1 M, pH 9.6.

– 3′end oligonucleotide labeling kit (Boehringer Mannheim, ref. 1362372) or FITC-coupled oligonucleotide (On CPG column, Clontech, ref. 5227).

– Alkaline-phosphatase-conjugated anti-digoxigenin antibody (Boehringer, ref. 1093274) or anti-FITC (Boehringer, ref. 1426338).

– Diethanolamin buffer. Stock solution 1 M: 96 ml diethanolamin (Sigma, ref. D2286), $MgCl_2$ (2 mM), sodium azide 0.2 g, $H_2O$ to 1 liter.

– PNP (Sigma, substrate 104–105) or CSPD (Tropix, ref. MC005).

*Making avidin-coated microtiter plates*: Avidin (0.1 mg/ml) in carbonate buffer, pH 9.6, 0.1 M (100 μl/well). Incubation: 2 hr at 37°C. Recover the avidin solution, which can be used four times. (Optional: three washes in PBS 0.1% Tween 20). Block the plates with BSA 1% in carbonate buffer (300 μl) for 2 hr at 37°C. Freeze the plate at −20°C.

*Amplicon capture and alkaline denaturation*: Thaw the plates, three washes with PBS, add 100 μl TE. Add 5 μl of PCR reaction. 1 hr incubation at 4°C. Add 100 μl/well NaOH 0.1 N, 10 min room temperature. Three washes with TE 0.1% Tween.

*Liquid hybridization*: Add 100 μl per well of the hybridization mix (0.5 X SSPE with 100 μg/ml herring sperm DNA, SDS 0.1%, and labeled probe 2 pmole/ml). Incubate 2 hr at 42°C. Three washes with PBS + NaCl 0.15 M, 0.1% Tween 20. One wash with PBS + NaCl 0.15 M + BSA 1%.

*Revelation*: Incubate for 1 hr with 100 μl/well of anti-digoxigenin antibody coupled to alkaline phosphatase (1/7500 in PBS + NaCl 0.15 M + BSA 1%). Four washes with PBS + NaCl 0.15 M, 0.1% Tween 20. Two washes with substrate buffer (0.1 M diethanolamin buffer, pH 10). Incubate 1–24 hr room temperature with substrate PNP (1 mg/ml in diethanolamin buffer). Read at 405 nm.

When using luminometry, white plates should be used and the substrate is CSPD (0.25 mM) with the enhancer Saphire II (Tropix, ref. LAX250). A microtiter plate luminometer is also required.

## *Processing the results of kinetic ELISA-PCR*

In parallel to the unknown samples, an external scale made of serial dilutions of purified amplified products is amplified. Every reaction is sampled at regular intervals during the PCR, and PCR products are quantified by ELISA. All the OD or RLU values are exported to a spreadsheet. For every reaction, one can draw a curve $S = f(\text{cycle number})$, where $S$ is the signal. We have used two methods to process the data to get absolute values.

### *Linear regression analysis*

This method is just a regular linear regression of ELISA data. Using signal ($S$) values above background and below saturation, for a given cycle, one can draw a straight line

$$\log_{10}\left(S\right) = a \log_{10}\left(C\right) + y_0,$$

where $C$ is the concentration of the standard, and $a$ and $y_0$ are the parameters (slope and ordinate intercept) of the straight line estimated by least-squares regression.

One can then compute the concentration of the unknown samples as

$$C = 10^{\left(\left(\log_{10}(S) - y_0\right)/a\right)}.$$

The estimation can be done at several samplings (cycle number), and the values are averaged. An example of linear regression analysis is shown in Figure 9.

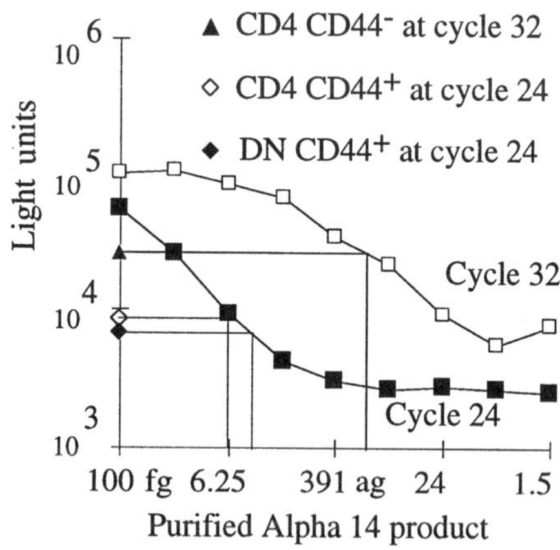


FIGURE 9. Analysis of ELISA-PCR data by linear regression analysis. In this example, the signal obtained after V$\alpha$14 sequence amplification in cDNA from the indicated subpopulations of mature thymocytes purified by FACS was compared to the signal obtained after amplification of serial dilution of purified V$\alpha$14 amplicon.

## *Logistic regression*

The principle of the method is to compute the "corrected cycle at half maximum" for every sample and for the scale. The values of the unknowns are then compared to those of the scale by straight linear regression. An example of such a scale is shown in Figure 10.

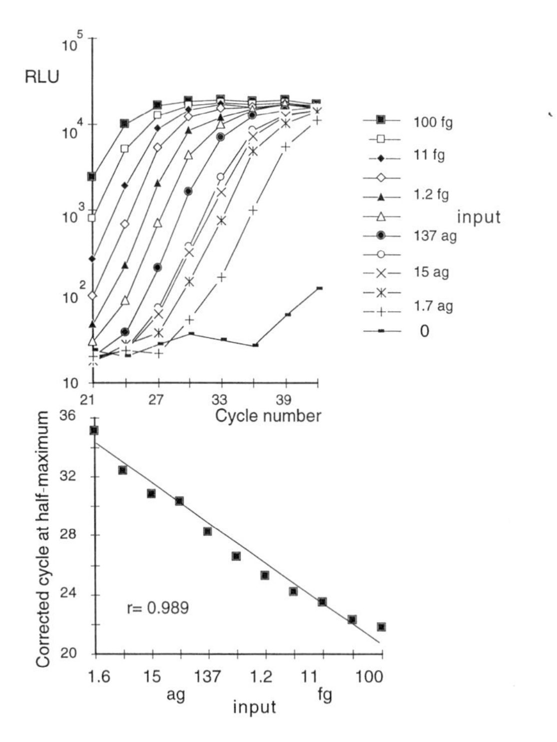


FIGURE 10. Logistic regression for analysis ELISA-PCR data (see text for explanations). Serial dilutions of IL-4 product were amplified, and amplicons were quantified by ELISA at the indicated cycle.

One can compute the four-parameter curve that fits the experimental values:

$$S = C + \left(\left(D - C\right) / \left(1 + \mathrm{Exp}\left(-A - BX\right)\right)\right)$$

*A* and *B* are curve parameters describing the intercept and slope of the line. *C* and *D* are the lower and upper plateaus of the logistic curve. The cycle ($X$) at the half-maximum occurs for a value $X^* = -B/A$. In Figure 10, values of $X^*$ are plotted against the corresponding input of the scale and the parameters of the straight lines are calculated by least-squares regression. By linear regression analysis, the values of the unknown can then be calculated.

*Troubleshooting guide (when using digoxigenin probe and colorimetry with alkaline phosphatase)*

If there is no signal in the ELISA-PCR assay:

1) Is it the PCR or the revelation assay?
   – Always keep the PCR plate and run an agarose gel if needed.
2) The PCR has not been successful. Is it the avidinated plate binding or the revelation step?
   – Use serial dilution of a 5′-biotinylated and 3′-digoxigeninated primer (from 10 pmole/well to 1 attomole/well).
   – Do an early reading to see the upper concentrations and estimate the binding capacity of the plate (it should not saturate at a concentration below 1 pmole).
   – Do a late reading to verify the sensitivity of the revelation step: it should be below 0.1 fmole/well.
3) Everything is nominal. It is the probe or the hybridizing step?
   – Pool PCR products, quantify on ethidium-bromide-stained agarose gel, make serial dilutions from 1 pmol/well to 10 attomole/well, and transfer to avidin-coated microplates. Use replicates and hybridize with several batches of the relevant probe.
   – If there has been a problem with a given batch of probe, one can wash the plates, redenaturate the PCR products with sodium hydroxide, and rehybridize with a new batch of probe.
4) Most of the time, the problem comes from the probe. For some not yet known reason, certain probes have problems being stored once labeled with digoxigenin.

*Constructing internal standard for ELISA-PCR*

In ELISA-PCR, specific hybridization with a probe is required either for capturing the amplicons or for measuring their amount. As seen above, we differentiate the target from the standard through the use of specific probes. We have chosen to avoid any cross-reactivity between the two probes by making the standard in such a way that the probing regions (18 nt) will be completly different in the standard and in the target. Figure 11 outlines the method used to construct such standards.

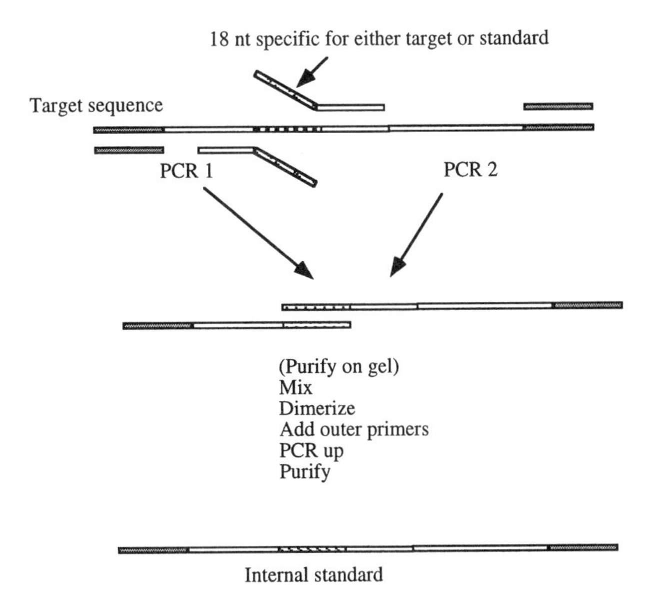


FIGURE 11. Construction of internal standard suitable for ELISA-PCR. Together with the two primers used for amplifying the cDNA, two overlapping primers corresponding to the probe sequence are used in two parallel PCR reactions. The two products are purified, mixed, allowed to dimerize by performing a few PCR cycles, and then the outer primers are added and a few PCR cycles are again carried out. The standard is identical to the target except for the 18nt sequence used for hybridizing either the cDNA or the standard probe.

# REFERENCES

Alard P, Lantz O, Sebagh M, Calvo CF, Weill D, Chavanel G, Senik A, and Charpentier B (1993): A versatile ELISA-PCR assay for mRNA quantitation from a few cells. *Biotechniques* 15:730–7.

Becker-Andre M and Hahlbrock K (1989): Absolute mRNA quantification using the polymerase chain reaction (PCR). A novel approach by a PCR aided transcript titration assay (PATTY). *Nucleic Acids Res* 17:9437–46.

Bouaboula M, Legoud P, Pessegué B, Delpech B, Dumont X, Piechaczyk M, Cassellas P, and Shire D (1992): Standardization of mRNA titration using a polymerase chain reaction method involving co-amplification with multispecific internal control. *J Biol Chem* 267:21830–8.

Chomczynski P and Sacchi N (1987): Single-step method of RNA isolation by acid guanidinium thiocyanate-phenol-chloroform extraction. *Anal Biochem* 162:156–9.

Dallman MJ, Larsen CP, and Morris PJ (1991): Cytokine gene transcription in vascularised organ grafts: analysis using semiquantitative polymerase chain reaction. *J Exp Med* 174:493–6.

Ferré F, Marchese A, Pezzoli S, Griffin S, Buxton E, and Boyer V (1994): Quantitative PCR: an overview. In: *PCR the polymerase chain reaction.* Mullis KB, Ferré F, and Gibbs RA, eds. Boston. Birkauser.

Gilliland G, Perrin S, Blanchard K, and Bunn HF (1990): Analysis of cytokine mRNA and DNA: Detection and quantitation by competitive polymerase chain reaction. *Proc Natl Acad Sci USA* 87:2725–9.

Grandchamp B (1995): Quantitative PCR with internal standard. In: *Quantitative PCR workshop*. Paris. Atelier Inserm.

Holodny M, Katzenstein DA, Sengupta S, Wang AM, Casipit C, Schwartz DH, Konrad M, Groves E, and Merigan T (1991): Detection and quantification of human immunodeficiency RNA in patient serum by use of the polymerase chain reaction. *J Infect Dis* 163:862–6.

Janezik A, Semper A, Holloway J, and Holgate S (1995): Detection of cytokine mRNA expression by a sensitive RT-PCR ELISA detection system. In: *Biochemica*. Boehringer Mannheim, ed., pp. 30–2.

Lantz O and Bendelac A (1994): An invariant T cell receptor a chain is used by a unique subset of MHC class I-specific CD4+ and CD4–8– T cells in mice and humans. *J Exp Med* 180:1097–106.

Martin CS, Butler L, and Bronstein I (1995): Quantitation of PCR products with chemiluminescence. *Biotechniques* 18:908–13.

Pannetier C, Delassus S, Darche S, Saucier C, and Kourilsky P (1993): Quantitative titration of nucleic acids by enzymatic amplification reactions run to saturation. *Nucleic Acids Res* 21:577–83.

Umlauf S, Beverly B, Lantz O, and Schwartz R (1995): Regulation of IL-2 gene expression by CD28 costimulation in mouse T cell clones. Both nuclear and cytoplasmic RNA are regulated with complex kinetics. *Mol Cel Biol* 15:3197–205.

Wang AM, Doyle MV, and Mark DF (1989): Quantification of mRNA by the polymerase chain reaction. *Proc Natl Acad Sci USA* 86:9717–21.

Zou W, Durand-Gasselin I, Dulioust A, Maillot MC, Galanaud P, and Emilie D (1995): Quantification of cytokine gene expression by competitive PCR using colorimetric assay. *Eur Cytokine Network* 6:257–64.

# PART I: METHODS/TECHNOLOGY ISSUES

## B. GENE QUANTITATION BASED ON OTHER TARGET AMPLIFICATION SYSTEMS

# Quantitation of RNA by NASBA™: Applications and Issues for HIV-1 and AIDS

Joseph Romano, Paul van de Wiel, and Stuart Geiger

## Introduction

The past 10 years have seen great advances in the fields of nucleic acid detection and quantitation. This can largely be attributed to the development of nucleic acid amplification methodologies. At the forefront of these technologies is the polymerase chain reaction (PCR; Saiki et al., 1985, 1988). However, a number of powerful alternative technologies have been extensively developed in recent years. These include the ligase chain reaction (Barany, 1991; Weidmann et al., 1994), Q-$\beta$ replicase amplification (Kramer et al., 1992), strand displacement amplification (Walker et al., 1992), and a number of transcription amplification systems (van Gemen et al., 1995a). There are specific features and individual advantages associated with each of these methods. Importantly, all of these technologies are capable of levels of sensitivity and specificity that were previously unattainable by conventional probe hybridization-based nucleic acid assays.

Several descriptions of transcription amplification systems (TAS) have appeared in the literature, including NASBA (Kievits et al., 1991; Compton, 1991) and 3SR (Guatelli et al., 1990), which are essentially the same technology. The primary advantage of TAS is its felicitousness for the amplification of an RNA analyte. There are several biologically-relevant reasons for specifically targeting RNA analytes. For example, several important viral pathogens utilize an RNA genome (e.g., retroviruses, enteroviruses, flaviviruses). Thus, detecting or quantifying these types of viruses can readily be achieved by targeting their genomes with a TAS-based assay. Further, RNA intermediates of any replicating virus are produced inside infected cells during virus replication. Therefore, these viral RNA transcripts can serve as markers for any replicating virus. The fact that RNA is not stable outside of viable cells makes it a highly suitable marker for other microbial pathogens as well. Although the DNA from such organisms (bacteria, protozoa, etc.) can be targeted with an amplification-based assay, the DNA may persist despite the fact that the organism is no longer viable. Thus, RNA serves as a general marker for viable microbial pathogens. The capacity to detect viable or replicating microorganisms is critical to the evaluation of an antimicrobial therapy. It is also useful to target RNA in genetic screening assays. The existence of a specific genetic sequence within a DNA coding region will be maintained in the RNA transcript of that gene. Typically, the transcript will be present in the cell at a higher copy number than the DNA version of the gene of interest, rendering the transcript more likely to be amplified. Last, the expression of specific transcripts frequently correlates with the occurrence of particular disease states. This has been clearly demonstrated in the field of oncology (e.g., Katz et al., 1994).

In this chapter, a detailed description of the transcription amplification system called NASBA will be provided. NASBA is actually a total system, comprised of three component technologies directed at nucleic acid isolation, target amplification, and product detection. Specifically, the application of NASBA in the quantitation of HIV-1 RNA will be presented. Finally, the use of HIV-1 RNA quantitation in the evaluation of antiviral therapy, the study of HIV-1 natural infection, and its utility in disease prognosis will be discussed.

## Quantitation with the NASBA System

Amplification-based nucleic acid assays all require the preliminary isolation of DNA and/or RNA, an actual amplification procedure, and the capacity to detect and quantify the amplified target. Different assay systems make use of a wide range of component technologies. The NASBA system utilizes a highly versatile extraction procedure, a unique isothermal method of amplification, and an electrochemiluminescence (ECL) based detection technology. The combination of these three component technologies makes the NASBA system highly appropriate for the detection and quantitation of HIV-1 RNA.

### *Nucleic acid extraction*

The NASBA system utilizes the guanidine isothiocyanate (GuSCN), acidified silica nucleic acid isolation procedure of Boom et al. (1990). Briefly, the procedure begins with the lysis of the clinical sample in nine volumes of lysis buffer (5.0M GuSCN, 1.0% Triton X-100, pH 6.2–7.2). This step will effectively solubilize a wide range of specimen types including whole blood, plasma, serum, cells, homogenized tissue, cerebrospinal fluid, and sputum.

Importantly, the lysis procedure will render infectious agents such as HIV-1 inactive. Since proteins are readily solubilized in this buffer, nucleases are denatured, thereby providing the nucleic acids with a stable environment for long-term storage. Typically, the HIV-1 NASBA assay involves the lysis of 100 μL of plasma (EDTA, citrate, or heparin) in 900 μL of lysis buffer.

After sample lysis, 50 μL of an acidified silica suspension (1.0 gm/mL, pH 2.0) are added. The association of the nucleic acids (both RNA and DNA) with the solid-phase silica allows for a series of pulse centrifugations and wash steps (twice with 5.0 M GuSCN, twice with 70% ethanol, once with acetone), which serves to eliminate the solubilized proteins, lipids, and carbohydrates. The final step in the procedure involves an elution of the nucleic acids from the surface of the silica in water or 10 mM Tris buffer. The association of the nucleic acids with the solid-phase silica helps to minimize the chance of crossover contamination between samples during the isolation procedure. Importantly, the procedure is very effective at eliminating potential inhibitors of the enzymatic amplification process, such as heparin. Heparinized plasma processed by this procedure yields a nucleic acid extract that is suitable for amplification. Further, the procedure can be applied to whole blood without interference by hemoglobin.

### *NASBA amplification*

Nucleic acid amplification by NASBA is possible through the action of three enzymes and two oligonucleotide primers (Kievits et al., 1991). The process is conducted entirely at 41°C; there is no requirement for thermal cycling. Although the procedure can be used to amplify RNA and DNA, the most significant advantage of the technology is its use in the amplification of RNA. This is due to the direct incorporation of reverse transcription into the amplification pathway. In the absence of heating steps, NASBA is entirely specific for RNA.

The NASBA pathway for the amplification of an RNA target is outlined in Figure 1. The initial event in the process is the annealing of the P1 primer to the RNA analyte. This primer is designed such that the 3′ half is complementary to a conserved sequence on the RNA analyte, while the 5′ (nonbinding) half encodes the T7 RNA polymerase promoter sequence. Therefore when P1 is annealed, the T7 RNA Pol promoter will exist as a free overhang. The first enzymatic step in the pathway is the extension from P1 of a DNA strand through the action of avian myeloblastosis virus reverse transcriptase (AMV-RT). This results in the production of a DNA:RNA hybrid. RNase H, the second enzyme to become involved, destroys the RNA portion of this hybrid. This action renders the second oligonucleotide primer, P2, able to anneal to the single-stranded cDNA at a site

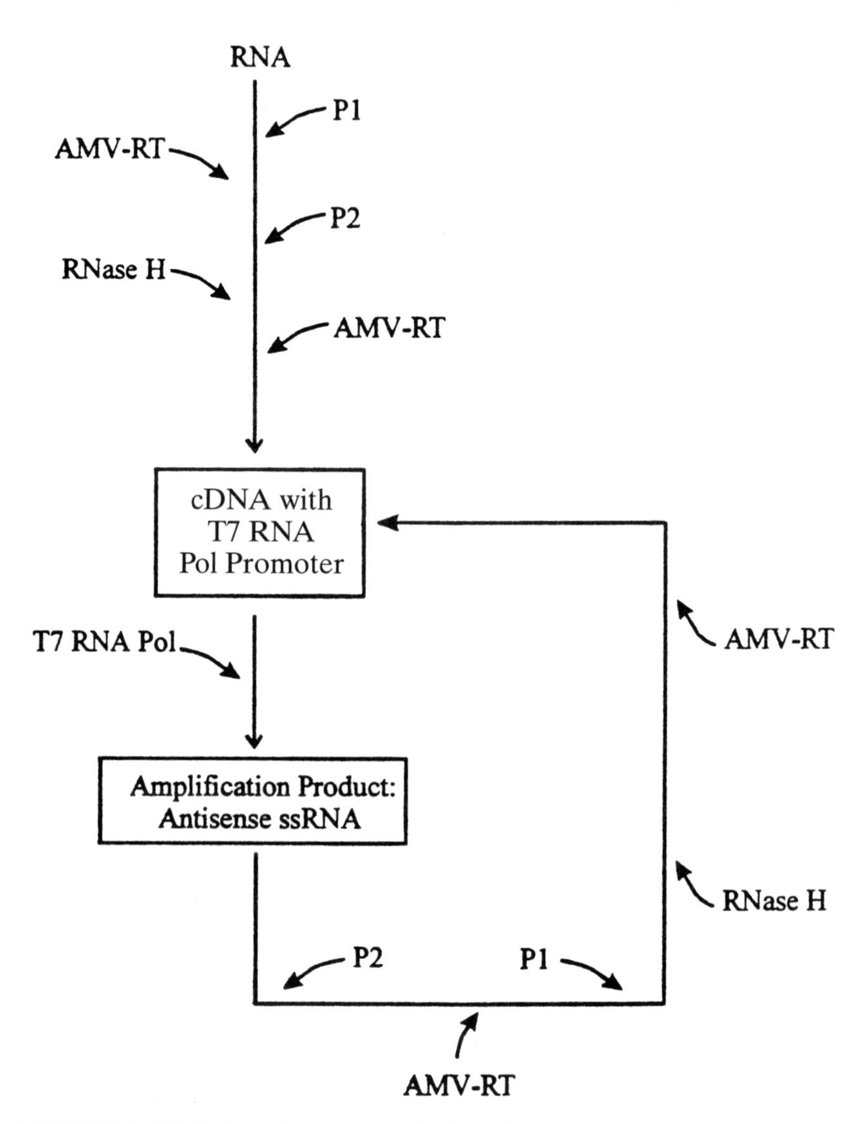


FIGURE 1. NASBA pathway. Amplification of a single-stranded RNA analyte by NASBA involves the initial production of a double-stranded cDNA copy of the original analyte, with an intact T7 RNA polymerase promoter at one end. This entity can be transcribed by the T7 RNA Pol, producing single-stranded RNA. The single-stranded RNA can serve as template for the production of more double-stranded cDNA in the cyclic phase of the process.

upstream from the P1 annealing site. The DNA-dependent DNA polymerase activity of the AMV-RT results in the extension downstream from the annealed P2, producing a double-stranded DNA copy of the original RNA analyte, with a complete T7 RNA polymerase promoter at one end. This promoter is recognized by the T7 RNA polymerase, which then begins to transcribe large amounts of antisense RNA corresponding to the original RNA analyte. This antisense RNA can then serve as a template for the production of additional double-stranded cDNA with the T7 RNA polymerase promoter; however, the P2 primer anneals first, followed later by P1.

Typically, the NASBA pathway results in a factor of amplification equal to approximately $10^9$. The entire process involves a single homogeneous reaction; all of the components are loaded into a single tube. Since the product of the amplification process is single-stranded RNA, the oligonucleotide primers are not incorporated into the product. Thus the antisense RNA amplification product can readily be subjected to a probe hy-bridization detection system without the requirement for a denaturing step. Interestingly, it is also possible to directly sequence the single-stranded RNA product of the NASBA reaction through the use of reverse transcriptase. Although cloning the double-stranded cDNA intermediate is possible, NASBA is typically not an appropriate method for obtaining fragments for cloning.

*Quantitative detection of amplified products*

Detection of NASBA products is achieved through probe hybridization methods. The use of an oligonucleotide probe, specific for the amplification product, brings an additional level of specificity to the system. Qualitative detection of NASBA products is readily achieved through conventional gel-based hybridization systems: e.g., Southern blot, liquid hybridization. Probes are typically labeled with a radioisotope ($^{32}P$) or horseradish peroxidase (HRP; van der Vliet et al., 1993). The HRP-labeled probe allows for a safe nonisotopic colorimetric system of detection.

Although isotopic- or colorimetric-based detection systems could be developed for quantitation of amplified products, the narrow dynamic ranges associated with these technologies are too limiting. For amplification-based quantitative assays, it is far more appropriate to utilize a detection system with a dynamic range large enough to enable quantitation across several orders of magnitude. This is achieved in the NASBA system through the application of electrochemiluminescence (ECL) chemistry. The use of ECL labels in nucleic acid probe assays has been described at length (Blackburn et al., 1991; Kenten et al., 1992). Basically, this technology makes use of a ruthenium compound label for the oligonucleotide probe. When the ruthenium is subjected to appropriate voltage at the surface of an electrode in the presence of tripropylamine (TPA), an oxidation-reduction reaction is induced, which results in the emission of light at 620 nm.

Actual quantitation is achieved with an ECL reader known as the NASBA QR System. This is an automated system that can measure the amount of ECL signal produced from the hybridization of a ruthenium-labeled probe to the RNA amplification product. The QR System requires a magnetic-bead-based hybridization format. After the amplification process, an aliquot of the product is diluted (typically by a factor of 50–100). The diluted amplificate is then mixed with the ruthenium-labeled detector probe that is specific for a sequence within the amplificate RNA. At the same time, a capture probe is included, which is immobilized onto the surface of a magnetic bead via a biotin:avidin linkage, and is complementary to a distinct sequence within the amplificate. The hybridization of the detector and capture probes to the amplificate is typically achieved at 41°C after 30 min. Next an oxidant (tripropylamine) is added to the hybridized product. At this point, the hybridization reaction is loaded into the QR System reader. The magnetic beads are captured onto the surface of an electrode inside the instrument cell through the action of a retractable magnet. Once they are captured, current is applied to the electrode. The ECL reaction is induced at the surface of the electrode wherever amplificate is bound to both the immobilized bead and the ruthenium- labeled capture probe. The amount of induced ECL is quantified through a photomultiplier tube that is positioned above the electrode. At the completion of the process, the retractable magnet is removed, freeing the magnetic beads, which are then washed from the ECL cell. The QR System process is summarized in Figure 2. This enclosed assay system is highly sensitive and specific, and allows for the generation of ECL signal that is quantifiable over six orders of magnitude (Blackburn et al., 1991).

*Quantitative configuration of the NASBA system*

Quantitation with the NASBA system is achieved through the use of three internal control, or calibrator, RNAs. These RNAs are produced in vitro from plasmid constructs that encode the analyte sequence of interest. The calibrator RNA sequences (also referred to as Q RNAs) are identical to the wild-type target RNA, except at the site encoding the detector probe target. For example, the HIV-1 calibrator RNAs consist of a 1500 bp sequence spanning a portion of the *gag* gene. This allows the Q RNAs to be amplified with the same primers used to amplify the wild-type target, with an identical level of efficiency. The detector probe site in each Q RNA (approximately 20 bases in length) is mutagenized, so that the A, C, G, and T content is maintained, but the base order is different. This allows

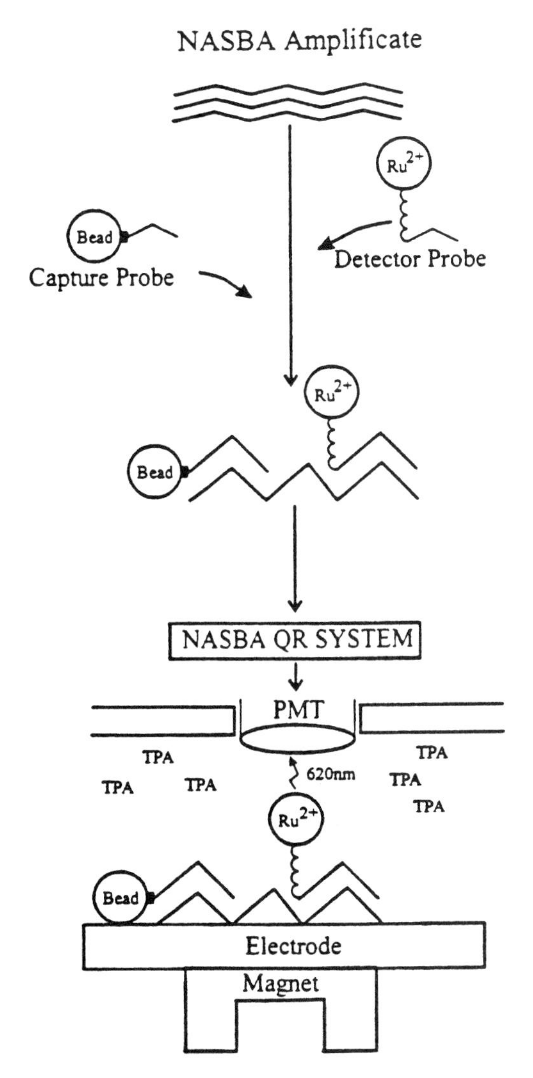


FIGURE 2. Detection of NASBA amplification product by ECL. After NASBA amplification, this product is hybridized with a ruthenium ($Ru^{2+}$) labeled detection probe and a capture probe immobilized onto the surface of a magnetic bead. This hybridization product is then loaded into the NASBA QR System, which will quantify the amount of ECL produced from captured amplificate, as shown (TPA = tripropylamine; PMT = photomultiplier tube).

for the specific detection of the wild-type and each Q RNA, with a distinct detector probe. Importantly, each of these probes will anneal with the same kinetics and melt at the same temperature.

Utilization of these Q RNAs for quantitation is based on the notion that their amplification is achieved with the same kinetics as the wild-type (WT) analyte RNA present in the sample. With complete homology between the primer annealing sites in the Q and WT RNAs, the priming of these different RNAs will occur in a similar manner. There will be no competitive advantage of one target RNA (Q or WT) over the other. Once they are annealed, the cDNA synthesis phase should again be kinetically equal, as should subsequent RNA synthesis, given the near identity in the sequences of the Q and WT RNAs. It is this nearly identical structural nature of the Q RNA to the WT analyte, and the subsequent equity in amplification and product detection, that permits the successful use of the internal standard strategy for quantitation. As long as the original input of Q RNA is known, and the amount of Q and WT amplification products can be determined, it is possible to construct mathematical algorithms to solve for WT input quantities.

The strategy for using these calibrator Q RNAs as internal quantitation standards is outlined in Figure 3 and has been described at length elsewhere (van Gemen et al., 1994). After the initial lysis of the clinical specimen, the three Q RNAs are added at known copy numbers ($Q_A = 10^6$, $Q_B = 10^5$, $Q_C = 10^4$) to the lysate. The Q RNAs are therefore coextracted with the wild-type RNA present in the clinical sample. This normalizes the quantitation standards relative to the wild-type RNA for any loss of material incurred during the extraction procedure. Importantly, the copy number for each control set will be determined empirically for each individual analyte, since all RNA targets will not require the same dynamic range. It is possible, for example, to more accurately quantify low-end HIV-1 RNA samples using $Q_A = 10^5$, $Q_B = 10^4$, and $Q_C = 10^3$. At the completion of the nucleic acid isolation step, 1/10th of the extract is used in the amplification reaction. Since this extract contains each Q RNA as well as the wild-type RNA, all four species of RNA will be co-amplified in the single reaction tube. This will normalize for the effect of anything present in the extract that could influence the amplification efficiency. The procedure ceases to be a single-tube assay at the point of detection, where a portion of the amplification reaction product is diluted and subjected to four independent hybridization detections, each with a specific ruthenium-labeled probe (WT, $Q_A$, $Q_B$, or $Q_C$). Using the NASBA QR System, the amount of ECL signal for the WT, $Q_A$, $Q_B$, and $Q_C$ is determined, and the ratio of these signals to each other is calculated. This ratio is then used to solve for the input quantity of WT RNA (which is the only unknown) by means of computerized calculations involving a curve fitting theory algorithm. The algorithm plots ECL signal relative to copy number for each Q calibrator RNA, and then confirms that the relationship among these points is statistically valid. The resulting line produced from these points can be used to solve for the WT copy number. Thus, this assay has the capacity to

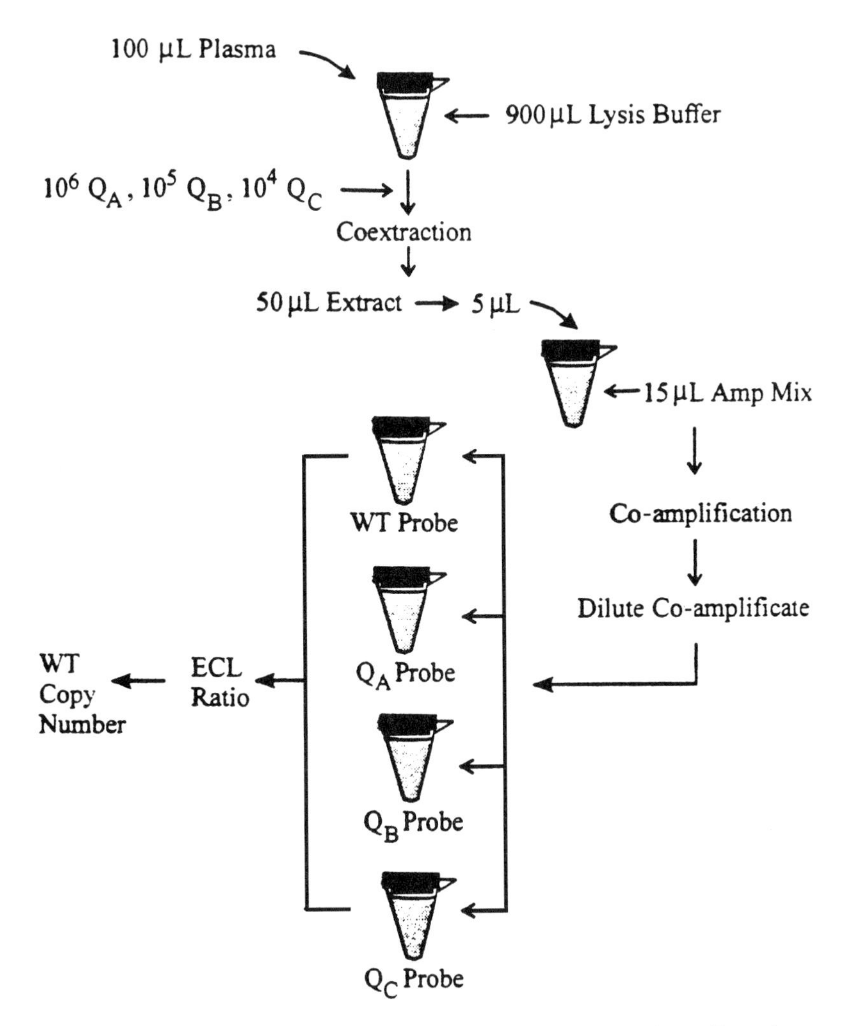


FIGURE 3. Quantitative NASBA assay strategy. This schematic outlines the strategy for quantitative NASBA assays that utilize the three internal control calibrator RNAs ($Q_A$, $Q_B$, $Q_C$).

provide absolute copy number quantitation of an RNA analyte by means of the three calibrator standards, which serve as internal controls for the entire procedure.

## The HIV-1 Quantitative NASBA Assay

The quantitative NASBA assay for genomic HIV-1 RNA was designed using precisely the strategy described in the previous section, involving primers that are homologous to a highly conserved region within the *gag* gene (van Gemen et al., 1993, 1994). A systematic evaluation of this single-tube NASBA assay for HIV-1 RNA required the availability of a highly characterized stock of HIV-1. Such a stock of the HIV-$1_{HXB3}$ isolate has been described (Layne et at., 1992). The particle content of this stock was determined by electron microscopy. The stock was also quantified at the level of p24, reverse transcriptase, gp120 content, and infectious dose. Interestingly, the RNA content of this stock was shown by the NASBA assay to correspond completely with the EM-determined particle count (particles per mL by EM = $2.9 \pm 1.0 \times 10^{10}$; particles per mL by RNA content = $2.2 \pm 1.6 \times 10^{10}$; i.e., particles = $0.5 \times$ RNA copies/mL) (Kievits and van Gemen, unpublished results). Once the RNA concentration of the stock was confirmed, the stock was usable in a series of experiments designed to evaluate performance.

A series of experiments using the HIV-$1_{HXB3}$ stock was designed to evaluate the HIV-1 quantitative NASBA assay for dynamic range, reproducibility, and accuracy (van Gemen et al., 1995b). For example, analysis of dilutions of the HIV-$1_{HXB3}$ stock in normal human plasma demonstratedthat the assay was accurate and reproducible within a factor of approximately 0.2 logs. This level of performance was found over a dynamic range extending from one log above the highest calibrator down to one log below the lowest calibrator. In the same study, it was shown that performance was not affected by artificially elevated levels of hemoglobin, lipid, or albumin. Importantly, it was also shown that the assay performed equally well with plasma obtained in sodium citrate, EDTA, or heparin.

The performance of the HIV-1 QT NASBA assay has also been evaluated using the ACTG Virology Quality Assurance (VQA) Laboratory proficiency panel samples. These panels are provided to all laboratories seeking certification for the quantitation of HIV-1 RNA in AIDS clinical trials. We have used the NASBA assay in two different configurations to analyze separate VQA panels. The proficiency panel 05 (PP05) was analyzed using the conventional configuration of the assay: $Q_A = 10^6$, $Q_B = 10^5$, $Q_C = 10^4$ copies. The proficiency panel 04 (PP04) was analyzed using a 1/10th dilution of the calibrators: $Q_A = 10^5$, $Q_B = 10^4$, $Q_C = 10^3$ copies. The undiluted calibrator configuration has a cut-off sensitivity of 400 copies RNA/input volume; the diluted calibrator format has a cut-off of 100 copies RNA/input volume. The results of these analyses are summarized in Tables 1 and 2. For PP04, there were two samples at an input value of $10^{2.38}$ copies/100 $\mu$L, five samples each at $10^{3.08}$, $10^{3.78}$, and $10^{4.48}$ copies per 100 $\mu$L, and two samples at $10^{5.18}$ copies/100 $\mu$L. In PP05, there were five samples at each of the five input values. In all cases, 100 $\mu$L of the provided sample were analyzed. The log mean and log SD for each sample set, in each assay format, are provided. Overall, the mean accuracy for the entire data set (PP04 and PP05), which is calculated as the average difference between each nominal and each NASBA measured value, is 0.12 logs (0.15 logs for the dilute format; 0.09 logs for the undilute format). The overall reproducibility for the entire data set (i.e., mean SD) is 0.11 logs (0.13 logs for the dilute format; 0.09 logs for the

TABLE 1. Reproducibility of the HIV-1 QT NASBA Assay: VAQ PP04.

| Log Nominal Input/100 $\mu$L | Log NASBA Values/100 $\mu$L | Mean Log | Log SD |
|---|---|---|---|
| 2.38 | 2.15 | 2.5 | 0.42 |
| | 2.74 | | |
| 3.08 | 2.90 | 3.05 | 0.11 |
| | 3.11 | | |
| | 2.96 | | |
| | 3.18 | | |
| | 3.08 | | |
| 3.78 | 3.60 | 3.67 | 0.08 |
| | 3.76 | | |
| | 3.59 | | |
| | 3.75 | | |
| | 3.67 | | |
| 4.48 | 4.41 | 4.34 | 0.08 |
| | 4.23 | | |
| | 4.28 | | |
| | 4.36 | | |
| | 4.41 | | |
| 5.18 | 4.97 | 4.89 | 0.12 |
| | 4.80 | | |

TABLE 2. Reproducibility of the HIV-1 QT NASBA Assay: VQA PP05.

| Log Nominal Input/100 $\mu$L | Log NASBA Values/100 $\mu$L | Mean Log | Log SD |
|---|---|---|---|
| 2.38 | 2.71* | NA | NA |
| 3.08 | 3.04 | 3.20 | 0.13 |
| | 3.15 | | |
| | 3.36 | | |
| | 3.15 | | |
| | 3.30 | | |
| 3.78 | 3.85 | 3.83 | 0.09 |
| | 3.95 | | |
| | 3.79 | | |
| | 3.71 | | |
| | 3.83 | | |
| 4.48 | 4.45 | 4.51 | 0.08 |
| | 4.60 | | |
| | 4.57 | | |
| | 4.43 | | |
| | 4.52 | | |
| 5.18 | 5.18 | 5.17 | 0.05 |
| | 5.23 | | |
| | 5.11 | | |
| | 5.20 | | |
| | 5.11 | | |

* Nominal value is below the cut-off for the assay with undiluted calibrators. One of five samples at this nominal input was successfully quantified.

undilute format). These analyses clearly demonstrate that the HIV-1 QT NASBA assay is sensitive, accurate, and reproducible over a dynamic range of four logs.

Performance of the HIV-1 QT NASBA assay was evaluated using an additional panel of HIV-1 RNA standards. This quantitation panel is produced by Boston Biomedica Inc. (W. Bridgewater, MA) and is used as a validation panel for quantitative HIV-1 RNA assays. The panel consists of nine dilutions of an HIV-1 stock in pooled normal human sera. The dilutions were quantified by the manufacturer using an in-house RT-PCR method. Quantitation of the panel using NASBA was achieved with 100 $\mu$L of each panel specimen.

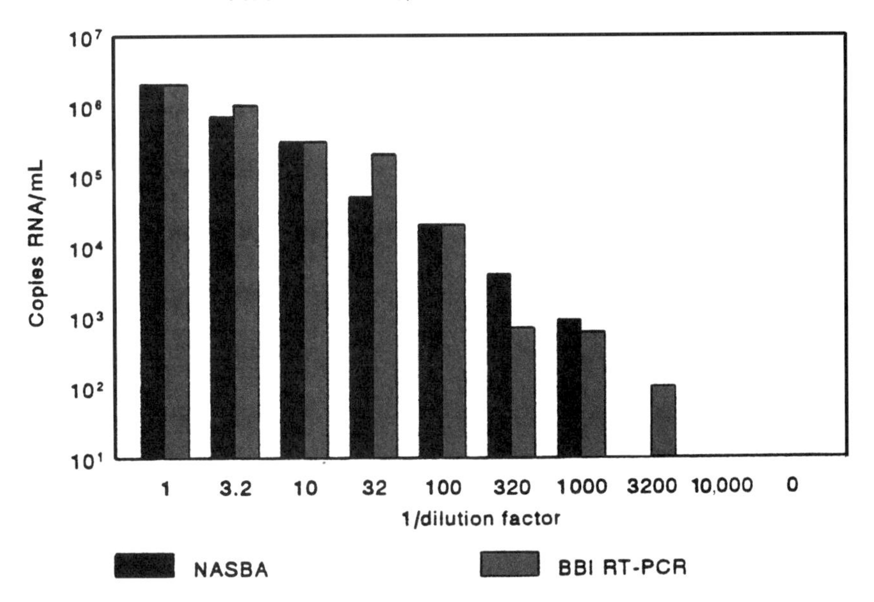


FIGURE 4. NASBA analysis of HIV-1 quantitation panel: BBI PRD801.

The results of this analysis are summarized in Figure 4, which compares the NASBA measured value to the stated nominal value calculated by the manufacturer. The NASBA values are reported per milliliter, and were obtained by multiplying the $100\,\mu$L results by a factor of 10. The results indicate that the HIV-1 QT NASBA assay produced essentially the same values as those obtained by the manufacturer. The 1/3200 and 1/10,000 dilutions were below cut-off for the NASBA assay. This is expected since $100\,\mu$L of sample were analyzed, and the manufacturer reported that concentration for the 1/3200 dilution was 100 copies/mL. In the case of each comparable sample, the NASBA result is essentially the same as that obtained by the manufacturer (mean log SD = 0.17).

The sensitivity of the HIV-1 QT NASBA assay is a function of the input volume of the sample. The assay can be readily applied to sample volumes ranging from $10\,\mu$L up to 1.0 mL. Assay results obtained with the NASBA QR System are provided as RNA copy number per input volume. Importantly, the same copy number of each of the three calibrator RNAs is added to the lysed sample, independent of the sample volume being used.

The amount of wild-type RNA in the input sample is solved from its ratio to the $Q_A$, $Q_B$, and $Q_C$ calibrators. A study designed to evaluate

TABLE 3. HIV-1 RNA Quantitation by NASBA: Effect of Ratio of Sample and Lysis Buffer Volume.

| Dilution | Sample Volume (mL) | Lysis Buffer Volume (mL) | Log Copies* | Log Copies/mL** |
|---|---|---|---|---|
| 1:10 | 0.01 | 0.9 | 3.18 | 6.18 |
| 1:10 | 0.10 | 0.9 | 3.94 | 5.94 |
| 1:10 | 0.10 | 9.0 | 4.10 | 6.10 |
| 1:10 | 1.00 | 9.0 | 4.98 | 5.98 |
| 1:30 | 0.01 | 0.9 | 2.79 | 6.34 |
| 1:30 | 0.10 | 0.9 | 3.44 | 5.92 |
| 1:100 | 0.01 | 0.9 | 2.19 | 6.19 |
| 1:100 | 0.10 | 0.9 | 3.59 | 6.59 |
| 1:100 | 0.10 | 9.0 | 3.08 | 6.08 |
| 1:100 | 1.00 | 9.0 | 4.05 | 6.05 |

* Values are the mean of two observations.
** Log copies/mL are calculated back from dilution factor and input volume.

this assay strategy on different sample volumes is summarized in Table 3. In this study, three dilutions (1:10, 1:30, and 1:100) of a patient plasma sample were prepared. Next, 10 μL, 100 μL, or 1.0 mL volumes of each dilution were analyzed with the assay. The 10 μL and 100 μL sample volumes were processed in 900 μL of lysis buffer; the 100 μL and 1.0 mL samples were processed in 9.0 mL of lysis buffer. The results of this analysis demonstrate that assay performance is independent of sample volume. The mean value of RNA copies per mL from the analysis of the different dilutions at different input volumes was $10^{6.14}$ copies/mL with a mean log SD of 0.16. Thus, the reproducibility of the assay is not affected by the size of the input sample.

## Biological Significance of HIV-1 RNA Plasma Load

The scientific literature is replete with reports addressing the biological significance of HIV-1 RNA levels in blood and plasma (Saag et al., 1996). The consensus of this significant body of literature can be summarized as follows: (1) HIV-1 RNA plasma levels can be used to assist in determining the prognosis of infected patients, (2) changes in plasma levels of HIV-1 RNA can be used to determine antiviral agent efficacy, and (3) HIV-1 RNA plasma levels can be used to study natural infection, as well as the impact on natural infection of therapeutic regimens. Interestingly, the HIV-1 QT NASBA assay has been used in several studies designed to address these issues. For example, in a retrospective natural history study, Jurriaans et al. (1994) demonstrated that from the time of seroconversion, a progressor patient population maintained serum levels of HIV-1 RNA that were significantly higher than serum levels of nonprogressor patients. Shortly after seroconversion, serum levels dropped in the nonprogressor populations who had nonsyncytia-inducing phenotype virus. Alternately, differences in RNA load between progressors and nonprogressors with syncytia-inducing phenotype were less pronounced.

More recently, the HIV-1 QT NASBA assay was used to demonstrate that determination of RNA plasma load at early time points after seroconversion (i.e., approximately 1 year) was the most useful means of predicting clinical outcome (Hogervorst et al., 1995). Further, this study confirmed the earlier findings of Jurriaans et al. (1994) that levels of HIV-1 RNA measured 5 years after seroconversion were decreased in nonprogressors; progressors were characterized by a lack of significant decrease by this time point.

The HIV-1 QT NASBA assay has also been used to evaluate therapy efficacy. In a retrospective study of 28 patients who were initially asymptomatic (16 who remained asymptomatic, 12 who progressed), RNA levels were determined before the initiation of AZT therapy, and for the first 8 weeks of therapy. Although RNA loads were reduced in the progressors and nonprogressors by week 4 of therapy, the reduction was significantly greater in the nonprogressor population (Jurriaans et al., 1995). Further, it was also demonstrated in this study that p24 antigen levels could not be used to predict clinical outcome, whereas the RNA levels were indicative. More recently, drastic reductions in HIV-1 RNA levels were determined by NASBA in HIV-1+ patients receiving Foscarnet treatment for CMV retinitis (Kaplan et al., 1995). The antiviral HIV-1 effect of Foscarnet was found to be very significant after as little as 3 weeks of therapy; cessation of the therapy due to resolution of the CMV infection resulted in the return to baseline of HIV-1 RNA levels in plasma. The assay has also been used to determine viral load in cerebrospinal fluid (CSF). Interestingly, CSF levels of HIV-1 RNA have proved useful in the differential diagnosis of AIDS-related encephalopathies (Ginocchio et al., 1995).

## Summary

The total NASBA system involves three component technologies: (1) a versatile nucleic acid isolation methodology, which can be applied to a wide array of specimen types and can be routinely utilized on sample volumes ranging from 10 $\mu$L to 1.0 mL, (2) a unique, isothermal amplification process that is highly suited for the amplification of RNA analytes, and (3) a powerful, semiautomated detection technology that is based in electrochemiluminescence chemistry. Quantitation with the NASBA system is achieved through the use of three internal calibrator RNAs, which serve as controls for the assay from the earliest stages of sample processing. The NASBA system has been highly developed for the quantitation of HIV-1 RNA, and has been shown to routinely perform in an accurate and reproducible manner, across a dynamic range of over four orders of magnitude, and with a maximum sensitivity of as little as 100 target molecules/mL. Importantly, the HIV-1 QT NASBA assay has been successfully applied in a number of studies designed to evaluate the natural history of HIV-1 infection, to determine antiviral agent efficacy, and to assist in determining the prognosis of HIV-1 infected patients.

*Acknowledgments*

The authors wish to thank the ACTG of the Division of AIDS, NIAID, NIH, for allowing the use of data from their proficiency panels. The authors are also grateful to Boston Biomedica, Inc., for allowing the use of their panel.

## REFERENCES

Barany F (1991): Genetic disease detection and DNA amplification using cloned thermostable ligase. *Proc Natl Acad Sci USA* 88:189.

Blackburn G, Shah H, Kenten J, et al. (1991): Electrochemiluminescence detection for development of immunoassays and DNA probe assays for clinical diagnostics. *Clin Chem* 37:1534.

Boom R, Sol C, Salimans M, et al. (1990): Rapid and simple method for purification of nucleic acids. *J Clin Microbiol* 28:495.

Compton J (1991): Nucleic acid sequence-based amplification. *Nature* 350:91.

Ginocchio C, Romano J, Kaplan M, et al. (1995): RNA plasma and CSF quantitation by NASBA technology for monitoring the efficacy of antiviral drug therapy and for the diagnosis of CMV and papovavirus encephalopathy. 95th General Meeting of the American Society for Microbiology, Washington, DC, Jan. 29–Feb.2.

Guatelli J, Whitfield K, Kwoh D, et al. (1990): Isothermal, in vitro amplification of nucleic acids by a multienzyme reaction modeled after retroviral replication. *Proc Natl Acad Sci USA* 87:7797.

Hogervorst E, Jurriaans S, de Wolf F, et al. (1995): Predictors for non- and slow progression in human immunodeficiency virus (HIV) type 1 infection: low viral RNA copy numbers in serum and maintenance of high HIV-1 p24-specific but not V3-specific antibody levels. *J Infect Dis* 171:811.

Jurriaans S, van Gemen B, Weverling G, et al. (1994): The natural history of HIV-1 infection: virus load and virus phenotype independent determinants of clinical course? *Virology* 204:223.

Jurriaans S, Weverling G, Goudsmit J, et al. (1995): Distinct changes in HIV type 1 RNA versus p24 antigen levels in serum during short-term zidovudine therapy in asymptomatic individuals with and without progression to AIDS. *AIDS Res Hum Retroviruses* 11:473.

Kaplan M, Ginocchio C, and Romano J (1995): The clinical utility of quantitative viral load: Implications about pathogenesis. *AIDS Res Hum Retroviruses* 11(S1):S83.

Katz A, Olsson C, Raffo A, et al. (1994): Molecular staging of prostate cancer with the use of an enhanced reverse transcriptase-PCR assay. *Urology* 43:765.

Kenten J, Gudibande S, Link J, et al. (1992): Improved electrochemiluminescent label for DNA probe assays: rapid quantitative assays of HIV-1 polymerase chain reaction products. *Clin Chem* 38:873.

Kievits T, van Gemen B, Van Strijp D, et al. (1991): NASBA isothermal enzymatic in vitro nucleic acid amplification optimized for the diagnosis of HIV-1 infection. *J Virol Meth* 35:273.

Kramer F, Lizardi P, and Tyagi S (1992): Q$\beta$ amplification assays. *Clin Chem* 38:456.

Layne S, Merges M, Dembo M, et al. (1992): Factors underlying spontaneous inactivation and susceptibility to neutralization of human immunodeficiency virus. *Virology* 189:695.

Saag MS, Holodniy M, Kuritzkes DR, et al. (1996): HIV-1 viral load markers in clinical practice. *Nature Medicine* 2:625.

Saiki R, Gelfard D, Stoffel S, et al. (1988): Primer directed enzymatic amplification of DNA with a thermostable DNA polymerase. *Science* 239:487.

Saiki R, Scharf S, Falcon F, et al. (1985): Enzymatic amplification of $\beta$-globin genomic sequences and restriction site analysis for diagnosis of sickle cell anemia. *Science* 230:1350.

van der Vliet G, Schukkink R, van Gemen B, et al. (1993): Nucleic acid sequence-based amplification (NASBA) for the identification of mycobacteria. *J Gen Microbiol* 139:2423.

van Gemen B, Kievits T, Schukkink R, et al. (1993): Quantification of HIV-1 RNA in plasma using NASBA during HIV-1 primary infection. *J Virol Meth* 43:177.

van Gemen B, van Beuningen R, Nabbe A, et al. (1994): A one-tube quantitative HIV-1 RNA NASBA nucleic acid amplification assay using electrochemiluminescent (ECL) labelled probes. *J Virol Meth* 49: 157.

van Gemen B, Kievits T, and Romano J (1995a): Transcription based nucleic acid amplification methods like NASBA and 3SR applied to viral diagnosis. *Rev Med Virol* 5:205.

van Gemen B, van de Wiel P, van Beuningen R, et al. (1995b): The one-tube quantitative HIV-1 RNA NASBA: precision, accuracy, and application. *PCR Meth Appl* 4:S177.

Walker G, Fraiser M, Schram J, et al. (1992): Strand displacement amplification — an isothermal, in vitro DNA amplification technique. *Nucleic Acids Res* 20:1691.

Wiedmann M, Wilson W, Czajka J, et al. (1994): Ligase chain reaction (LCR) — overview and applications. *PCR Meth Appl* 3:S51.

# APPENDIX I
## GENERIC NASBA METHODS

The HIV-1 quantitative NASBA assay described in this report is available from Organon Teknika as a commercial kit. The procedures described in this appendix are similar to those utilized in the HIV-1 Quantitative NASBA kit, however, they will allow for the generic amplification of any RNA analyte. These procedures are intended for use in the basic research laboratory, which has a need for isothermal RNA amplification and detection. It is important to remember that the product of the NASBA amplification reaction is RNA, which is highly susceptible to degradation by contaminating RNAse. Strict control of the plasticware and reagents is required to ensure the integrity of the assay. We routinely use commercially available molecular-biology-grade water (without DEPC) to prepare our reagents. However, we have not found it necessary to resort to autoclaving. Chemicals used in these methods are restricted for use in the NASBA procedures. Finally, NASBA is an amplification procedure, and it is therefore susceptible to template contamination. Consequently, great care is taken to separate activities (i.e., isolation, amplification, and detection).

A. Isolation of Nucleic Acids (Boom et al., 1990)

1. Into a 1.5 mL screw-cap microcentrifuge tube containing 900 μL of lysis buffer (50 mM Tris, pH 6.4, 20 mM EDTA, 1.3% w/v Triton-X 100, 5.25 M guanidine isothiocyanate [GuSCN]), add 100 μL of sample (i.e., whole blood, plasma, serum, cerebral spinal fluid, sputum, $10^5$ cells). Incubate at ambient temperature, 5 min, vortexing once per min.
2. Add 50 μL of activated silica suspension (1 gm silica [Sigma]/mL 0.1 M HCl), and incubate at ambient temperature for 5 min, vortexing once per min.
3. Centrifuge 10 sec at 10,000× g. Aspirate off the supernatant and resuspend the silica pellet in 1.0 mL of wash solution (50 mM Tris, pH 6.4, 20 mM EDTA, 5.25 M GuSCN). Vortex silica into an even suspension.
4. Repeat step 3.
5. Repeat step 3 two additional times; however, substitute 70% ethanol for wash solution.
6. Repeat this washing procedure one final time with acetone (Fisher Ultra Pure).
7. After removing the acetone supernatant, dry the silica approximately 10 min by incubating the tubes in a 56°C heating block, with the caps off the tubes.

8. When the silica has been completely dried, add 100 μL of sterile water, and vortex the silica into an even suspension. Incubate the suspension for 10 min in the 56°C heating block (caps on).
9. Pellet the silica by centrifuging for 2 min at 10,000×g. Remove 90 μL of the supernatant to a fresh tube and store at –70°C until needed for the amplification reaction. Typically 5 μL of this eluted nucleic acid solution are used in the amplification step.

B. Standard NASBA Amplification (van Gemen et al., 1993)

1. Using stock solutions, prepare a quantity of 2.5× reaction buffer, sufficient for the desired number of reactions, which is: 100 mM Tris, pH 8.5, 30 mM $MgCl_2$, 100 mM KCl, 12.5 mM DTT, 2.5 mM each dNTP, 5.0 mM each NTP.
2. Prepare an independent 5× primer solution, which is: 75% DMSO, 1.0 μM each primer (note: the standard HIV-1 *gag* region NASBA primers used are: P1 = 5′*AATTCTAATACGACTCACTATAGGG*TGCTATGTCACTTCCCCTTGGTTCTCTCA3′; P2 = 5′AGTGGGGGGACATCAAGCAGCCATGCAAA3′. The italicized sequence is the T7 RNA polymerase promoter region).
3. On ice, prepare a master mix enzyme solution that consists of: 8 units/μL T7 RNA polymerase (Pharmacia), 1.6 unit/μL AMV-RT (Seikagaku), 0.02 units/μL RNase H (Pharmacia), and 0.5 μg/μL BSA.
4. Thaw the nucleic acid extract and transfer 5 μL to a fresh tube. Add 5 μL of the 5× primer solution and 10 μL of the 2.5× buffer solution. Incubate the tubes at 65°C for 5 min.
5. Cool the tubes at 41°C for 5 min. Add 5 μL of the enzyme master mix to each tube and incubate at 41°C for 90 min. Store the completed reactions at –20°C until subjecting to a detection procedure.

C. Detection of Amplified Products

Detection of amplified products can be achieved by standard hybridization-based procedures (e.g., Southern Blot, liquid hybridization, etc.). Probes can be labeled with a radioisotope, or with a reporter molecule (horseradish peroxidase, biotin, etc.). Standardized techniques can be used to apply any of these strategies to the detection of a NASBA reaction product. Importantly, the probe used to detect the product must be the same strand as the original RNA analyte (i.e., the sense strand), since antisense RNA is the product of the NASBA process.

# Application of Transcription-Mediated Amplification to Quantification of Gene Sequences

Frank Gonzales and Sherrol H. McDonough

## Introduction

To be useful in clinical settings, hybridization-based assays must have high specificity and adequate sensitivity to detect clinically relevant concentrations of nucleic acid, and must be easy to use. Processing of patient specimens for these assays should involve a minimal number of steps so that the entire assay procedure will fit easily into the clinical laboratory.

We have developed commercial products for routine clinical use that utilize chemiluminescent DNA probes to directly detect nucleic acids in patient specimens (Limberger et al., 1992; Stockman et al., 1993). The sensitivity and ease-of-use of these assays are based on the hybridization protection assay, a homogeneous, sensitive detection format that simplifies hybrid detection by eliminating binding and washing steps commonly associated with hybridization-based formats (Arnold et al., 1989). These assays are adequate for situations in which the target sequences are found in relative abundance and correlate well with reference methods such as culture.

Assays with more stringent sensitivity requirements incorporate an enzymatic amplification step prior to the hybridization assay. Target amplification technology has been especially useful in clinical assays in which the pathogen is present in limiting quantities and is difficult to detect by conventional means such as culture (Jonas et al., 1993). The amplification method used in our assays, transcription-mediated amplification or TMA, is a versatile, rapid, and sensitive target amplification system that requires no thermal cycling. The system is capable of amplifying RNA or DNA targets and can amplify more than one nucleic acid at a time. Amplification-based systems have been refined beyond qualitative outputs to quantitative modes whereby estimates of the initial concentration of target nucleic acid can be made. In TMA, conditions can be adapted to detect fewer than ten copies of target if required, and also to allow amplified product accumulation and detection in proportion to the input target over at least a four-log range. The utility of the information gained by the clinician from these assays is affected by assay accuracy, precision, and reproducibility. These factors, in turn, are dependent on the consistency of the sample preparation, amplification, and detection steps.

## Assay Description

A sample assay protocol is shown in Figure 1. We have adapted our technology to allow not only sensitive detection of RNA or DNA targets, but quantitative measurements of these analytes as well. The assays under development follow the three main steps of sample processing, target amplification using transcription-mediated amplification, and specific amplicon detection using the hybridization protection assay. All assay steps were designed to integrate into an easy-to-perform, batch format that requires no thermocycling, no sample transfers, and no amplicon dilutions. Up to 80 specimens are processed at a time in a manual mode, and all assay steps are easy to automate for improved throughput.

### *Step One — Sample Preparation*

This step secures an adequate number of cells or virions to provide sensitivity for a given application. In addition, this step releases and stabilizes the nucleic acids of interest, eliminates nuclease activity, and removes or inactivates potential

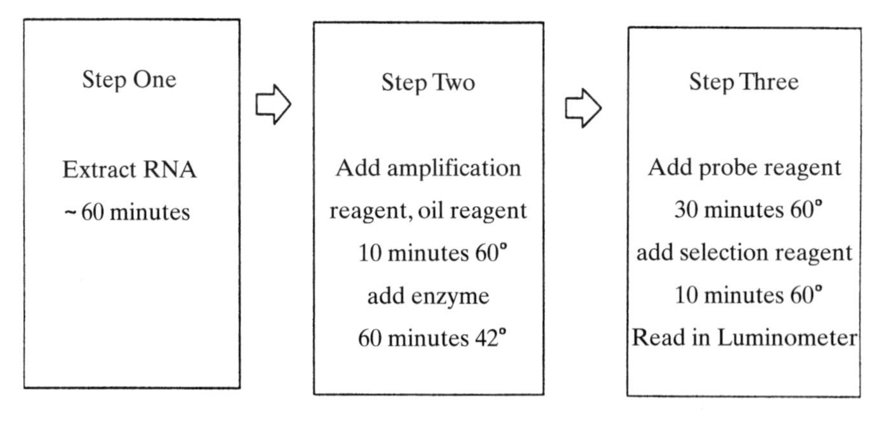


FIGURE 1. Representative TMA assay protocol.

inhibitors of amplification and detection. As the initial step of an assay, sample preparation is critical to the success of subsequent steps and has obvious effects on the accuracy and precision of the quantitative result. We have worked to develop methods that maximize sensitivity, maximize reproducibility, and are user-friendly (batch format, no centrifugations, no toxic chemicals).

### *Step Two — Amplification by TMA*

Target amplification is performed with the RNA transcription system referred to as transcription-mediated amplification, or TMA. This process uses two enzymes, reverse transcriptase and T7 RNA polymerase, to autocatalytically cycle between RNA and DNA intermediates without thermocycling. The system is versatile and unique in that it amplifies RNA and DNA targets with the same number of steps and enzymes. The reaction kinetics are extremely rapid, resulting in excess of one billion-fold amplification of target sequences in less than an hour. The reactions produce single-stranded RNA amplicons that are readily detected in a hybridization detection format. The principle behind TMA is the conversion of target sequences into promoter-containing DNA intermediates that are transcribed by RNA polymerase to produce hundreds of copies of RNA. In contrast to other amplification assay formats, this system allows amplification and detection to be performed in the same tube, omitting postamplification sample transfers, and reducing the potential for carryover contamination. A reaction scheme for an RNA target is shown in Figure 2. The reaction begins when a promoter-primer hybridizes to a specific RNA template and is extended by reverse transcriptase, forming an RNA:DNA intermediate. RNAse H and other activities of reverse transcriptase make the new cDNA available to hybridize to a second primer, primer 2, which can be extended by the DNA-dependent DNA polymerase activity of reverse transcriptase to give a DNA intermediate with a double-stranded DNA promoter. This DNA intermediate is recognized by T7 RNA polymerase, and multiple copies of single-stranded RNA transcript, or amplicon, are produced. Primer 2 hybridizes to this transcript and is extended, forming a new RNA:DNA intermediate. The RNAse H activities of the reverse transcriptase again make the cDNA available for binding of the promoter-primer. The promoter-primer hybridizes to the cDNA and is extended and copied by reverse transcriptase to form a new DNA intermediate. This new DNA intermediate is recognized by T7 RNA polymerase, producing more RNA amplicon, which continues the amplification cycle. Primer access to the DNA strand of the RNA:DNA hybrids (see Figure 2, steps 3 and 10) is accomplished by the activities of reverse transcriptase; thus, in contrast to PCR, no thermocycling is required. For amplification of a DNA target,

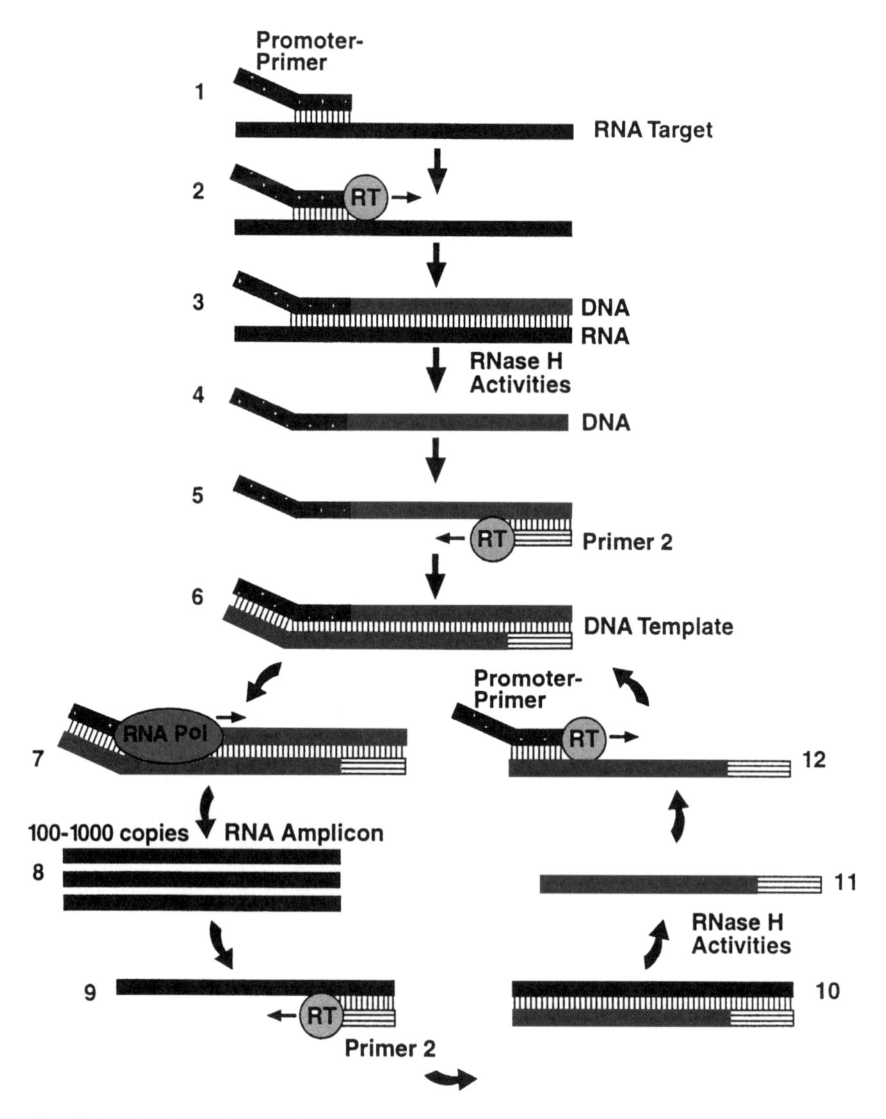


FIGURE 2. Reaction scheme for amplification of an RNA target using transcription-mediated amplification.

the unwinding and strand-displacing activities of the reverse transcriptase may replace the RNase H activity, as shown in Figure 2, steps 3 and 4, in separating the first cDNA strand from the target strand.

### *Step Three — Detection by HPA*

Specific detection of the product of an amplification reaction is always preferred, as nonspecific side products can contribute to false signals. Specific detection of the TMA RNA amplicon is performed by HPA with acridinium-ester-labeled DNA probes. HPA is unique in that distinction of hybridized from unhybridized probe is performed by chemical hydrolysis in solution, rather than by physical separation. This detection format involves a hybridization step followed by a selection step. In the hybridization step, an excess of chemiluminescent acridinium-ester-labeled probe is allowed to anneal to the single-stranded RNA amplicon. In the selection part of the assay, the label on any remaining unhybridized probe is rendered nonchemiluminescent by addition of an alkaline selection reagent. This alkaline reagent creates conditions under which the chemiluminescent label on unhybridized probe is hydrolyzed, leaving chemiluminescent label on hybridized probe intact (Figure 3). Chemiluminescence from the acridinium ester retained within the hybrids is measured in a luminometer, and signals are expressed numerically in relative light units or RLU. HPA has at least a five log-linear range, and as little as 0.01 fmole of product can be detected.

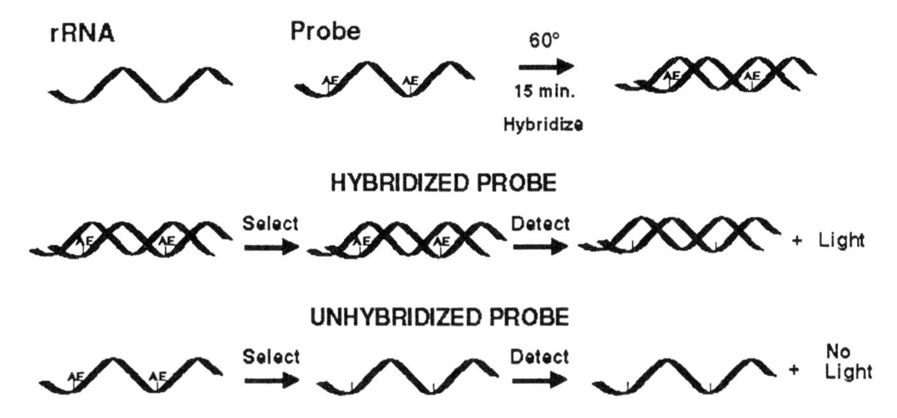


FIGURE 3. Mechanism of the hybridization protection assay. RNA amplicon is allowed to hybridize to an excess of acridinium ester (AE) labeled probe for 15–30 min at 60°C. Following the hybridization step, a selection reagent is added that hydrolyzes label on unhybridized probe to a nonchemiluminescent form. Label on hybridized probe is protected from hydrolysis and retains its chemiluminescent properties. Chemiluminescence remaining after the selection step is a direct measure of the amount of hybrid formed.

### *Controls*

The use of accurate and universally accepted standards is important for general applicability of quantitative methods. The examples shown in this chapter use external standard curves to quantify the amount of target in test reactions. The standard curves were prepared with *in vitro* transcribed RNA for CML and HIV, or from cloned DNA for HBV. The HIV RNA was quantified by absorbance, by hybridization with a number of probes in the HPA format, and by comparison to an externally calibrated source of HIV RNA, the VQA virion standard obtained from the AIDS Clinical Trials Group Virology Quality Assurance Reference Laboratory. Cloned HBV DNA was quantified by absorbance.

## Results

### *Detection of cellular RNA*

This first example shows relative quantification of a cellular RNA using chronic myelogenous leukemia or CML as a model system. This leukemia is characterized by the *t*(9;22) chromosomal translocation where the *bcr* and *abl* genes, normally found on separate chromosomes, are brought into juxtaposition. This fusion gene is expressed as a fusion mRNA that can be detected by molecular methods (Hochhaus et al., 1996; Kantarjian et al., 1993). Patients with CML usually die within 1 year of diagnosis without treatment, while appropriate treatment can place CML patients in clinical remission. In clinical remission, the white blood cell count returns to normal, and monitoring of chromosome smears (usually 20 smears are evaluated) reveals no translocations.

These cytogenetic methods are imprecise and have low sensitivity; therefore, we have applied our technology to relative quantification of the fusion RNA as a potentially more specific and sensitive method of CML patient monitoring. The steps of the assay are 1) isolate mRNA from whole blood or bone marrow, 2) amplify with TMA, and 3) detect the specific amplicon with HPA. Up to 80 specimens can be processed in under 4 hours. To perform the assay, 0.05–0.5 ml of whole blood or bone marrow was processed, and RNA from 10–40 $\mu$l of original specimen was amplified. The quantitative character of the TMA reaction is shown in Figure 4. Dilutions of *in vitro* transcribed RNA containing the *bcr-abl* translocation, or RNA from K562 cells that express the *bcr-abl* translocation, showed a linear relationship between input RNA and signal output.

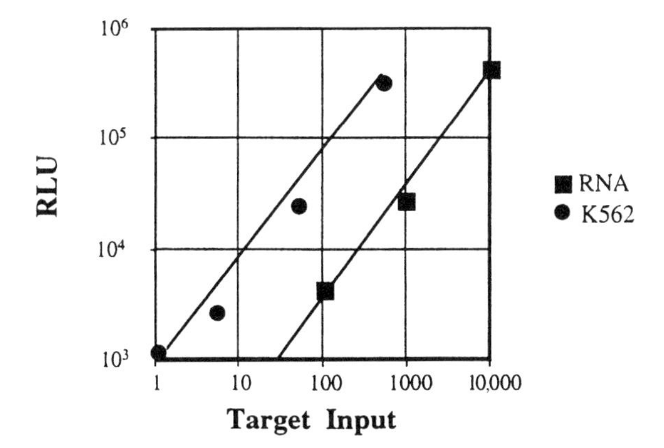


FIGURE 4. Quantitative character of the TMA assay. *In vitro* transcribed RNA or RNA from K562 cells was amplified and analyzed by HPA. Increasing amounts of *in vitro* transcript (squares) or RNA from K562 cells (circles) gave increasing RLU signals over the range analyzed. A portion of the reaction was analyzed and RLU extrapolated.

To determine the assay utility in clinical specimens, coded whole blood or bone marrow specimens from M. D. Anderson Cancer Center were evaluated for white cell count in a Coulter Counter and for *bcr-abl* RNA in our assay. A standard curve of *in vitro* transcript was run with each set of clinical specimens. As shown in Table 1, the fusion RNA was detected at high levels in patients with active CML disease, and no RNA was detected in eight of eight normal patients or patients without CML disease. Of particular interest were those patients in the remission phase of CML.

TABLE 1. Detection of *bcr/abl* RNA in Patient Specimens.

| Status of patient | Copies of *bcr/abl* RNA per $10^5$ WBCs | Number of patients (total tested) |
|---|---|---|
| Normal/no CML disease | 0 | 8 (8) |
| Active CML disease | >>1000 | 6 (6) |
| Early remission (<1 year) | >1000 | 1 (1) |
| Long-term remission (3–10 years) | 101–1000 | 1 (19) |
| " | 11–100 | 3 (19) |
| " | >0, <10 | 15 (19) |

These patients showed no cytologic or hematologic evidence of leukemia. One specimen from a patient in early remission showed fairly high levels of *bcr-abl* RNA (>$10^3$ copies/$10^5$ WBCs), while the remaining specimens had very low levels of fusion RNA. Specimens from the majority, 15 of 19 patients, showed signals above background but below the ten-copy standard. These results show that this assay can be adapted to allow relative quantification of a specific RNA in a complex mixture of cellular RNAs. Sensitivity of at least ten copies of RNA standard per 20 $\mu$l of blood was easily achieved, and the assay was capable of detecting fusion RNA in blood or bone marrow from patients at all stages of CML disease.

## *Quantification of a viral DNA*

One advantage of TMA compared to other amplification systems is the ability to amplify either RNA or DNA targets with the same number of enzymes and assay steps. To demonstrate the ability of TMA to quantify a DNA target, we applied the Gen-Probe assay to quantification of hepatitis B virus (HBV) DNA. HBV is a highly infectious virus that is usually transmitted through contact with infected blood or blood products, or through sexual contact, or is passed from mother to infant during the perinatal period. Although the virus replicates in the liver, virus is also found in peripheral blood. Acute hepatitis infections can resolve, or in about 5% of cases, proceed to a chronic phase. Chronic infection can lead to more serious conditions such as cirrhosis or hepatocellular carcinoma. Interferon therapy is the only FDA-approved treatment for chronic hepatitis B, but only about 30% of eligible patients show a sustained response. Viral load may play a role in response to therapy, and viral monitoring may provide a way to predict as well as assess treatment efficacy (Perrillo et al., 1990; Zaaijer et al., 1994). The approach used to monitor HBV DNA in serum was to lyse the virus directly in a patient specimen, amplify with TMA in the presence of the lysate, and detect with HPA. Figure 5 shows the results obtained when cloned HBV DNA was spiked into HBV-negative serum. Samples were amplified in duplicate and analyzed with two concentrations of probe. A greater than four log dynamic range was seen with cloned DNA or virus in serum.

The ability of the assay to detect changes in viral load was assessed using samples from chronic HBV patients undergoing interferon therapy. Samples were kindly provided by Dr. Robert Perrillo of the Oschner Clinic. Figure 6A shows that two pretreatment specimens gave consistent HBV DNA levels of about 6700 pg/ml. Following initiation of interferon treatment at week 7, a significant drop in DNA concentration was observed by the DNA assay, dropping below assay threshold (<250 copies) by week 18.

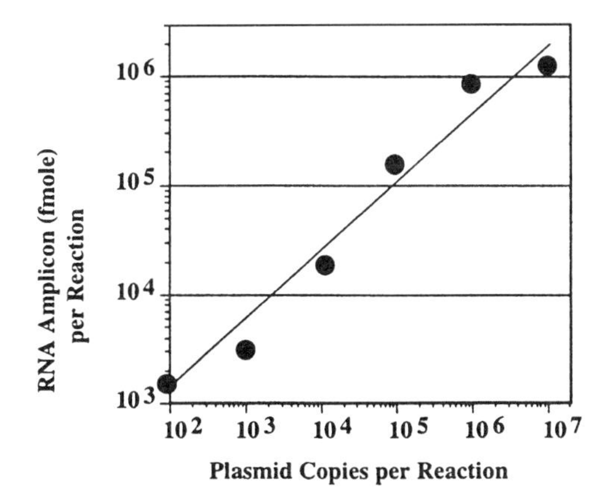


FIGURE 5. Dynamic range of TMA assay for HBV DNA. Cloned HBV DNA was spiked into normal serum and amplified by TMA and analyzed by HPA. Duplicate reactions were run and analyzed with two concentrations of probe.

This patient showed a small rebound in HBV DNA levels at week 31; however, this level represents a small fraction of the pretreatment viral burden. The figure also shows the concomitant drop in HBeAg reactivity following treatment. The coincident drop in both DNA and HBeAg markers indicated the initial effectiveness of interferon treatment in lowering peripheral blood viral load in this patient. A different type of treatment response is illustrated in Figure 6B. The two pretreatment samples for this patient also showed reproducible DNA levels of about 7100 pg/ml. A drop in DNA concentration was also observed in this patient following initiation of interferon treatment at weeks 13 and 27; however, the DNA level in this patient never dropped below the assay threshold, and a return to pretreatment concentrations was observed by week 47. This is believed to signal treatment failure in this individual.

These data show that viral DNA can be quantified in crude serum lysate using this assay and that at least a four log dynamic range was obtained by amplifying in duplicate and analyzing with two concentrations of probe. The changes in HBV DNA closely mirror those for another marker of infection, HBeAg, and therefore the results appear to be useful for patient monitoring.

### *Quantification of a viral RNA*

The final example demonstrates the ability of this technology to quantify a viral RNA target in plasma using HIV RNA as the target nucleic acid. This

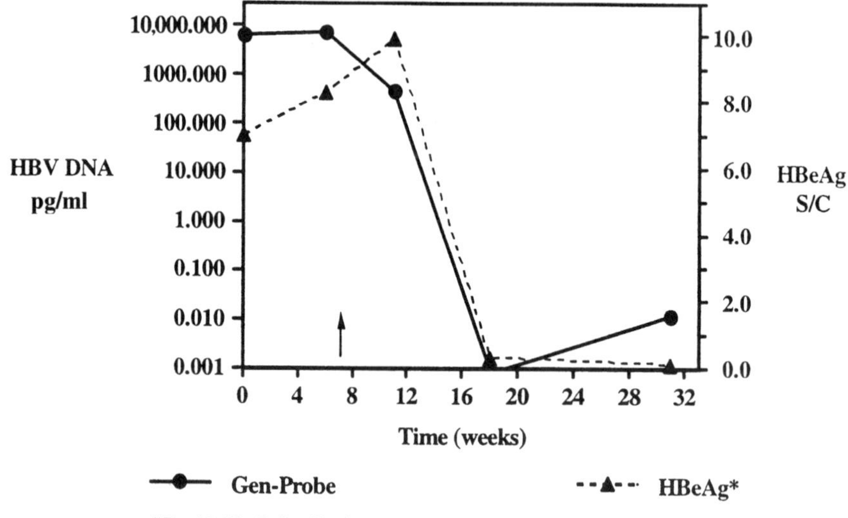


a

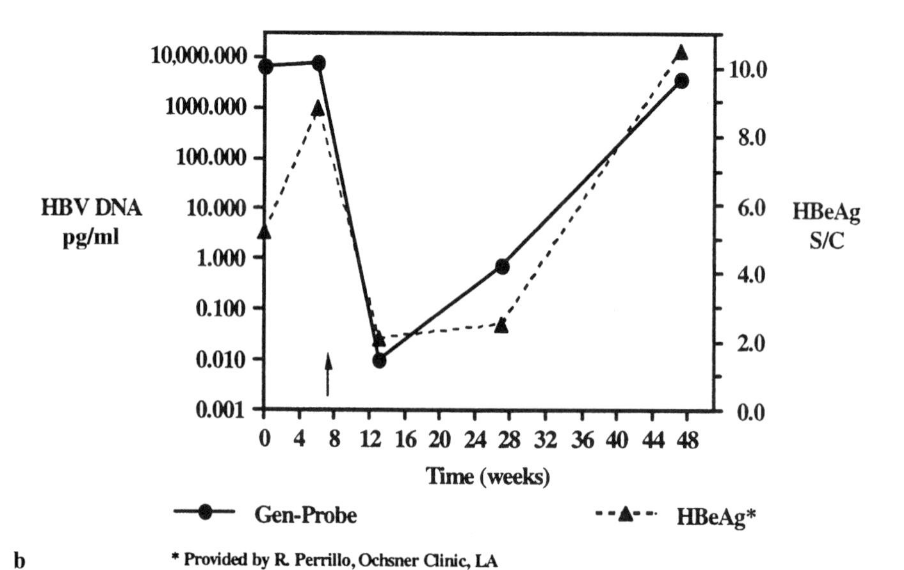


b

FIGURE 6. Analysis of specimens from a chronic hepatitis patient before and during interferon treatment. HBV DNA levels were determined using the TMA and HPA format. HBeAg levels, expressed as signal:cut-off, were determined at the Oschner Clinic. The patients were given interferon at week 7.

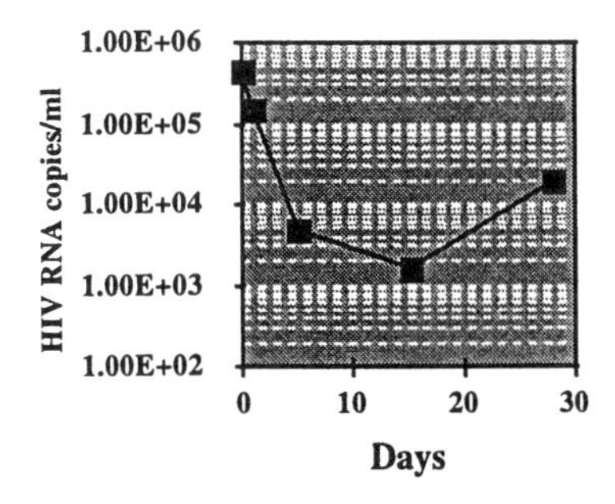


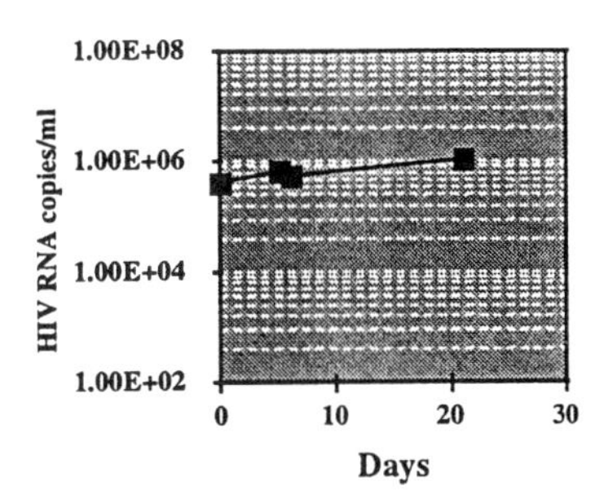


FIGURE 7. Analysis of specimens from HIV-infected patients with (left panel) and without (right panel) Nevirapine therapy. ACD plasma samples were stored and analyzed retrospectively. Samples were kindly provided by Dr. Douglas Richman, University of California, San Diego.

virus is found in plasma following infection, and a drop in plasma virus RNA concentration has been observed to follow some drug treatments (Kappes et al., 1995). We developed an assay for HIV RNA in plasma that involved an RNA isolation step, amplification with TMA, and detection with acridinium-ester-labeled probe. All steps of this assay were performed in a single tube to eliminate carryover due to pipetting of amplicon, and to simplify the procedure. A dynamic range spanning 20–200,000 copies was obtained when *in vitro* transcript or HIV virions were spiked into 100 $\mu$l of plasma. To assess the assay's ability to detect changes in plasma HIV RNA levels, we assayed specimens provided by Dr. Douglas Richman at the University of California, San Diego Medical School, from patients undergoing treatment with the nonnucleoside reverse transcriptase inhibitor Nevirapine. Figure 7 shows that the assay identified changes in RNA level following initiation of drug therapy, and showed consistent RNA levels in an untreated control individual over four time points. The rise in viral RNA concentration following the initial drop is believed to represent the outgrowth of drug-resistant variants.

## Conclusions

In summary, quantification of nucleic acids is useful for many applications. Adaptation of TMA and HPA technologies to a quantitative mode has been demonstrated with cellular and viral targets in many specimen types. Excellent sensitivity and specificity have been observed with the assay formats under development, and the greater than three log dynamic range allows potential application of this technology to many systems. The assay allows for high throughput in a manual mode, and automation of these assays is under development.

## REFERENCES

Arnold LJ, Hammond PW, Wiese WA, and Nelson NC (1989): Assay formats involving acridinium-ester-labeled DNA probes. *Clin Chem* 35:1588–94.

Hochhaus A, Lin F, Reiter A, Skladny H, Mason PJ, van Rhee F, Shepherd PCA, Allan NC, Hehlmann R, Goldman JM, and Cross NCP (1996): Quantification of residual disease in chronic myelogenous leukemia patients on interferon-$\alpha$ therapy by competitive polymerase chain reaction. *Blood* 87:1549–55.

Jonas V, Alden MJ, Curry JI, Kamisango K, Knott CA, Lankford R, Wolfe JM, and Moore DF (1993): Detection and identification of *Mycobacterium tuberculosis* directly from sputum sediments by amplification of rRNA. *J Clin Microbiol* 31:2410–16.

Kantarjian HM, Deisseroth A, Kurzrock R, Estrov Z, and Talpaz M (1993): Chronic myelogenous leukemia: A concise update. *Blood* 82:691–703.

Kappes JC, Saag MS, Shaw GM, Hahn BH, Chopra P, Chen S, Emini EA, McFarland R, Yang LC, Piatak Jr. M, and Lifson JD (1995): Assessment of antiretroviral therapy by plasma viral load testing: standard and ICD HIV-1 p24 antigen and viral RNA (QC-PCR) assays compared. *J Acquir Immune Defic Syndr* 10:139–49.

Limberger RJ, Biega R, Evancoe A, McCarthy L, Slivienski L, and Kirkwood M (1992): Evaluation of culture and the Gen-Probe PACE2 assay for detection of *Neisseria gonorrhoeae* and *Chlamydia trachomatis* in endocervical specimens transported to a state health laboratory. *J Clin Microbiol* 30:1162–6.

Perrillo RP, Schiff ER, Davis GL, Bodenheimer Jr. HC, Lindsay K, Payne J, Dienstag JL, O'Brien C, Tamburro C, Jacobson IM, Sampliner R, Feit D, Lefkowitch J, Kuhns M, Meschievitz C, Sanghvi B, Albrecht J, and Gibas A

(1990): A randomized, controlled trial of interferon $\alpha$-2b alone and after prednisone withdrawal for the treatment of chronic hepatitis B. *New Engl J Med* 323:295–301.

Stockman L, Clark KA, Hunt JM, and Roberts GD (1993): Evaluation of commercially available acridinium ester-labeled chemiluminescent DNA probes for culture identification of *Blastomyces dermatitidis*, *Coccidioides immitis*, *Cryptococcus neoformans* and *Histoplasma capsulatum*. *J Clin Microbiol* 31:845–50.

Zaaijer HL, ter Borg F, Cuypers HTM, Hermus MCAH, and Lelie PN (1994): Comparison of methods for detection of hepatits B virus DNA. *J Clin Microbiol* 32:2088–91.

# PART I: METHODS/TECHNOLOGY ISSUES

## C. GENE QUANTITATION BASED ON SIGNAL AMPLIFICATION

# Branched DNA (bDNA) Technology for Direct Quantification of Nucleic Acids: Design and Performance

Mark L. Collins, Peter J. Dailey, Lu-Ping Shen, Mickey S. Urdea, Linda J. Wuestehube, and Janice A. Kolberg

## Introduction

The branched DNA (bDNA) assay represents a significant advance in the direct quantification of nucleic acid molecules for research and clinical applications. Other methods for the detection and quantification of nucleic acid molecules, such as polymerase chain reaction (PCR) and nucleic acid sequence-based amplification (NASBA), involve molecular amplification of target nucleic acids followed by detection of amplified products. While these target amplification methods can be extremely sensitive, they suffer from two major disadvantages. First, it is difficult to obtain reproducible quantification since the target nucleic acid must be extracted and amplified in order to measure it. Second, numerous precautions must be taken to avoid false positive results caused by contamination of samples with PCR products and carryover from other specimens. A new departure from target amplification methods, the bDNA assay, directly measures nucleic acid molecules by boosting the reporter signal, rather than by replicating target sequences as the means of detection (Figure 1), and hence it is not subject to the errors inherent in the extraction and amplification of target nucleic acids. An ideal tool for nucleic acid quantification, the bDNA assay detects nucleic acid molecules at their physiological concentration, yields highly reproducible quantification, and eliminates false positives due to contamination.

The bDNA assay has been applied successfully in a number of clinical research areas, including the prognosis and therapeutic monitoring of patients with viral diseases. As a reliable method for quantification of viral load, bDNA assays have been developed to measure human immunodeficiency virus type 1 (HIV-1) RNA (Kern et al., in press; Pachl et al., 1995; Todd et al., 1995), hepatitis B virus DNA (Hendricks et al., 1995),

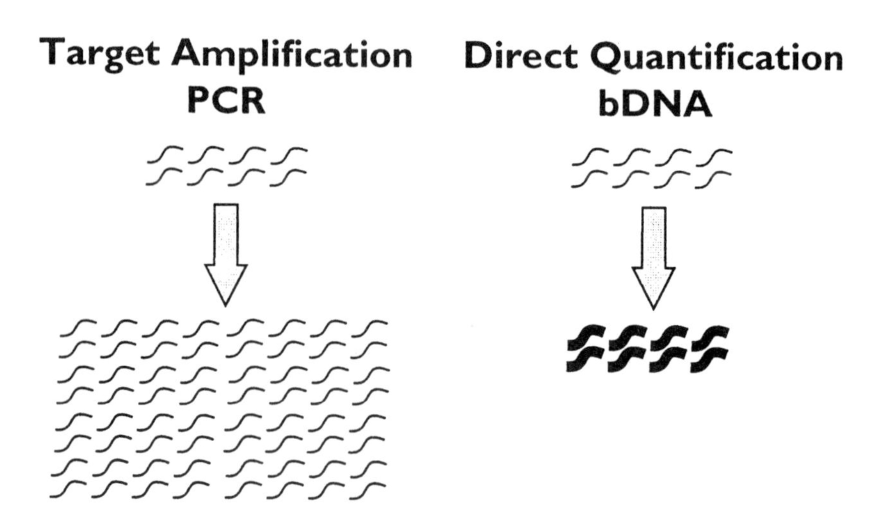


FIGURE 1. Comparison of target amplification methods to the bDNA assay for detection and quantification of nucleic acid molecules. Target amplification methods such as PCR first increase the number of target nucleic acid molecules to facilitate detection, then back-calculate to derive the original number of target nucleic acid molecules present. By contrast, the bDNA assay directly labels all target nucleic acid molecules and then counts them.

hepatitis C virus RNA (Davis et al., 1994; Detmer et al., 1996), and cytomegalovirus DNA (Chernoff et al., submitted; Flood et al., submitted). Yet, the potential applications of the bDNA assay extend far beyond viral nucleic acid quantification. Recent efforts have focused on quantification of cellular mRNAs. For example, the bDNA assay has been used for investigations into the concentration and intracellular location of specific cellular messages, splicing of intracellular messages, and expression of stress-induced genes for toxicology applications. Specific examples of novel applications of the bDNA assay are described in the chapter "Branched DNA (bDNA) Technology for Direct Quantification of Nucleic Acids: Research and Clinical Applications" in this book. With the development of bDNA assays for custom use, it is possible for researchers to use bDNA technology for a wide variety of specific research applications. This chapter describes the design of the bDNA assay for quantification of nucleic acid molecules and details the protocols and procedures for its use in a research or clinical laboratory. The development of reference standards for the bDNA assay is also presented, as well as specific examples illustrating the performance of the bDNA assay with regard to accuracy, sensitivity, and reproducibility.

# bDNA Assay Design

## *bDNA assay format*

Amenable to routine use in a clinical setting, the bDNA assay is similar to an ELISA in its basic approach and uses a 96-well microplate format (Figure 2). Each well has "capture probes" attached to its surface that contain a specific nucleotide sequence (Running and Urdea, 1990). These capture probes bind to an overhang sequence on a subset of "target probes," which are bound to specific sequences of the target nucleic acid molecule, thereby anchoring the target nucleic acid molecule to the microplate surface. The use of several target probes for capture increases the specificity of the capture process. Another set of target probes contains an overhang sequence that is designed to bind to the bDNA amplifier molecules (Horn and Urdea, 1989), each of which contains a maximum of 45 binding sites for alkaline-phosphatase-conjugated "label probes" (Urdea et al., 1988). At the end of the hybridization steps, the target nucleic acid molecule may contain as many as 1700 separate alkaline-phosphatase-conjugated label probes, thus providing significant enhancement of the signal. After introduction of a chemiluminescent dioxetane substrate, which is activated by the alkaline

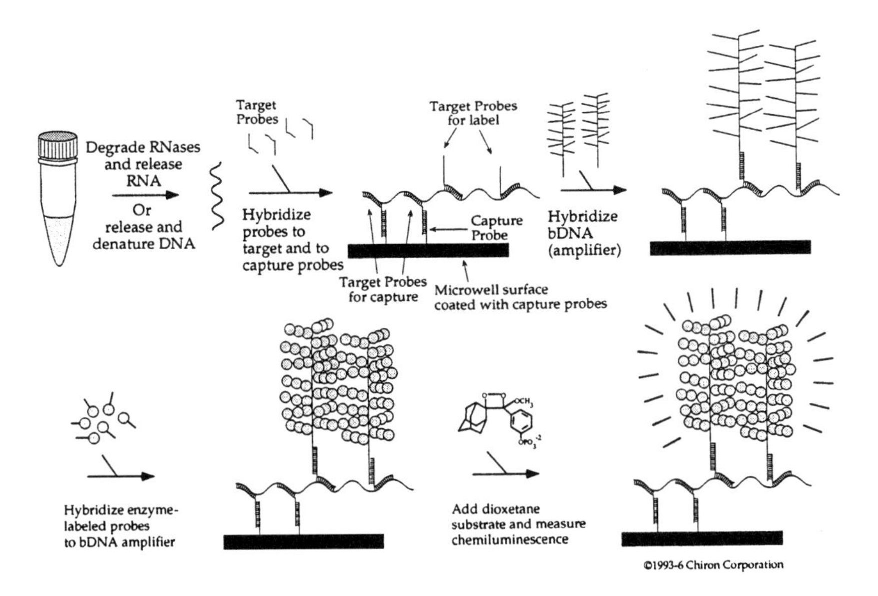


FIGURE 2. Schematic representation of the bDNA assay for direct quantification of nucleic acid molecules.

phosphatase, the signal is quantified easily by photon counting in a luminometer. The bDNA assay is inherently quantitative since the number of photons emitted is directly related to the amount of target nucleic acid molecules in the specimen.

With current technology, the bDNA assay can accurately and precisely measure between approximately 500 and 10,000,000 molecules (Kern et al., 1996). New advances in bDNA technology include the addition of preamplifier molecules and the incorporation of the novel nucleotides isoC and isoG into oligonucleotide probe sequences to further enhance signal and reduce noise caused by nonspecific hybridization of bDNA assay components (Kern et al., 1996; Collins et al., submitted). These improvements have extended the quantitative detection limit of the bDNA assay to as low as 50 molecules.

The practical features of the bDNA assay facilitate ease of use in a clinical or research laboratory. Sample preparation usually is minimal, and no special handling procedures are required for containment of potential contaminants. Also, since a chemiluminescent detection system is used, the bDNA assay does not suffer from the limited shelf-life of radioactive assays. With the 96-well microplate format, up to 42 samples in duplicate plus controls and standards (optional) can be included in each bDNA assay run, allowing many specimens to be processed by one individual in less than 24 hours. Basic reagents and a typical protocol for the bDNA assay are described in Appendix I.

## *Specimen preparation*

One of the advantages of the bDNA assay is that it does not require highly purified nucleic acid preparations, and hence it can be applied to a number of specimen types. Reliable measurement of nucleic acid molecules using the bDNA assay has been achieved with both crude serum and plasma specimens as well as extracts from cells and biopsy specimens. Serum and plasma specimens collected using standard procedures have been used directly and after storage at –20°C or below. Measurement of relatively abundant mRNAs has been accomplished by processing cells ($<10^6$) directly in the bDNA assay. However, for some applications, such as measuring very low-abundance mRNA transcripts, prior extraction of RNA is preferable, since chromosomal DNA may interfere with assay performance if present in large amounts. Isolation of RNA and DNA from the same specimens also allows for determination of an RNA : DNA ratio, an index that is useful as a measure of gene expression for comparison between samples. Protocols for processing cells and biopsy tissue specimens for use in the bDNA assay are provided in Appendix II.

### *Oligonucleotide probe design*

In addition to the bDNA assays for quantification of viral nucleic acids commercially available for research and the services offered through the Chiron Reference Testing Laboratory, custom design of the oligonucleotide probes for use in the bDNA assay is possible. Given the number of probe components in a bDNA assay, the design of the oligonucleotide probes is critical to bDNA assay performance. Chiron has developed and intends to introduce a number of novel software applications allowing for the computer design of oligonucleotide probes for the bDNA assay that can greatly reduce the time spent optimizing assay conditions (Bushnell et al., in preparation). Custom design of bDNA probe sets will be performed in conjunction with Chiron-certified technical support personnel.

The basic methodology for bDNA assay probe design involves a series of steps, including selection of a target nucleic acid sequence, generation of a set of candidate probes, cross-hybridization screening, and final probe assignment. A number of critical parameters must be considered in the selection of a target nucleic acid sequence. For example, the degree of exclusivity or inclusivity of the target probes must be determined according to the particular research application. Some applications require probes that are specific to unique sequences in the nucleic acid molecule of interest, while other applications need probes that recognize highly conserved or homologous sequences. Probes also can be designed for contiguous coverage of the target nucleic acid sequence to offer protection against possible nuclease degradation. In the probe generation step, a variety of program options is available for the design of potential oligonucleotide probes with constant length, constant Tm, or constant Td. In cross-hybridization screening, sequences of the candidate oligonucleotide probes are compared to sequences in regularly updated databases, such as GenBank and European Molecular Biology Laboratory (EMBL), to search for possible similarities to other known nucleotide sequences. Probe sequences also are screened for potential cross-hybridization with other bDNA assay components. These screening steps for cross-hybridization reduce nonspecific hybridization. Finally, in the probe assignment step, only those probes that are matched for the desired exclusivity/inclusivity, appropriate hybridization properties, and minimal nonspecific hybridization are chosen for use in the bDNA assay. Although the total number of probes will depend upon the specific application, a typical probe set for the bDNA assay includes 3–5 target probes for capture, and 10–15 target probes for labeling.

## Performance of the bDNA Assay

The high level of performance exhibited by the bDNA assay is maintained by rigorous testing of several parameters, including accuracy, sensitivity, and reproducibility. Carefully prepared reference standards, extensively characterized by independent quantification methods, can be included in the bDNA assay to ensure accurate results. In evaluating the sensitivity of the bDNA assay, two limits can be defined — one that is appropriate for use in clinical applications (clinical quantification limit), and one that reflects the absolute sensitivity of the bDNA assay (molecular detection limit). The clinical quantification limit is based on analysis of clinical specimens and is necessary to avoid unacceptable false positive results. The molecular detection limit is based on dilution series analysis and is less than or equal to the clinical quantification limit. To ensure reproducibility, the bDNA assay is thoroughly tested for both within-lot and between-lot precision, so that the variations introduced by testing specimens at different times, by different operators, or at different sites are minimal. The following sections further describe the issues important for maintaining the high performance characteristics of the bDNA assay with regards to accuracy, sensitivity, and reproducibility.

### *Accuracy*

Accuracy is the measure of how close the quantification values generated by an assay for a sample are to the true quantity of nucleic acid molecules present in the sample, and it is the key to the prognostic value of nucleic acid quantification. The prediction of disease progression and the likelihood of response to therapy are based upon clinical studies involving populations of patients in which nucleic acid concentrations are measured (for example, Mellors et al., 1996). Thus, in order to be clinically meaningful, it is imperative that accurate results be obtained. To ensure the accuracy of assay results, the bDNA assay includes reference standards characterized by well-established, independent methods (Collins et al., 1995).

A schematic diagram of the procedures used to characterize the reference RNA standards that are included in bDNA assays for measurement of viral nucleic acids is shown in Figure 3. Purified transcripts are analyzed by agarose gel electrophoresis to determine the percent of the preparation that is full-length. Four independent quantification methods are then applied, including optical density, incorporation of $^{32}$P-GTP, total phosphate

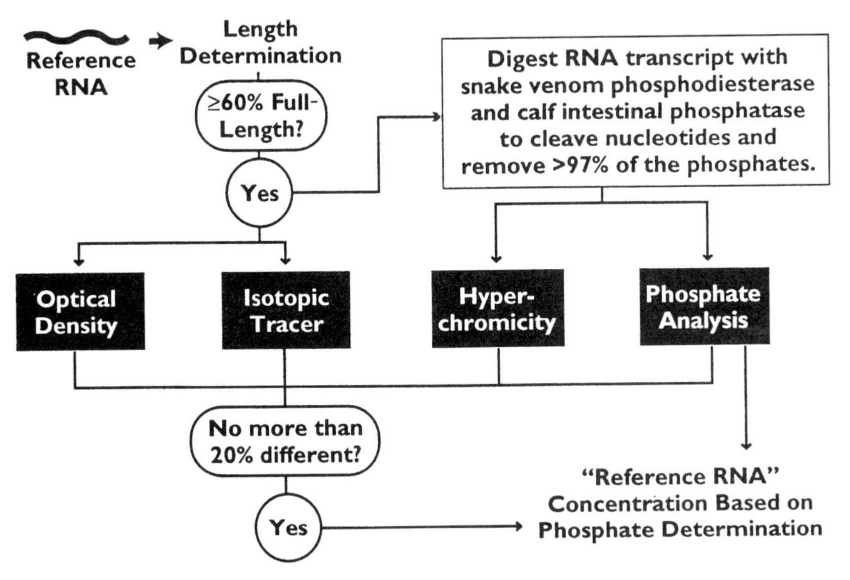


FIGURE 3. Flow chart of the procedures used for characterization of reference RNA standards by multiple independent quantification methods.

determination, and hyperchromicity. To qualify for use as reference RNA standards, no less than 60% of the transcripts in each preparation must be full-length and results from the four analytical methods must agree within 20%. If these criteria are met, the final concentration of RNA is based on the phosphate determination, since the standard curve for this assay is derived from a phosphate standard from the National Institute of Standards and Technology. RNA transcripts derived following these methods serve as the Chiron Gold Standard. This Chiron Gold Standard is then used for quantification of a DNA-based Chiron Silver Standard, a secondary standard against which all kit standards are matched.

It is important to note that absolute quantification using reference standards is not necessary for all applications of the bDNA assay. Although some clinical applications, such as measurement of viral load, require absolute nucleic acid quantification using reference standards, other applications need only measure relative changes in nucleic acid concentrations. For example, researchers may wish to evaluate changes in gene expression that may be biologically significant. One possibility would be to use changes in relative light units generated in the bDNA assay as a means to monitor changes in gene expression under different conditions. For this type of research application, the inclusion of reference standards in the bDNA assay would not be essential.

TABLE 1. Quantification Range and Specificity of bDNA Assays for Measurement of Viral Nucleic Acids

| bDNA Assay | Quantification Range* | Specificity | Reference |
|---|---|---|---|
| Quantiplex® HBV DNA | 0.7–5000 MEq | 99.7% | Hendricks et al., 1995 |
| Quantiplex® HCV RNA 2.0 | 0.2–120 MEq | ≥95% | Detmer et al., 1996 |
| Quantiplex® HIV RNA 1.0 | 10–1600 kEq | 100% | Pachl et al., 1995<br>Todd et al., 1995 |
| Quantiplex® CMV DNA | 4.4–5600 kEq | 94% | Chernoff et al., submitted |

* 1 MEq = 1,000,000 equivalents; 1 kEq = 1000 equivalents.

## *Sensitivity*

In evaluating the sensitivity of the bDNA assay, it often is useful to consider two limits—the molecular detection limit and the clinical quantification limit. The molecular detection limit is the lowest number of nucleic acid molecules that can be distinguished from zero. Since the molecular detection limit is influenced by the design of the oligonucleotide probes, it is important that the probe design be optimized (see above). The molecular detection limit always is less than or equal to the clinical quantification limit, an experimentally established limit that is necessary to achieve a high degree of specificity for nucleic acid quantification in clinical specimens. The clinical quantification limit is defined by screening large numbers of negative specimens and setting a desired specificity for quantification. In practice, the clinical quantification limit can have more relevance than the molecular detection limit, which may be far below what is required for patient management. For example, although the molecular detection limit of the Quantiplex® HCV RNA 2.0 assay is below 0.2 MEq/ml, the clinical quantification limit has been set at 0.2 MEq/ml to maintain a ≥95% specificity.

Even so, the Quantiplex® HCV RNA 2.0 assay can measure HCV RNA in 96% (95% confidence interval, 91–98%) of specimens from patients chronically infected with HCV and not undergoing therapy (Detmer et al., 1996). The clinical quantification range and corresponding specificities for several bDNA assays for viral nucleic acid quantification are listed in Table 1. Similar molecular sensitivity can be achieved with bDNA assays for quantification of cellular mRNAs. The number of cells required for mRNA quantification depends on the level of expression of particular genes. For highly expressed genes such as TNF$\alpha$, mRNA can be measured in a few cells (50–100 tissue culture cells stimulated with lipopolysaccharide). With ongoing improvements in the chemistry of the bDNA assay (Kern et al., 1996; Collins et al., submitted), even greater sensitivity for the bDNA assay can be achieved.

### *Reproducibility*

In clinical practice, it is common for specimens to be tested at different times, by different operators, and possibly at different sites. Given the variables that arise from the practical application of nucleic acid quantification assays, a high level of assay reproducibility is necessary to distinguish true differences among quantification values. To ensure reproducible results, the bDNA assay is thoroughly tested for both within-lot and between-lot precision. Studies have shown that the reproducibility of the bDNA assay is superior to that of RT-PCR, allowing detection of even relatively small changes in nucleic acid concentrations. For example, whereas two- to threefold changes in HIV-1 RNA concentrations were discerned as statistically significant by the bDNA assay, 3.7- to 5.8-fold changes in HIV-1 RNA levels were required for detection by RT-PCR to be considered clinically meaningful (Todd et al., 1994). Indeed, the bDNA assay for HIV-1 RNA quantification is considered one of the most reproducible quantification assays ready for use in clinical trials (Lin et al., 1994). With the high degree of reproducibility exhibited by the bDNA assay, subtle differences in nucleic acid quantification values that may be clinically meaningful can readily be measured.

## Concluding Remarks

The bDNA assay provides a powerful approach for the quantification of nucleic acid molecules. Fundamentally different from target amplification methods, the bDNA assay avoids the errors inherent in the extraction and amplification of target sequences by boosting the reporter signal as the means of detection, rather than amplifying target nucleic acid sequences. Standardization is an important feature of the bDNA assay, and procedures have been developed for rigorous characterization of reference standards using several independent quantification methods. The bDNA assay maintains high levels of performance with regard to accuracy, sensitivity, and reproducibility, and has been used in numerous studies to measure viral nucleic acid molecules as well as cellular mRNAs. With the recent developments in technology, the bDNA assay should also be useful in other hybridization applications. For example, *in situ* bDNA assays may be useful for detection of nucleic acid targets in cells. With the appropriate selection of internal standards, it should be possible to achieve relative quantification with an *in situ* bDNA assay, both in determining the number of infected cells, and in measuring the number of target nucleic acid molecules per cell.

Amenable to routine use in a clinical or research setting, the bDNA assay can be custom-designed for specific research applications. Custom kits for bDNA assays will allow researchers to investigate a number of important questions that require highly reproducible quantification of nucleic acids. For example, with reproducible quantification of nucleic acids, investigators can evaluate gene expression in sorted cell populations and thereby identify which cells contain the specific mRNA molecule(s) of interest. Also, with custom design of probe sets that distinguish between different spliced forms of mRNA, researchers can investigate the mechanisms and regulation of splicing events. In addition, by measuring nucleic acid molecules in cells exposed to various stimuli *in vitro*, investigators can explore the molecular mechanisms of cellular response. In fact, for any research question where an easy-to-use method for reproducible quantification of nucleic acid is desired, the bDNA assay could be applied. As a unique and powerful tool for reliable quantification of nucleic acid molecules, the bDNA assay provides researchers and clinicians with an effective means to address fundamental questions in biomedical research.

## REFERENCES

Chernoff DN, Hoo BS, Shen L-P, Kelso RJ, Kolberg JA, Drew WL, Miner RC, Lalezari JP, and Urdea MS: Quantification of CMV DNA in peripheral blood leukocytes using a branched DNA (bDNA) signal amplification assay. Submitted.

Collins ML, Irvine B, Tyner D, Fine E, Zayati C, Chang C, Horn T, Ahle D, Detmer J, Shen L-P, Kolberg J, Bushnell S, Ho DD, and Urdea MS: System 8 bDNA hybridization assay quantitates targets below 100 molecules per mL. Submitted.

Collins ML, Zayati C, Detmer JJ, Daly B, Kolberg JA, Cha TA, Irvine BD, Tucker J, and Urdea MS (1995): Preparation and characterization of RNA standards for use in quantitative branched DNA hybridization assays. *Anal Biochem* 226:120–9.

Davis GL, Lau JY, Urdea MS, Neuwald PD, Wilber JC, Lindsay K, Perrillo RP, and Albrecht J (1994): Quantitative detection of hepatitis C virus RNA with a solid-phase signal amplification method: definition of optimal conditions for specimen collection and clinical application in interferon-treated patients. *Hepatology* 19:1337–41.

Detmer J, Lagier R, Flynn J, Zayati C, Kolberg J, Collins M, Urdea M, and Sánchez-Pescador R (1996): Accurate quantification of HCV RNA from all HCV genotypes using branched DNA (bDNA) technology. *J Clin Microbiol* 34:901–7.

Flood J, Drew WL, Miner R, Jekic-McMullen D, Shen L-P, Kolberg J, Garvey J, Follansbee S, and Poscher M: Detection of cytomegalovirus (CMV) polyradiculopathy and documentation of in vivo anti-CMV activity in cerebrospinal fluid using branched DNA signal amplification and CMV antigen assays. Submitted.

Hendricks DA, Stowe BS, Hoo BS, Kolberg J, Irvine BS, Neuwald PD, Urdea MS, and Perillo RP (1995): Quantitation of HBV DNA in human serum using a branched DNA (bDNA) signal amplification assay. *Am J Clin Pathol* 104:537–46.

Horn T and Urdea MS (1989): Forks and combs and DNA: The synthesis of branched oligodeoxyribonucleotides. *Nucleic Acids Res* 17:6959–67.

Kern D, Collins M, Fultz T, Detmer J, Hamren S, Peterkin JJ, Sheridan P, Urdea M, White R, Yeghiazurian T, and Todd J: An enhanced sensitivity branched DNA (ES bDNA) assay for the quantification of HIV-1 RNA in plasma. *J Clin Microbiol* 34:3196–202.

Lin HJ, Myers LE, Yen LB, Hollinger FB, Henrard D, Hooper CJ, Kokka R, Kwok S, Rasheed S, Vahey M, Winters MA, McQuay LJ, Nara PL, Reichelderfer P, Coombs RW, and Jackson JB (1994): Multicenter evaluation of quantification methods for plasma human immunodeficiency virus type 1 RNA. *J Infect Dis* 170:553–62.

Mellors JW, Rinaldo CRJ, Gupta P, White RM, Todd JA, and Kingsley LA (1996): Prognosis in HIV-1 infection predicted by the quantity of virus in plasma. *Science* 272:1167–70.

Pachl C, Todd JA, Kern DG, Sheridan PJ, Fong S-F, Stempien M, Hoo B, Besemer D, Yeghiazarian T, Irvine B, Kolberg J, Kokka R, Neuwald P, and Urdea MS (1995): Rapid and precise quantification of HIV-1 RNA in plasma using a branched DNA (bDNA) signal amplification assay. *J Acquir Immune Defic Syndr Hum Retrovirol* 8:446–54.

Running JA and Urdea MS (1990): A procedure for productive coupling of synthetic oligonucleotides to polystyrene microtiter wells for hybridization capture. *BioTechniques* 8:276–9.

Todd J, Pachl C, White R, Yeghiazarian T, Johnson P, Taylor B, Holodniy M, Kern D, Hamren S, Chernoff D, and Urdea M (1995): Performance characteristics for the quantitation of plasma HIV-1 RNA using branched DNA signal amplification technology. *J Acquir Immune Defic Syndr Hum Retrovirol* 10 (supplement 2):S35–S44.

Todd J, Yeghiazarian T, Hoo B, Detmer J, Kolberg J, White R, Wilber J, and Urdea M (1994): Quantitation of human immunodeficiency virus plasma RNA by branched DNA and reverse transcription coupled polymerase chain reaction assay methods: A critical evaluation of accuracy and reproducibility. *Serodiagn Immunother Infect Dis* 6:1–7.

Urdea MS, Warner BD, Running JA, Stempien M, Clyne J, and Horn T (1988): A comparison of non-radioisotopic hybridization assay methods using fluorescent, chemiluminescent and enzyme labeled synthetic oligodeoxyribonucleotide probes. *Nucleic Acids Res* 16:4937–56.

# APPENDIX I

## bDNA ASSAY PROCEDURE FOR RNA QUANTIFICATION

1. Materials
   1.1 Reagents:
      A) Day 1: Lysis Diluent
         Lysis Reagent (Proteinase K)
         Target Probes for capture
         Target Probes for label
         Oligonucleotide-modified microwells
         Standards Diluent
         Controls and Standards
      B) Day 2: Amplifier Concentrate
         Label Diluent
         Label Probe Concentrate
         LumiphosPlus
      10% sodium dodecyl sulfate (SDS)
      Wash A (0.1X standard saline citrate[SSC; 1X SSC is 0.15 M NaCl plus 0.015 M sodium citrate], 0.1% SDS, 0.05% sodium azide, 0.05% Proclin 300 [Supelco, Inc., Bellefonte, PA])
      Wash B (0.1X SSC, 0.05% sodium azide, 0.05% Proclin 300)

      Note: bDNA assays for custom use include all reagents except Target Probes for capture and Target Probes for label.
   1.2 Equipment: Chiron plate heater
      Plate-reading luminometer

2. Plate set-up (Day 1)
   2.1 Samples
      A) Prepare specimen working reagent
         For example: 6 ml Lysis Diluent
         600 μl Lysis Reagent
         1.32 μl Target Probes for capture at 500 fm/μl (final concentration: 20 fm/200 μl)
         3.96 μl Target Probes for label at 500 fm/μl (final concentration: 60 fm/200 μl)

      Note: The concentrations of Target Probes for capture and Target Probes for label will vary depending on the target nucleic acid molecule. The concentrations given above are an example.

B) Add 450 μl Specimen Working Reagent to each RNA pellet specimen tube. Vortex vigorously for 10 sec and incubate tubes in a 63°C heat block incubator with shaking for a total of 30 min. The RNA pellet should be solubilized by the end of the incubation.
Note: The procedure for preparing RNA pellets is described in Appendix II. Prior RNA extraction is not necessary for measurement of highly expressed cellular mRNAs. If the specimen contains fewer than 1 million cells, the cellular lysate can be used directly in the bDNA assay.

C) Centrifuge briefly (23,500 × g, 30 sec, room temperature) and immediately remove the tubes from the centrifuge.

D) Carefully pipette 200 μl into duplicate oligonucleotide-modified microwells from each RNA sample.

2.2 Controls and Standards (Optional)

A) For specificity purposes, each bDNA assay run should contain positive and negative controls run in duplicate.

B) If absolute quantification is desired, a standard curve should be included. Start by pipetting 150 μl Standards Working Reagent, then add 50 μl of the appropriate standard curve member. Run in duplicate.

C) Incubate the microwells overnight at 53°C or 63°C in the Chiron plate heater, depending on the defined optimal temperature for the specific assay.

3. Plate Processing (Day 2)

3.1 Take the plate from the Chiron plate heater and let stand 10 min at room temperature. While the plate is cooling down, prepare the amplifier solution by diluting the amplifier concentrate at 200 fm/μl in Label Diluent. The final concentration should be 10 fm/50 μl.

3.2 Wash the microwells twice with Wash A, then pipette 50 μl of diluted amplifier into each well. Incubate the microwells for 30 min at 53°C in the Chiron plate heater.

3.3 Take the plate from the Chiron plate heater and let it stand 10 min at room temperature. While the plate is cooling down, prepare the Label Working Solution by diluting the Label Probe at 500 fm/μl into Label Diluent. The final concentration should be 20 fm/50 μl.

3.4 Wash the microwells twice with Wash A, then pipette 50 μl of the Label Working Solution into each well. Incubate the microwells for 15 min at 53°C in the Chiron plate heater.

3.5 Take the plate from the Chiron plate heater and let it stand 10 min at room temperature. During this time, prepare the Substrate Solution made of 99.7% v/v LumiphosPlus and 0.3% v/v 10% SDS (for example: 3 ml LumiphosPlus, 9 $\mu$l 10% SDS).
3.6 Wash the microwells twice with Wash A and then three times with Wash B. Add 50 $\mu$l of the Substrate Solution into each well.
3.7 Measure the chemiluminescence in the plate-reading luminometer. (This includes incubation of the microwells for 30 min at 37°C prior to measurement to reach steady-state kinetics).

## APPENDIX II

### TISSUE–PROCESSING PROTOCOL FOR THE bDNA ASSAY

1. Precautions: The homogenization of peripheral blood mononuclear cells (PBMCs), tissue culture pellets, and biopsies should be performed in a certified Biology Safety Cabinet Class II, because this procedure is likely to produce aerosols. Other parts of this protocol and the bDNA assay can be performed on an open bench covered with plastic-backed absorbent paper to contain possible spills. Lab coats, disposable gloves, and face shields should be worn.

2. Materials:
   2.1 Reagents:
      A) 8 M Guanidine-HCl solution, Sequenal Grade (Pierce Chemical Co., Rockford IL, #24115 G)
      B) 3 M Sodium Acetate
      C) N-Lauryl-Sarcosine, sodium salt (Sarcosyl), (Sigma Chemical Co., St. Louis, MO, #L5125)
      D) 100% Ethanol, Gold Shield
      E) Polyadenilic Acid (5′) (polyA), potassium salt (Sigma Chemical Co., St. Louis, MO, #P-9403)
      F) Phosphate Buffered Saline (PBS) solution
   2.2 Equipment:
      A) Homogenization equipment
         • Disposable pellet pestle mixer and matched polypropylene 1.5 ml microtubes (Fisher Scientific, Pittsburgh, PA, #K749520-0000). The matched pestle and microfuge tubes are appropriate for homogenizing PBMC samples, tissue culture

samples, and biopsy samples (<25 mg). Both the pellet mixer and microtubes may be autoclaved.

**Critical parameter:** Abrasion of the plastic pestle with coarse sandpaper before use assists in rapid homogenization of tissue samples.

- Disposable mixer for 1.5 ml tubes (Phenix Research Products, Hayward, CA, #PM-449). When specimens are received in microfuge tubes other than the Fisher Scientific tubes (see above), the Phenix Research Products pestles can be used. These fit many conical 1.5 ml microfuge tubes with O-rings and external threads.
- Cordless motor for disposable pellet pestle mixers (distributed by Fisher Scientific, #K749540-0000).
- Sterile disposable tissue grinder (Sage Products Inc., Crystal Lake, IL, #3505 small, #3500 large; also available from Baxter S/P). These are useful for larger tissue specimens (i.e., >25 mg).

B) Other Equipment:

- 2 ml microcentrifuge tubes.

  **Critical parameter:** For ethanol precipitation of PBMC, tissue culture pellets, or tissue homogenized in Guanidine-HCl solution, it is necessary to use larger volume microcentrifuge tubes with screw caps and O-rings (such as Sarstedt, Newton, NC, #72.693).
- A microfuge tube that is able to accommodate the 2-ml tubes. Centrifugation should be performed at 2–8°C and approximately 10,000 to 12,000 × g.
- Pipettors and large-orifice (micropipette) tips, 1–200 $\mu$l and 100–1000 $\mu$l.

3. Reagent Preparation:

A) 3 M Sodium Acetate: Dissolve 40.81 g Sodium Acetate· $3H_2O$ 80 ml distilled water. Adjust the pH to 5.2 with glacial Acetic Acid. Adjust the volume to 100 ml with distilled water.

B) Guanidine-HCl homogenization solution: Mix 280 ml 8 M Guanidine-HCl solution with 20 ml 3 M Sodium Acetate. Mix *very well* to avoid precipitation, and store at 2–8°C.

C) PolyA 10 mg/ml: Add polyA to distilled water, mix well, and aliquot. Store at –20°C.

D) 10% Sarcosyl: Add N-Lauryl-Sarcosine, sodium salt, to distilled water. Store at room temperature.

E) 100% Ethanol: Store at –20°C.
F) 70% Ethanol: Add distilled water to 100% Ethanol, store at –20°C.

4. Specimen Collection and Storage:
4.1 PBMCs and tissue culture material have been successfully used with this method. Specimens of at least 5 million and preferably 10 million cells are strongly recommended. Isolate PBMCs from whole blood as quickly as possible after collection (within 4 hours) using either Ficoll-Hypaque or leukoprep methods. Harvest tissue culture cells at designated time points. After purification of PBMCs and/or harvest of tissue culture cells, hold on wet ice until they can be frozen (e.g., while counting cells).
A) Frozen "dry" PBMC or tissue culture material pellets: To prepare the PBMCs or tissue culture pellets, centrifuge them 5 min in a refrigerated microcentrifuge (10,000–12,000 × g). Aspirate as much as possible of the supernatant, leaving only a small amount of fluid (10–20 μl). Wash with 500 μl PBS. Centrifuge again for 5 min. Aspirate the supernatant leaving only a small amount of fluid (10–20 μl).
**Critical parameter:** Use the appropriate Fisher microfuge tubes (see homogenization equipment above) to pellet and freeze the cells in. These tubes are specially designed to match the pellet pestle mixers.
B) Cryopreserved PBMCs or tissue cultures: Alternately, PBMCs or tissue culture material can be cryopreserved in DMSO and tissue culture medium.
**Critical parameter:** It is important to freeze the biopsy specimen without placing it in any type of preservative or diluent, such as formalin or saline.

5. Specimen Preparation:
5.1 Biopsies: Determination of tissue specimen mass.
A) Select an appropriate tissue grinder
- For small tissue specimens (<25 mg), including most needle biopsies, use the pellet pestle mixers and matched centrifuge tubes.
- For larger tissue specimens use the sterile disposable tissue grinders from Sage Products, Inc. Specimens up to 1 g can be homogenized in the small grinder; larger specimens require the larger grinder.

B) Determine the mass of the tube for grinding. A pan balance accurate to 1 mg would be a good choice (Mettler PM460 Delta Range, Mettler Instruments Corp., Hightstown, NJ).

C) Place the frozen tissue in the appropriate tube and keep on dry ice before and after weighing. Do not allow sample to thaw.

D) Determine the mass of the tube and tissue and calculate mass of tissue.

5.2 Cryopreserved PBMCs or tissue culture: Rapidly thaw the cells at 37°C, and then pellet the cells in one of the Fisher microfuge tubes (which match the pellet pestle mixer, see homogenization equipment above). Remove supernatant and wash with 500 μl PBS. Centrifuge again, remove supernatant and keep on wet ice until the pellets can be homogenized in Guanidine-HCl homogenization solution as described below.

5.3 Frozen "dry" PBMCs or tissue culture pellets: Do not allow these samples to thaw before homogenization. Keep on dry ice.

**Critical Note:** Frozen dry biopsies, PBMCs and tissue culture pellets should not be allowed to thaw, but should be removed from the freezer or dry ice container, one at a time, and homogenized immediately. For maximum nucleic acid recovery, do not allow these specimens to thaw before homogenization. This step should be performed in a certified Biological Safety Cabinet. After homogenization, keep the specimens on ice or in the refrigerator.

6. Homogenization of biopsies, PBMCs, and tissue culture material, and isolation of RNA:

6.1 *Quickly* add cold Guanidine-HCl homogenization solution (which contains sodium acetate) to the frozen tissue, frozen "dry" pellets or pellet of cryopreserved PBMCs or tissue culture. Homogenize as quickly as possible with the pellet pestle mixer, using the cordless motor. This many take a few minutes.

A) For most small biopsy specimens (<25 mg) and PBMC or tissue culture specimens (10 million cells or less), 250 μl Guanidine-HCl homogenization solution is sufficient for initial homogenization. After thorough homogenization in this small volume, add an additional 250 μl Guanidine-HCl homogenization solution; homogenize again with the pellet pestle mixer. Finally, add an additional 500μl Guanidine-HCl homogenization solution for a total of 1 ml per 25 mg of tissue or 10 million cells. Vortex gently. If the tissue specimen is less than 25 mg (most cases) or less than 10 million cells, add Guanidine-HCl homogenization solution to *make a final volume of 1 ml.*

B) For larger biopsy specimens (>25 mg) and PBMC or tissue culture specimens (>10 million cells), add proportionally more

Guanidine-HCl homogenization solution so that the final volume is 1 ml per 25 mg of tissue or 1 ml per 10 million cells.
**Critical parameter:** Be sure to homogenize samples thoroughly; keep homogenized samples on wet ice or in the refrigerator (2–8°C).

6.2 To each tube containing 1 ml of homogenized sample, add 50 μl 10% Sarcosyl (final concentration 0.5%), vortex gently and hold for 5 min at 2–8°C.

6.3 Centrifuge biopsy specimens for 1 min in a refrigerated microcentrifuge at approximately 12,000 × g to sediment particulates (this step is not necessary for PBMC or tissue culture specimens).

6.4 Transfer 1 ml of supernatant from each homogenized biopsy, PBMC, or tissue culture specimen to a large volume (2 ml) tube from Sarstedt.
**Critical parameter:** It is essential to use these large-volume tubes to ensure mixing efficiency.

6.5 To each tube add 10 μl polyA (10 mg/ml). Vortex.

6.6 Add 500 μl 100% EtOH to each tube and vortex thoroughly.
**Critical parameter:** Look carefully at each tube to ensure thorough mixing of EtOH and Guanidine-HCl homogenization solutions. These solutions have very different densities and must be mixed well. It is helpful to invert the tube several times in addition to vortexing.

6.7 Hold at –20°C overnight (approximately 12–18 hours).

6.8 Take the samples from the –20°C freezer, centrifuge the tubes in a microfuge (~12,000 × g, 20 min, 2–8°C) and carefully remove the supernatants without disturbing the pellets (important).
Note: If desired, DNA from the sample can be saved at this stage by separating the supernatant into three tubes and adding 2 volumes 100% EtOH. This step precipitates the DNA, which may be run in the DNA assay.

6.9 Add 500 μl 70% EtOH to each tube. Vortex to mix, and centrifuge again (12,000 × g, 20 min, 2–8°C).
**Critical parameter:** Remember to keep tubes on ice during manipulations.

6.10 Aspirate the supernatants and dry down the pellets in a speedvac rotary vacuum device or equivalent. Use the speedvac at room temperature until visible liquid is removed.
**Critical parameter:** Do not evaporate to complete dryness.

# Hybrid Capture™ — A Sensitive Signal-Amplified Test for the Detection and Quantitation of Human Viral and Bacterial Pathogens

Attila T. Lörincz, Mariana G. Meijide, James G. Lazar, and Abel De La Rosa

## Introduction

Since the early 1970s, researchers have used nucleic acid detection techniques to identify either specific genes or organisms. These early nucleic acid detection methods were very time-consuming, cumbersome, isotopic, technically elaborate, and only moderately sensitive, and thus not suitable for routine clinical use (Southern, 1975). The first nonisotopic nucleic acid detection systems, developed in the late 1970s, were fairly specific but not as sensitive as their radioactive counterparts (Matthews and Kricka, 1988). To increase overall sensitivity, amplification methodologies have been developed and adapted to nonisotopic nucleic acid detection assays. The two most common amplification methods are target amplification (Mullis and Faloona, 1987) and signal amplification employing chemiluminescence (Pollard-Knight et al., 1990; Bronstein and Olesen, 1995).

Target amplification protocols increase the total number of nucleic acid molecules, thus permitting the use of less sensitive signal detection methods. Although amplification techniques such as the polymerase chain reaction (PCR) are remarkably sensitive, extreme care must be taken to avoid cross-contamination during sample processing. In addition, quantitation of the original target is a challenge for the user and is usually accomplished by the inclusion of multiple internal and external controls. Some steps involved in target amplification have unascertained or variable efficiencies, especially in the presence of crude clinical material, making it difficult to accurately correlate the final copy numbers to the original amounts of target.

Signal amplification methods increase the signal of the system without changing the total number of target molecules. In methods involving signal amplification, the nucleic acid of interest is measured directly, and the resulting signal is directly proportional to the concentration of the specific nucleic acid in the sample. This type of direct detection method does not require complex and processive enzymatic reactions, and, therefore, accurate quantitation is preserved in the presence of relatively crude specimens.

With the advent of sophisticated amplification techniques, the sensitivity, ease of use, and turnaround times of nonisotopic nucleic-acid-based detection assays have been dramatically improved. Signal-amplified tests are generally more user-friendly than assays like PCR, but improvements are still needed in all amplified methods. Enzyme immunoassay methods, originally developed for antibody and antigen detection, are easily automated, rapid, easy to perform, and familiar to those working in clinical laboratories. To make it possible for more laboratories to utilize nucleic acid diagnostic assays, test systems must be endowed with the ease of handling, simplicity, and versatility for automation found in traditional enzyme immunoassay and clinical chemistry methods.

The Hybrid Capture™ System (HCS) (Digene Corporation, Silver Spring, MD) combines well-known nucleic acid hybridization techniques with the ease of sample manipulation shared by traditional immunoassays. HCS is a sensitive, nonradioactive, nucleic-acid-based signal amplification method. The technology offers researchers and clinicians an effective, standardized, and accurate tool for pathogen detection. This chapter will cover the use of Hybrid Capture signal amplification technology for the detection of human papillomavirus (HPV), cytomegalovirus (CMV), hepatitis B virus (HBV), and human immunodeficiency virus (HIV).

## Hybrid Capture Technology

HCS is a universal nucleic acid detection technology (Figure 1). Following a simple specimen-processing method, in which the sample quickly becomes a uniform nucleic acid suspension, DNA or RNA probes are hybridized in solution t o the specific target. The hybridization reaction is transferred to an antibody-coated tube (HCT) or microtiter plate (HCM), where the RNA:DNA hybrids are captured. The capture antibody is a polyclonal universal reagent that specifically recognizes all RNA:DNA hybrids, regardless of their sequence. This antibody does not react with single-stranded DNA, RNA, or other hybrids (such as DNA:DNA or RNA:RNA). Because of multiple recognition moieties on the hybrids, the capture antibody molecules are able to rapidly, specifically, and efficiently immobilize RNA:DNA hybrids. An alternate method of capture involves the use of biotinylated DNA probes that hybridize to RNA targets and are captured by streptavidin-coated microwells; the Hybrid Capture HIV

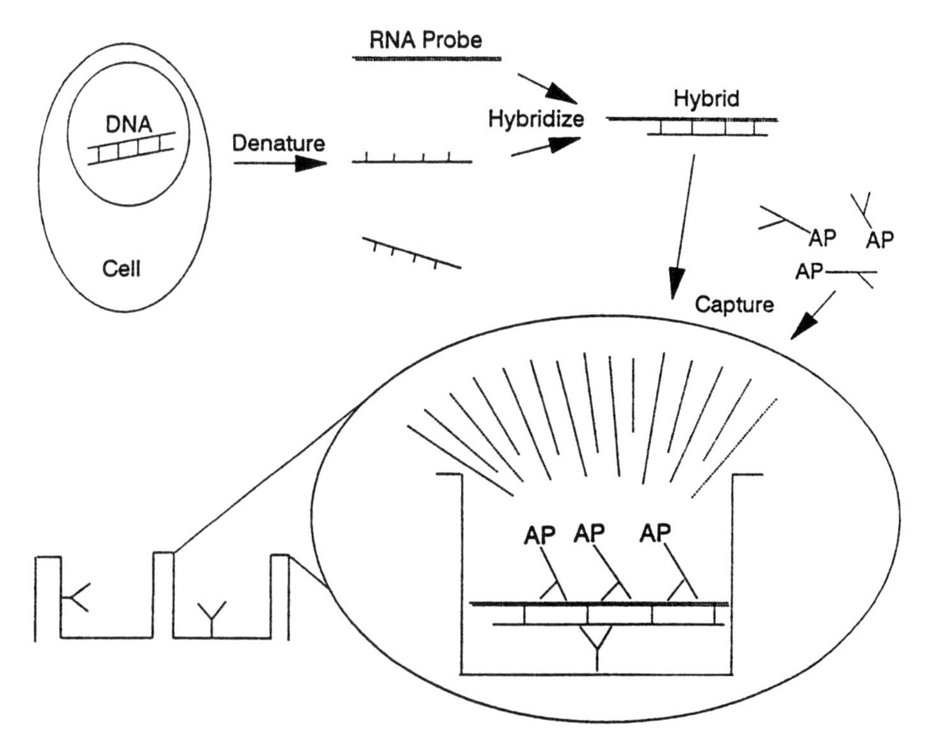


FIGURE 1. Principle of the Hybrid Capture™ System. Full genomic RNA probes are hybridized in solution with denatured target DNA from specimens, forming RNA:DNA hybrids. Hybrids are captured onto an antibody-coated solid phase and reacted with an alkaline-phosphatase-labeled conjugate. The alkaline phosphatase with a dioxetane substrate generates light that is measured in a luminometer. (Reprinted with permission of Leeds Medical Information from Lörincz, 1996.)

assay uses this format. The immobilized hybrids are subsequently reacted with a second antibody. This antibody is an alkaline-phosphatase-labeled antibody that is also specific for RNA:DNA hybrids. Multiple alkaline-phosphatase-conjugated antibodies bind to each captured hybrid, proportionately increasing the number of enzyme labels present on the complex. These properties are largely responsible for the sensitivity of the Hybrid Capture System. The hybrids are then detected by chemiluminescence using dioxetane-based substrates. Multiple substrate molecules are cleaved by the conjugated alkaline phosphatase molecules, and the light emitted is a long-lived glow reaction, which promotes fast results without timing constraints. The reaction is quantitated using a luminometer such that the relative light units (RLU) emitted over time are proportional to the amount of target nucleic acid in the original specimen.

## Human Papillomavirus

High-risk and intermediate-risk (carcinogenic) types of HPV are responsible for over 90% of cervical cancers worldwide (Morrison et al., 1991; Lörincz et al., 1992; Bosch et al., 1995). Most HPV infections and their induced lesions are self-limiting and spontaneously disappear within months or a few years. Persistent infections, however, are cause for concern (Schneider et al., 1992; Ho et al., 1995a). The Pap smear is a valuable cancer prevention aid but is error-prone and difficult to maintain reproducibly, because of its subjective nature and the need for highly trained cytologists and pathologists. Even among experts, the Pap smear can miss about 20% of cancers and high-grade cervical intraepithelial neoplasms (CIN).

Industrialized nations have seen 70% or greater decreases in cervical cancer rates during the last 50 years; however, recent trends in the U.S. are worrisome, insofar as cancer rates have started to increase (Boring et al., 1991; Wingo et al., 1995). After hovering at about 13,000 cases per year for several years in the late 1980s and early 1990s, cervical cancer estimates from the US National Cancer Institute's Surveillance, Epidemiology, and End Results (SEER) program have grown to almost 16,000 cases in 1995 (a 17% increase normalized for population growth), with a disproportionate increase among younger women (Wingo et al., 1995). Considering that the U.S. spends about $5 billion per year on cervical cancer prevention while the malignancy remains relatively prevalent and is now increasing, the situation appears to need some remedies. Thus, HPV DNA testing has been proposed as an effective adjunct to the Pap smear in cervical cancer prevention programs.

To evaluate the performance of the first-generation HCT, aliquots of cervical specimens from 199 women were tested by Southern blot, PCR, and HCT in three independent laboratories. These clinical specimens were derived from a study described earlier (Sherman et al., 1994). HCT was shown to be a simple, reliable test suitable for routine use (Schiffman et al., 1995). Pairwise interlaboratory agreement ranged from 87% to 94%, with no clearly superior results from any given laboratory.

Five clinical studies of patient management using HPV DNA testing by HCT have been completed (Cox et al., 1995; Wright et al., 1995; Hatch et al., 1995; Hall et al., in press; Ferenczy et al., in press). The studies examined the utility of HPV DNA detection as an adjunct to the Pap smear for managing women with previous ASCUS (atypical squamous cells of undetermined significance) or LSIL (low-grade squamous intraepithelial lesion) Pap smears. The general study design was to obtain specimens of exfoliated cervical cells for HPV DNA testing, and perform a repeat Pap smear just prior to colposcopy. Histopathology was used as the reference standard definition of prevalent cervical disease. Three studies employed the recommended specimen collection devices (swabs or Cone Brush) for HPV DNA testing. The fourth study (Hall et al.) compared the use of a cervicovaginal lavage to the use of a cytobrush. The fifth study (Ferenczy et al.) examined the use of a liquid cytology transport medium (PreservCyt, Cytyc Corporation, Boxborough, MA) that is compatible with both HPV DNA testing and a Pap slide from the same specimen. This latter method is particularly interesting, as it permits HPV DNA testing to be performed from routine Pap screening specimens, thus eliminating an additional office visit.

Selected results of the studies are shown in Table 1. The Pap smear and HCT appear equivalent in sensitivity, with some studies showing a slight advantage for HCT and others showing a slight advantage for the Pap smear. Interestingly, in four of the five studies, HCT was better at predicting women with CIN 2/3 than the Pap smear. In all studies, the combination of HCT and the Pap smear was better than either alone, generally reaching a sensitivity of over 95% for important cervical disease.

Clinical data on HCM are still quite limited. From an analytical perspective, the test can detect and distinguish among a wide array of genital high-risk HPVs in the range of $1 \times 10^3$ to $5 \times 10^7$ HPV genomes per assay, using about 5% of the total specimen.

Analytical specificity of the HCM test is good and has improved over the HCT test, as the result of the formulation of a new conjugate reagent that is ultrapure and contains a completely synthetic base matrix. The microplate format is easier to manufacture reproducibly on automated machinery and is easier to run in clinical laboratories. Analytical specificity of HCM is better than 97% for novice users and exceeds 99% in well-trained hands. There is no detectable cross-reactivity of HCM to a panel of over 30 different bacteria tested at a level of $10^7$ organisms per ml, or to any of 15 different human viruses tested at a level of $10^8$ genomes per ml. Cross-reactivity among most HPVs is not observed, but in certain cases (e.g., type 18 vs. type 45, or type 16 vs. type 35) additional controls must be included in each test run to confidently distinguish multiply-infected specimens from low-level cross-reactivity.

Two preliminary clinical studies have been performed with HCM. The first study (Chesebro et al., personal communication) was similar in design to the Hatch and Hall studies of Table 1, except that concurrent

TABLE 1. HCT appears to be a useful tool for managing women with ASCUS Pap smears: results of five recent studies. (Reprinted with permission of Leeds Medical Information from Lörincz, 1996)

| Study (CIN 2/3/total)[a] | Protocol | % Sensitivity | Colposcopies per CIN 2/3 | Referral Pap[b] |
|---|---|---|---|---|
| Cox et al., 1995 (15/217) | Colposcopy | 100 | 14.5 | ASCUS |
| | Pap smear | 73 | 6.1 | |
| | HCT | 93[c] | 5.3 | |
| | Pap + HCT | 100 | 7.1 | |
| Wright et al., 1995 (50/398) | Colposcopy | 100 | 8.0 | ASCUS or LSIL |
| | Pap smear | 80 | 5.1 | |
| | HCT | 78 | 5.3 | |
| | Pap + HCT | 96 | 5.9 | |
| Hatch et al., 1995[d] (126/311) | Colposcopy | 100 | 2.5 | SIL |
| | Pap smear | 75 | 2.3 | |
| | HCT | 74 | 1.8 | |
| | Pap + HCT | 91 | 2.2 | |
| Hall et al., 1996[e] (15/75) (in press) | Colposcopy | 100 | 5.0 | ASCUS or SIL |
| | Pap smear | 87 | 4.2 | |
| | HCT | 93 | 4.0 | |
| | Pap + HCT | 100 | 4.1 | |
| Ferenczy et al., 1996 (47/364) (in press) | Colposcopy | 100 | 8.3 | ASCUS or SIL |
| | Pap smear | 87 | 5.1 | |
| | HCT | 77 | 4.8 | |
| | Pap + HCT | 95 | 5.4 | |

[a] Although a few lesions were cancers, these cases have only a minimal impact on the numbers in this table.
[b] This Pap smear was the basis of the original referral of the women for colposcopy. ASCUS = atypical squamous cells of undetermined significance. LSIL = low-grade squamous intraepithelial lesion.
[c] The sensitivity of HCT was 100% when HPVs 39 and 58 were added to the probe mix, indicating the importance of an expanded probe set for HPV. The other studies in the table did not have HPVs 39 or 58 as part of the probe test.
[d] Endpoint was based on cervical intraepithelial neoplasia grades 2 and 3 (CIN 2/3).
[e] Data are for high-risk HPV types only.

Pap smears were not taken. Sensitivity of HCM for CIN 2/3 was 92% (36/39) and required 3.5 colposcopies per high-grade CIN confirmed on histology, while 4.1 colposcopies were required to reach 100% sensitivity for CIN 2/3. In a second ongoing study (Cox et al., personal communication), only sensitivity data are currently available for HCM: of CIN 1s, 91% (88/97) were detected; of CIN 2s, 96% (24/25) were detected; and of CIN 3s, 100% (8/8) were detected. Vaginal and vulvar high-grade lesions were also detected at 95% or better sensitivity in this study.

It appears that HCM has consistently high sensitivity for CIN 2/3 in a fairly large group of specimens from two different study sites in the U.S., with an average of 95% sensitivity among a combined set of 72 histologically confirmed high-grade cervical lesions.

## Human Cytomegalovirus

Cytomegalovirus (CMV) related diseases are a serious and growing international healthcare problem. CMV is a herpes virus that infects approximately 50% of the world's population. Primary CMV infection in healthy individuals is usually mild or asymptomatic and is generally followed by a persistent latent (nonreplicating) infection that is controlled by the immune system. However, when the immune system is compromised, either by drugs or disease, the latent CMV can become active (replicating) and cause serious disease. Antiviral drugs are available for CMV, but they are only effective if treatment is initiated in the early stages of infection. In addition, these drugs are both toxic and expensive. Currently, there is no rapid, reliable CMV detection method available to detect CMV infection accurately and quickly, and to help physicians assess therapeutic efficacy.

CMV is the single most important microbial pathogen affecting organ transplant recipients and is a major cause of morbidity and mortality following transplantation. CMV disease usually occurs 4–12 weeks after transplantation in approximately 60% of organ recipients (Rubin, 1994). Prior to transplantation, organ donors and recipients are tested by serology for CMV to determine the presence of latent infection. Recipients who test positive or are receiving organs from seropositive donors often undergo prophylactic antiviral treatment independent of clinical findings. Under this treatment strategy most patients are treated, many unnecessarily. Transplant recipients are usually treated for CMV with ganciclovir at a cost per patient for a full treatment course of approximately $20,000. Unfortunately, ganciclovir causes marrow suppression, which is a cause of great concern in the management of bone marrow transplant patients. Foscarnet, another anti-CMV drug, has good clinical efficacy against CMV without marrow suppression but is extremely toxic (Rubin, 1994).

After transplantation, patients are monitored weekly for active CMV on an inpatient basis and biweekly for up to 14 weeks as out-patients. Current strategies to deal with CMV infection include therapy of established disease and preemptive therapy of patients with active but asymptomatic infection. The latter process restricts treatments to those at highest risk of developing CMV disease.

There are an estimated one million HIV-positive people in the United States and another one million in Europe. Most if not all of these people will eventually develop clinical AIDS. CMV infects approximately 95% of all AIDS patients and is the leading cause of AIDS-related blindness (Ho, 1991; Gozlan et al., 1993; Young, 1996). In approximately 20–40% of AIDS patients, CMV causes retinitis, a condition that can progress rapidly to blindness if left untreated. Ophthalmological detection of CMV retinitis is a reactive diagnostic strategy that does not provide sufficient protection from disease, since permanent retinal damage has occurred once ophthalmological symptoms are observed. The strategy then becomes to control further damage to slow the rate of visual impairment. Other end-organ or systemic CMV disease is often associated with CMV retinitis (Ho, 1991).

The two FDA-approved drugs for the treatment of CMV retinitis in AIDS patients, ganciclovir and foscarnet, are not considered cures but improve symptoms and delay disease progression in about 75% of AIDS-related CMV infections. Because these drugs are most effective when administered early in the course of infection, the ability to detect active CMV disease earlier should result in more effective treatment, by allowing intervention before symptoms appear. Both ganciclovir and foscarnet are associated with serious side effects and therapeutic limitations. A study at the University of California at San Francisco demonstrated that the combination of side-effect monitoring and infusion pump expenses resulted in annual total costs between $37,000 and $115,000, depending upon the therapeutic regimen employed (The Pink Sheet, 12/20/93). The recent availability of oral ganciclovir will add a new dimension to these strategies, reinforcing the need for early detection of CMV disease to implement therapy as early as possible.

Currently, detection of CMV infection is accomplished by rapid culture in shell vials (SV) and by traditional cell culture in tubes (TC). With the SV method, a purified specimen is added to a vial containing a monolayer of cells growing on a 12-mm circular cover slip and is incubated for 24–48 hours. The cover slip is washed, then stained with monoclonal antibodies to immediate early antigens specific for CMV. The cover slip is mounted on a microscope slide, and the positive cells are read under a fluorescent microscope. Cell culture requires a tube containing a cell monolayer, such as MRC-5 cells. The monolayer is inoculated with the specimen, and the TC is rotated in the presence of media for 5–28 days. If more than 5 days is required, the primary culture is passed to another tube and TC is continued. The endpoint is the detection of a cytopathic effect (CPE), which appears as enlarged and frequently elliptical cells lying in compact groups of 6 to 8. These assays, however, lack sensitivity, are labor-intensive, and can take up to a month to generate results.

Antigenemia assays are designed to detect the presence of an early structural CMV protein called pp65. This protein is believed to be associated with actual CMV DNA viral replication. To perform an antigenemia assay, blood white cells (WC) are purified from whole blood, washed, and counted. Usually 100,000 cells are centrifuged onto the slide, and duplicate slides are prepared for each specimen. The cells are stained with a primary antibody specific for the CMV pp65 protein. A secondary antibody conjugated with fluorescein is added, and the slide is read using a fluorescent microscope. Positive cells are indicated by nuclear fluorescence, and results can be generated in 4 to 5 hours. Antigenemia assays are labor-intensive, subjective, and poorly standardized. Moreover, for acceptable results using the antigenemia assay, specimens must be processed within 6 hours of collection, making the timing of this test difficult for routine clinical laboratories.

Most importantly, none of these methods can be used to assess viral load accurately, an important feature of monitoring disease progression and assessing therapeutic efficacy. The Hybrid Capture System CMV DNA assay (HCS CMV) was developed to address the limitations of traditional techniques by providing a reliable tool for early CMV diagnosis that provides accurate, reliable results with same-day turnaround. In addition to early detection, HCS CMV is expected to offer effective monitoring of transplant and AIDS patients to assist in determination of timely and appropriate therapy.

HCS CMV is designed to provide significant performance benefits compared to other currently available methods. Specimen processing is simple, rapid, and convenient. WC are collected in EDTA or ACD, and can be stored for up to 6 days at 4°C before processing. A lysis/centrifugation procedure is used to separate WC from other blood components and to prepare a WC pellet. No gradient separations are required. WC pellets can be stored at –20°C for up to 6 weeks, and the assay has a flexible format that allows testing of 1 to 54 specimens in each assay run. The HCS CMV probe mix contains a cocktail of approximately 40,000 bases of single-stranded RNA that constitute approximately 17% of the CMV genome. Because certain regions of the CMV genome are known to cross-react with other herpes viruses and human DNA sequences, extensive cross-reactivity studies were performed to choose appropriate probe sequences. The current probe cocktail has been tested against herpes simplex virus types 1 and 2, Epstein-Barr virus, varicella zoster virus, and human herpes virus type 6 at concentrations up to $10^{10}$ viral particles per ml, and no cross-reactivity was observed. The CMV

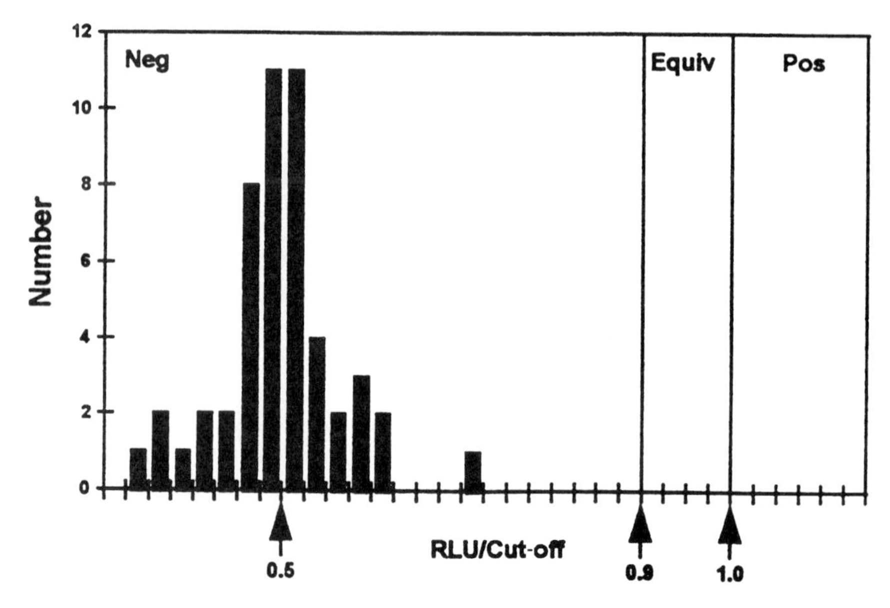


FIGURE 2. HCT CMV results on normal specimens

probe cocktail was also tested against human genomic DNA as high as 200 $\mu$g/ml, and again no cross-reactivity was observed.

To demonstrate the specificity of the assay in a normal population, specimens from 50 normal, healthy subjects were tested in the HCT CMV DNA assay. A frequency distribution of the results is shown in Figure 2, and demonstrates the high specificity of the assay.

The analytical sensitivity of the assay was determined by testing a set of positive calibrators of known CMV DNA concentration. The results of the assay were analyzed by regression, as shown in Figure 3. The analytical sensitivity of the assay is demonstrated by the intersection of the regression line with the assay cut-off and yields a sensitivity of 5000 genomes per ml of blood. A second-generation assay based on a 96-well microplate format was developed recently; this new test is named HCM CMV and is described later in this review.

The HCT CMV DNA assay has been compared to SV and TC in several clinical trials. The results of two of these studies are shown in Tables 2 and 3, and both demonstrate the relatively low sensitivity of culture techniques for detection of CMV viremia. Additional studies have confirmed these results (Isada et al., 1996; Mazzulli et al., 1996; Miller et al., 1996).

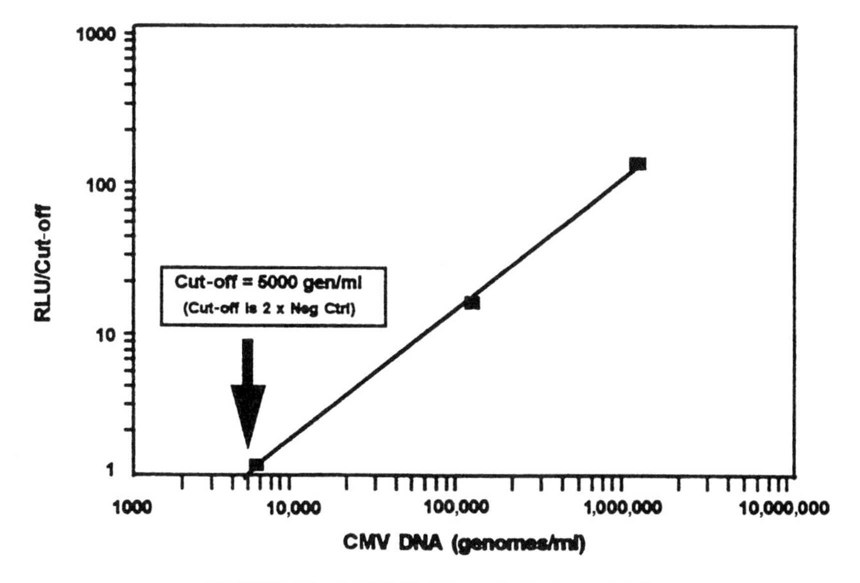


FIGURE 3. HCT CMV analytical sensitivity

TABLE 2. Hybrid Capture System – CMV DNA assay performance. Results of the HCT CMV alpha trial. Specimen processing and culture testing were done by an outside lab, but Hybrid Capture testing was performed at Digene. N = 302.

| | HCS CMV DNA | Culture |
|---|---|---|
| Sensitivity | 93% | 47% |
| Specificity | 99% | 100% |
| Agreement | 98% | 97% |

TABLE 3. Hybrid Capture System – CMV DNA assay performance. Results of the HCT CMV beta trial. Specimen processing and all testing were performed by an outside laboratory. N = 114.

| | HCS CMV DNA | Culture |
|---|---|---|
| Sensitivity | 94% | 31% |
| Specificity | 94% | 100% |
| Agreement | 94% | 76% |

The utility of the HCT CMV DNA assay for patient monitoring has also been demonstrated in several studies. Figures 4–7 show the results of the HCT CMV DNA assay, TC, SV, an antigenemia assay, and PCR in longitudinal studies of AIDS patients. These studies demonstrate that HCT CMV detects active infection earlier than culture methods, and at approximately the same time as antigenemia and PCR. Furthermore, the data show that the HCT CMV can be used to monitor the effectiveness of drug therapy.

A further utility of the Hybrid Capture System assays is the application of these assays for drug screening or drug susceptibility testing in clinical isolates (Isada et al., 1996). The standard method for drug susceptibility testing is a culture method called the plaque reduction assay (PRA). In this assay, viral isolates are grown until they generate approximately 100 plaques in culture. The culture is then passaged and split into different cultures containing variable concentrations of the drug being tested. Drug susceptibility is then determined by counting the reduction in the number of plaques as the ganciclovir concentration is increased. To compare this method to the HCT CMV DNA assay, cells from a plaque reduction assay were lysed after the plaques were counted and run in HCT. As

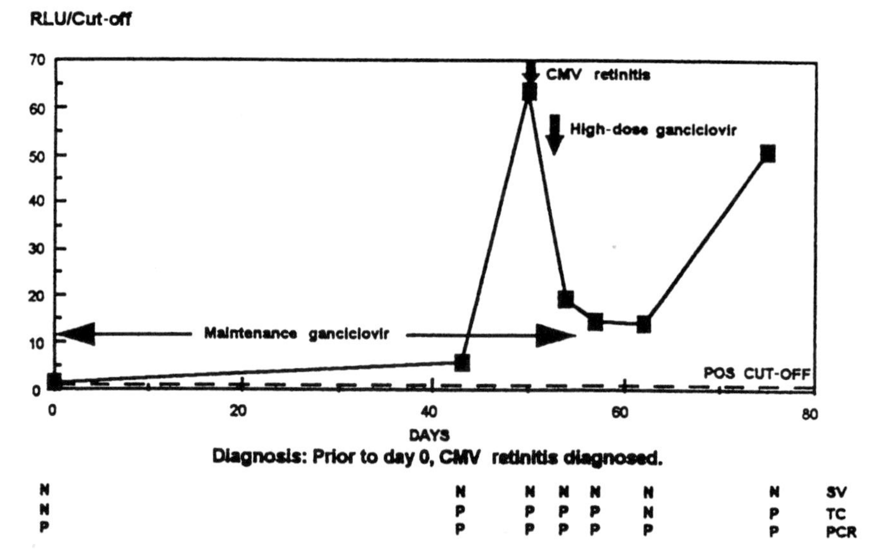


FIGURE 4. Longitudinal monitoring of a patient with CMV retinitis (P = positive; N = negative). HCT CMV DNA levels fall after administration of high-dose ganciclovir. However, rising levels indicate the possible emergence of a drug-resistant strain. (Provided courtesy of Dr. David Parker, Murex Biotech, Ltd., Dartford, UK.)

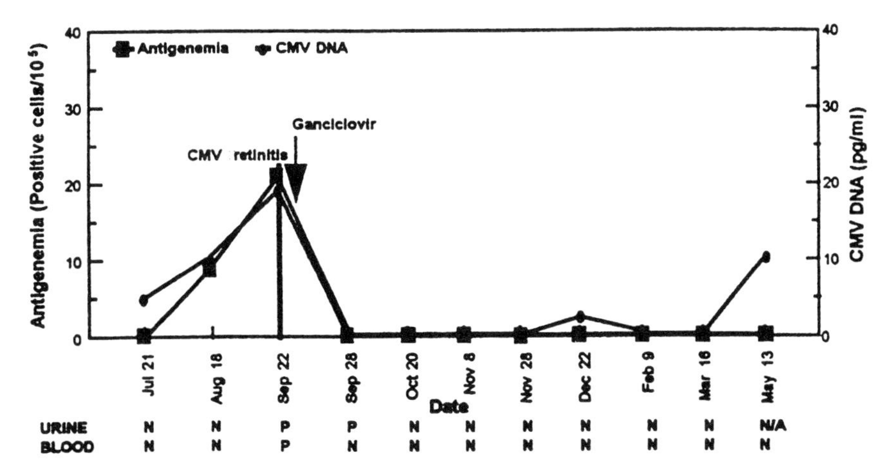


FIGURE 5. Longitudinal study of an AIDS patient comparing HCT CMV to antigenemia, and urine and blood culture (P = positive; N = negative; N/A = not available). Both antigenemia and HCT CMV detect CMV earlier than culture. Response to ganciclovir therapy is demonstrated by both antigenemia and HCT CMV. (Provided courtesy of Dr. Tony Mazzulli, Mt. Sinai Hospital, Toronto, Ontario, Canada.)

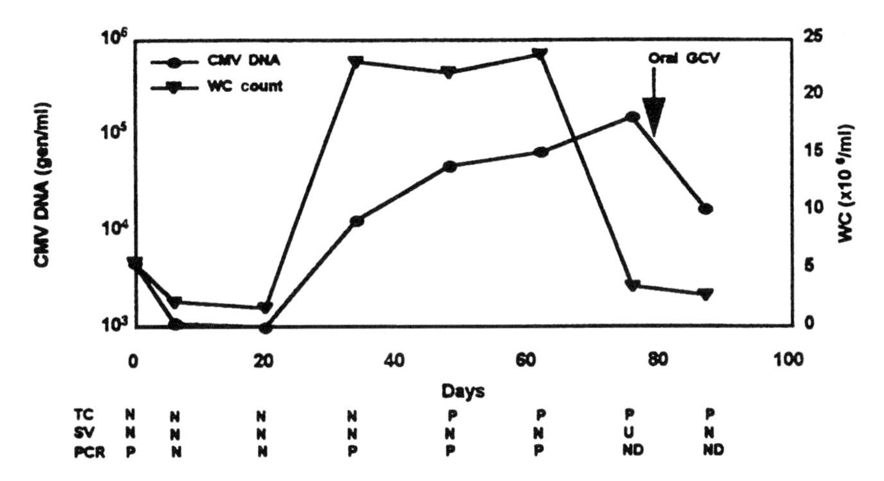


FIGURE 6. Longitudinal study of an AIDS patient with CMV retinitis (P = positive; N = negative; ND = not done; GCV = ganciclovir; WC = white cell). HCT and PCR detect active CMV simultaneously. SV is negative throughout. TC is positive 10 days after HCT CMV (Miller et al., 1996).

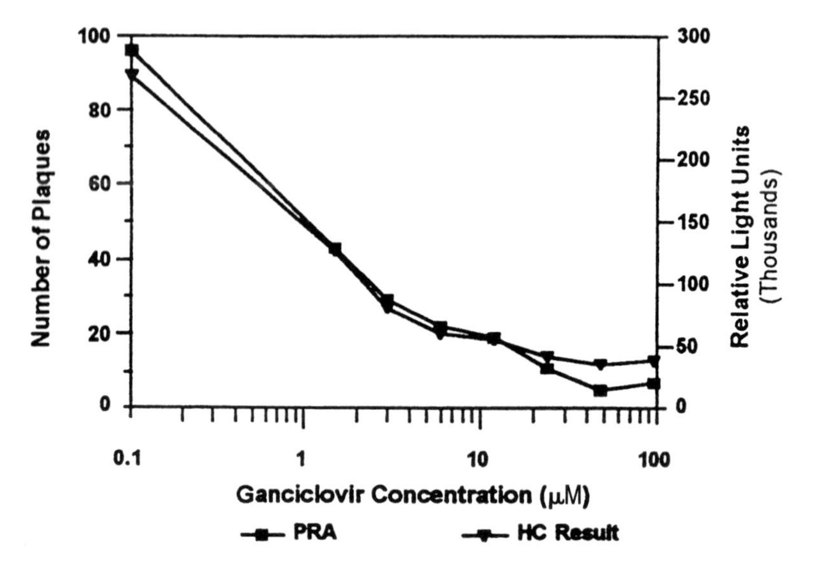


FIGURE 7. Comparison of the Plaque Reduction Assay (PRA) and the HCT CMV assay for measuring drug susceptibility. (Data provided courtesy of Dr. Carlos Isada, Cleveland Clinic Foundation, Cleveland, OH.)

shown in Figure 7, the number of plaques correlated very well with the RLU values.

## Hepatitis B Virus

Infection with HBV in humans is common and results in many cases of acute and chronic hepatitis. The high incidence of acute cases as well as chronic infections with a potential to develop liver cirrhosis and primary hepatocellular carcinoma has given this virus an important place in the arena of diagnostic assays. Traditionally, HBV disease has been diagnosed with serological assays for HBsAg, anti-HBs, HBeAg, anti-HBeAg, anti-HBc, IgM anti-HBc, and DNA polymerase (Hoofnagle and Di Bisceglie, 1991). With the advent of molecular biology techniques, it is possible to detect HBV DNA levels in serum. The most important issues in reference to HBV DNA testing are: the utility of monitoring DNA levels during the course of antiviral drug treatment (Kuhns et al., 1989), eligibility of patients for treatment (Hoofnagle, 1990), and assessing patient response and eligibility of patients for liver transplant (Samuel et al., 1993; Perrillo and Mason, 1993). In the case of acute and chronically infected patients, resolution of active viral replication is demonstrated by loss of HBV DNA levels. The newly emerging and promising field of viral hepatitis therapeutics would not be possible if accurate, sensitive, reproducible, and quantitative assays for HBV DNA detection were not available.

The Hybrid Capture HBV DNA assay was developed for detection and quantitation of HBV DNA, as a diagnostic tool as well as a prognostic and therapeutic guide. The specimen type required for the assay is serum, and the volume tested is 50 $\mu$l. After a minimal sample preparation step, the specimen is hybridized with a specific full-length genomic HBV RNA, and signal is produced and detected as described in Figure 1.

To accurately quantitate the amount of HBV DNA in a clinical sample, we chose a 4-point calibration curve as part of the first-generation HCT HBV assay procedure (Figure 8). The HBV calibrators contain the specified concentration of HBV DNA (5, 10, 200, and 2000 pg/ml) and carrier DNA in a transport medium that prevents degradation and guarantees long-term stability. The upper calibrator at 2000 pg/ml is at the approximate "clinically significant threshold level" referred to by investigators or clinicians involved in trials of antiviral therapy (Hoofnagle, 1990). The HBV DNA used in the calibrators is quantitated and tested using rigorous methods and quality control specifications.

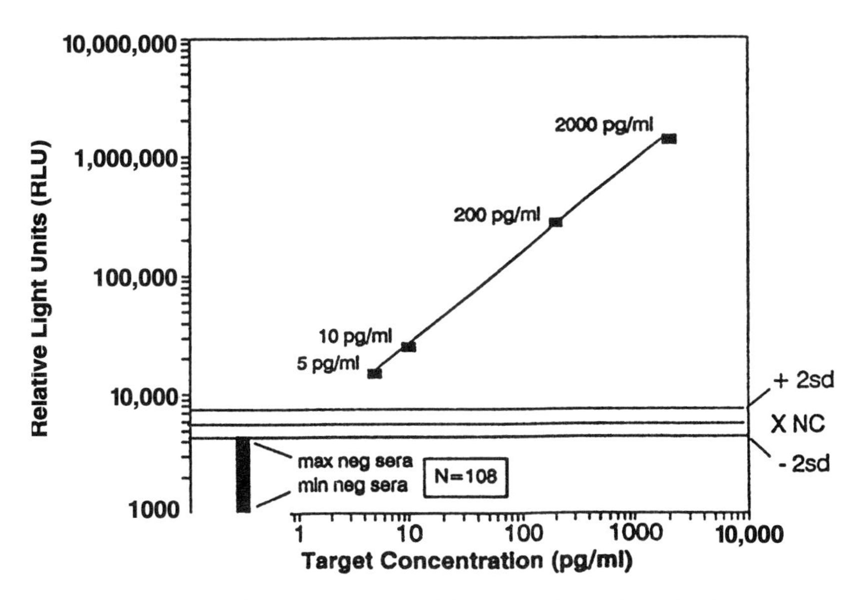


FIGURE 8. Dynamic range and linearity of the HCT HBV assay

In Figure 8, the values obtained with 108 negative sera are indicated by the solid bar (maximum and minimum). There is good separation between the cut-off and the values obtained with the negative sera and the negative control. None of the negative sera tested had values ±2 SD from the negative control of the assay.

The rate of success of antiviral drug treatment for chronically infected HBV patients has been reported to be tightly linked to the initial HBV DNA level prior to commencement of treatment. Thus, the importance of a sensitive and precise baseline measurement is crucial (Perrillo et al., 1990; Hoofnagle and Di Bisceglie, 1991).

The effectiveness of HCT in following HBV DNA levels in patients undergoing antiviral treatment has been evaluated extensively. Figure 9 represents the profile of a responder patient monitored for 9 months who underwent interferon treatment at the Liver Unit of the National Institutes of Health. HBV DNA levels were compared with DNA polymerase activity (the traditional marker of active viral replication). The quantitative changes in HBV DNA levels closely paralleled results with the DNA polymerase assay. From the patients evaluated in this study, it was possible to discriminate responders from nonresponders.

HCT HBV has been shown to be a sensitive and reliable test, superior to other commercially available molecular HBV DNA hybridization assays. It has been described as a "simple immunoassay-like procedure

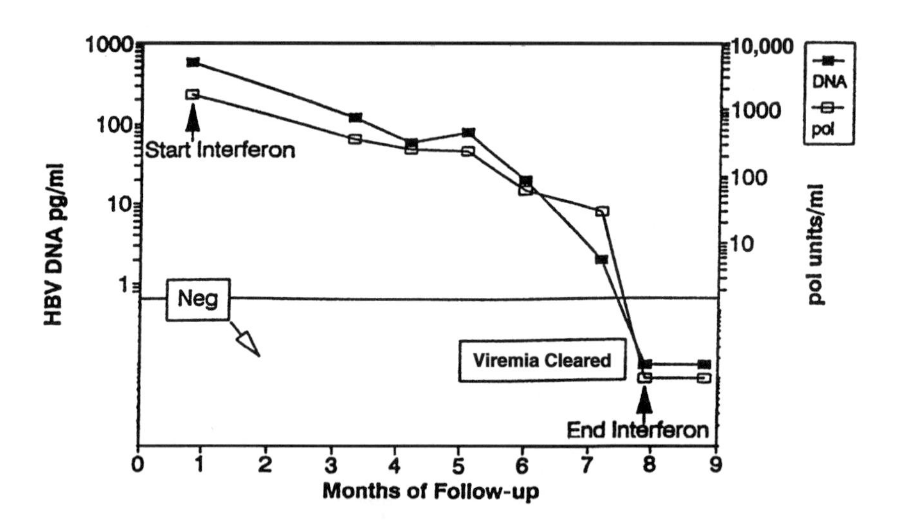


FIGURE 9. Typical course of a case of chronic hepatitis B responding to antiviral treatment, monitored for HBV DNA levels with the HCT HBV assay ("DNA") and for DNA polymerase activity ("pol").

offering nonisotopic direct detection of HBV DNA" (Aspinall et al., 1995).

## Human Immunodeficiency Virus

Direct HIV RNA quantitation has recently become a useful method for clinically gauging HIV disease progression (O'Brien et al., 1996). The concentration of cell-free genomic RNA in plasma directly reflects the titer of HIV viral particles during infection; thus, circulating levels of HIV RNA are perhaps the best measure of *in vivo* viral replication (Pantaleo et al., 1993; Ho et al., 1995; O'Brien et al., 1996). Viral load, as measured by quantitative plasma RNA, predicts disease progression independently of CD4+ counts, and levels of 10,000 RNA copies/ml of plasma (or lower) have been associated with good prognosis, whereas levels above 100,000 copies/ml may predict an enhanced risk of progression (Mellors et al., 1995; Mellors, 1996). There is also a strong correlation between maternal plasma HIV RNA levels and vertical transmission. At levels of 20,000 RNA copies/ml of plasma (or lower) transmission was unlikely (0/63 mothers transmitted), but at 50,000 copies/ml (or higher) transmission occurred at a higher rate (15/20 mothers transmitted) (Dickover et al., 1996). Consequently, quantitative plasma HIV RNA measurements may be the most useful method of assessing HIV disease stage, disease progression, and risk of transmission.

Another potential application of HIV RNA quantitation is to help clinicians evaluate antiviral therapy. HIV RNA levels usually decline in response to antiretroviral therapeutic intervention and increase when drug resistance develops (Piatak et al., 1993; Dickover et al., 1996; O'Brien et al., 1996; Ho et al., 1995). In this context, HIV RNA levels can be used as a sensitive marker to determine baseline levels and to monitor the effect of therapeutic strategies in the case of a change in therapy.

Using Hybrid Capture technology, we have developed a quantitative, microtiter-based, HIV RNA detection assay. The assay employs biotin-labeled DNA probes that cover 92% of the HIV RNA genomic sequence. The standard curve is $10^3$ to $5 \times 10^6$ HIV RNA equivalents. The HIV RNA detection assay procedure is summarized in Figure 10. Including sample preparation, the time to results is 7 hours with 1.5 hours of total hands-on time. After hybridization in the presence of complete HIV RNA molecules, the hybrids are captured on streptavidin-coated microtiter plates. An alkaline-phosphatase-labeled antibody, specific for RNA:DNA hybrids, is then added to the immobilized hybrids. Bound alkaline phosphatase is detected by the enzymatic dephosphorylation of a dioxetane chemiluminescent sub-

HYBRID CAPTURE HIV RNA DETECTION ASSAY

Add 50 μl of precipitation buffer to 1 ml of plasma or serum sample.
↓
Centrifuge for 1 hour at 25,500 X*g*.
↓
Aspirate and discard the supernatant.
↓
Resuspend the pelleted virus in 50 μl of lysis buffer.
↓
Incubate at 37°C for 10 minutes.
↓
Add 50 μl of hybridization buffer + probe.
↓
Hybridize at 65°C for 2 hours.
↓
Transfer samples to microtiter plate wells.
↓
Capture for 1 hour at room temperature.
↓
Aspirate.
↓
React with 100 μl of conjugate at room temperature for 1 hour.
↓
Aspirate.
↓
Wash 6X with buffer.
↓
Add 100 μl of detection reagent.
↓
Incubate at room temperature for 15 minutes.
↓
Signal detected using a luminometer.

Time to results = 7 Hours
Hands-on time = 1.5 Hours

FIGURE 10. Hybrid Capture HIV RNA detection assay

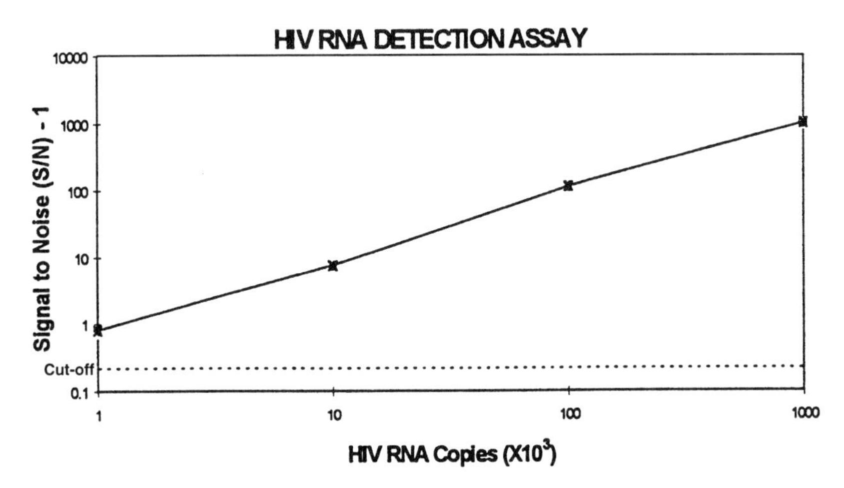


FIGURE 11. Hybrid Capture HIV RNA detection assay. In this prototype assay, the detection range was $10^3$ to $10^6$ HIV genomes per ml of plasma. Using an extra calibrator, it is possible to accurately quantitate to $5 \times 10^6$/ml or higher.

strate, and quantitated using a luminometer. RLUs are converted to signal-to-noise ratios and plotted against HIV RNA equivalents as depicted in Figure 11. (Cut-off values were determined by measuring the signal generated by multiple negative samples and determining the level above which the assay could reproducibly discern a negative from a positive sample.) The detection assay is very sensitive (can detect less than $10^3$ HIV RNA copies/ml of plasma without target amplification), extremely flexible and easy to automate.

## Future Developments

A new generation of Hybrid Capture assays is currently being developed that will increase the usefulness and scope of these signal-amplified tests. These more sensitive new assays, named the HCM family of tests, are based in a microtiter plate format to enhance assay throughput and ease of use, and to enable automation of the tests. Furthermore, refinements and improvements made to the probes and other reagents have minimized the volume of specimen required and increased the sensitivity of the HCM assays by 20- to 50-fold over the first-generation HCT. In addition to the HPV, HIV, and CMV HCM tests described above, additional tests for HBV, HSV-1, and HSV-2, and a chlamydia/gonorrhea test are also under

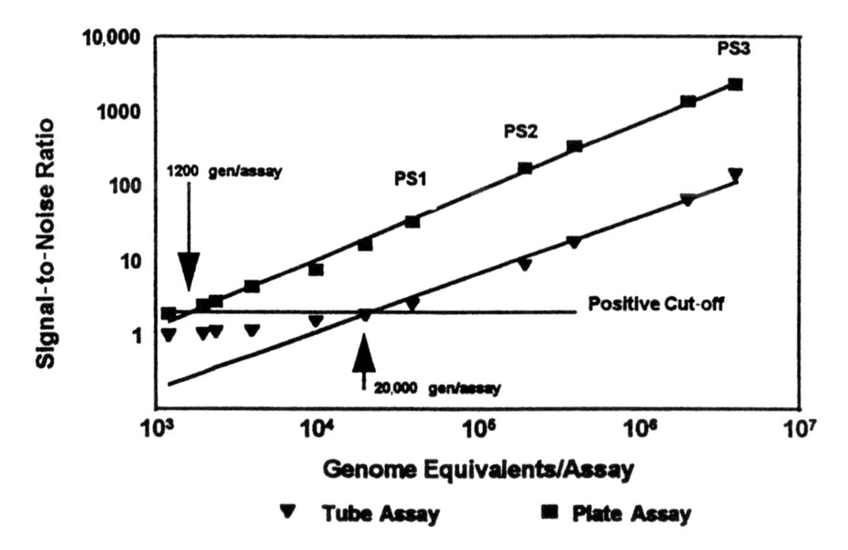


FIGURE 12. A comparison of the HCT and HCM formats, showing that the plate prototype is approximately 20 times more sensitive than the tube assay.

development. A comparison of the tube and prototype plate version of the Hybrid Capture CMV assays is shown in Figure 12.

## REFERENCES

Aspinall S, Steele AD, Peenze I, and Mphahlele MJ (1995): Detection and quantitation of hepatitis B virus DNA: comparison of two commercial hybridization assays with polymerase chain reaction. *J Viral Hepatitis* 2:107–11.

Boring CC, Squires TS, and Tong T (1991): Cancer statistics, 1991. *CA Cancer J Clin* 41:19–36.

Bosch FX, Manos MM, Munoz N, Sherman M, Jansen AM, Peto J, Schiffman MH, Moreno V, Kurman R, and Shah KV (1995): Prevalence of human papillomavirus in cervical cancer: A worldwide perspective. *JNCI* 87:796–802.

Bronstein I and Olesen CEM (1995): Detection methods using chemiluminescence. In: *Molecular Methods for Virus Detection* (pp. 147–74). Wiedbrauk DL, Farkas DH, eds. San Diego: Academic Press.

Cox JT, Lorincz AT, Schiffman MH, Sherman ME, Cullen A, and Kurman RJ (1995): Human papillomavirus testing by hybrid capture appears to be useful in triaging women with a cytologic diagnosis of atypical squamous cells of undetermined significance. *Am J Obstet Gynecol* 172:946–54.

Dickover RE, Garratty EM, Herman SA, Sim MS, Plaeger S, Boyer PJ, Keller M, Deveikis A, Stiehm R, and Bryson YJ (1996): Identification of levels of maternal HIV-1 RNA associated with risk of perinatal transmission. *JAMA* 272:599–605.

Ferenczy A, Franco E, Arseneau J, Wright TC, and Richart RM (1996): Diagnostic performance of Hybrid Capture™ HPV DNA Assay combined with liquid-based cytology. *Am J Obstet Gynecol* (in press).

Gozlan J, Salord JM, Chouaid C, Duvivier C, Picard O, Meyohas MC, and Petit JC (1993): Human cytomegalovirus (HCMV) late-mRNA detection in peripheral blood of AIDS patients: Diagnostic value for HCMV disease compared with those of viral culture and HCMV DNA detection. *J Clin Microbiol* 31:1943–5.

Hall S, Lörincz A, Shah F, Sherman ME, Abbas F, Kurman RJ, and Shah KV (1996): HPV DNA detection in cervical specimens by Hybrid Capture™: correlation with cytologic and histologic diagnoses of squamous intraepithelial lesions of the cervix. *Gynecol Oncol* (in press).

Hatch KD, Schneider A, and Abdel-Nour MW (1995): An evaluation of human papillomavirus testing for intermediate- and high-risk types as triage before colposcopy. *Am J Obstet Gynecol* 172:1150–7.

Ho M (1991): *Cytomegalovirus: Biology and Infection.* 2nd Ed. New York: Plenum Publishing Corporation.

Ho DD, Neumann AU, Perelson AS, Chen W, Leonard JM, and Markowitz M (1995): Rapid turnover of plasma virions and CD4 lymphocytes in HIV-1 infection. *Nature* 373:123–6.

Ho GYF, Burk RD, Klein S, Kadish AS, Chang CJ, Palan P, Basu J, Tachezy R, Lewis R, and Romney S (1995a): Persistent genital human papillomavirus infection as a risk factor for persistent cervical dysplasia. *J Natl Cancer Inst* 87:1365–71.

Hoofnagle JH (1990): α-Interferon therapy of chronic hepatitis B. Current status and recommendations. *J Hepatol* 11:S100–S107.

Hoofnagle JH and Di Bisceglie AM (1991): Serologic diagnosis of acute and chronic viral hepatitis. *Sem Liver Dis* 11:73–83.

Isada C, Kohn D, Lazar JG, Lorincz AT, Salim H, and Yen-Lieberman B (1996): Rapid diagnosis of cytomegalovirus (CMV) using the Digene Hybrid Capture System: A comparison with tissue culture and polymerase chain reaction (PCR). *12th Annual Clinical Virology Symposium and Annual Meeting, Pan American Society for Clinical Virology.* Clearwater Beach, FL, April 28-May 1. Abstr S19.

Kuhns MC, McNamara AL, Perrillo RP, Cabal CM, and Campbell CR (1989): Quantitation of hepatitis B viral DNA by solution hybridization: Comparison with DNA polymerase and hepatitis B e antigen during antiviral therapy. *J Med Virol* 27:274–81.

Lörincz A, Reid R, Jenson AB, Greenberg MD, Lancaster W, and Kurman RJ (1992): Human papillomavirus infection of the cervix: relative risk associations of 15 common anogenital types. *Obstet Gynecol* 79:328–37.

Lörincz A (1996): Hybrid Capture™ method for detection of human papillomavirus DNA in clinical specimens. *Pap Report* 7:1–5.

Matthews JA and Kricka LJ (1988): Analytical strategies for the use of DNA probes. *Anal Biochem* 169:1–25.

Mazzulli T, Wood S, Chua R, Clark J, and Walmsley S (1996): Clinical evaluation of the Digene Hybrid Capture System (DHCS) assay for the quantitative detection and monitoring of cytomegalovirus (CMV) in HIV infected patients. *12th Annual Clinical Virology Symposium and Annual Meeting, Pan American Society for Clinical Virology.* Clearwater Beach, FL, April 28-May 1. Abstr S10.

Mellors JW, Kingsley LA, Rinaldo CR Jr, Todd JA, Hoo BS, Kokka RP, and Gupta P (1995): Quantitation of HIV-1 RNA in plasma predicts outcome after seroconversion. *Ann Intern Med* 122:573–9.

Mellors JW (1996): Prognostic value of viral load determinations. *3rd Conference on Retroviruses and Opportunistic Infections*. Washington DC, January 28-February 1. Abstr S22.

Miller MJ, Lazar J, Moe AA, and Tufail A (1996): Evaluation of Hybrid Capture™ System for the detection of CMV viremia. *12th Annual Clinical Virology Symposium and Annual Meeting, Pan American Society for Clinical Virology*. Clearwater Beach, FL, April 28-May 1. Abstr S8.

Morrison EAB, Ho GYF, Vermund SH, Goldberg GL, Kadish AS, Kelley KF, and Burk RD (1991): Human papillomavirus infection and other risk factors for cervical neoplasia: a case-control study. *Int J Cancer* 49:6–13.

Mullis KB and Faloona FA (1987): Specific synthesis of DNA *in vitro* via a polymerase-catalyzed chain reaction. *Meth Enzymol* 155:335–50.

O'Brien WA, Hartigan PM, Martin D, Esinhart J, Hill A, Benoit S, Rubin M, Simberkoff MS, and Hamilton JD (1996): Changes in plasma HIV-1 RNA and CD4+ lymphocyte counts and the risk of progression to AIDS. *N Engl J Med* 334:426–31.

Pantaleo G, Graziosi C, Demarest JF, et al. (1993): HIV infection is active and progressive in lymphoid tissue during the clinically latent stage of disease. *Nature* 362:355–8.

Perrillo RP and Mason AL (1993): Hepatitis B and liver transplantation: Problems and promises. *N Engl J Med* 329:1885–7.

Perrillo RP, Schiff ER, Davis GL, Bodenheimer HC Jr, Lindsay K, Payne J, Dienstag JL, O'Brien C, Tamburro C, Jacobson IM, Sampliner R, Feit D, Lefkowitch J, Kuhns M, Meschievitz C, Sanghvi B, Albrecht J, Gibas A, and the Hepatitis Interventional Therapy Group (1990): A randomized, controlled trial of interferon alfa-2b alone and after prednisone withdrawal for the treatment of chronic hepatitis B. *N Engl J Med* 323: 295–301.

Piatak M, Jr., Saag MS, Yang LC, Clark SJ, Kappes JC, and Luk K (1993): High levels of HIV-1 in plasma during all stages of infection determined by competitive PCR. *Science* 259:1749–54.

*The Pink Sheet* (1993): Cytovene induction, Foscavir maintenance for CMV retinitis may be "clever way" to minimize toxicities – UCSF's Drew; Gilead's GS 504 entering Phase II/III for CMV. 55(51):4–5; 20 Dec 93. F-D-C Reports Inc.

Pollard-Knight D, Simmonds AC, Schaap AP, Akhavan H, and Brady MAW (1990): Nonradioactive DNA detection on Southern blots by enzymatically triggered chemiluminescence. *Anal Biochem* 185:353–8.

Rubin RH (1994): Infection in organ transplant recipients. In: *Clinical Approach to Infection in the Compromised Host.* 3rd Ed. Rubin RH, Young LS, eds. New York: Plenum Publishing.

Samuel D, Muller R, Alexander G, Fassati L, Ducot B, Benhamou JP, and Bismuth H (1993): Liver transplantation in European patients with the hepatitis B surface antigen. *N Engl J Med* 329:1842–7.

Schiffman MH, Kiviat NB, Burk RD, Shah KV, Daniel RW, Lewis R, Kuypers J, Manos MM, Scott DR, Sherman ME, Kurman RJ, and Stoler MH (1995): Accuracy and interlaboratory reliability of human papillomavirus DNA testing by Hybrid Capture. *J Clin Microbiol* 33:545–50.

Schneider A, Kirchhoff T, Meinhardt G, and Gissmann L (1992): Repeated evaluation of human papillomavirus 16 status in cervical swabs of young women with a history of normal Papanicolaou smears. *Obstet Gynecol* 79:683–8.

Sherman ME, Schiffman MH, Lorincz AT, Manos MM, Scott DR, Kurman RJ, Kiviat NB, Stoler M, Glass AG, and Rush BB (1994): Toward objective quality assurance in cervical cytopathology: Correlation of cytopathologic diagnoses with detection of high-risk human papillomavirus types. *Am J Clin Pathol* 102:182–7.

Southern EM (1975): Detection of specific sequences among DNA fragments separated by gel electrophoresis. *J Mol Biol* 98:503–17.

Wingo PA, Tong T, and Bolden S (1995): Cancer statistics, 1995. *CA Cancer J Clin* 45:8–30.

Wright TC, Sun XW, and Koulos J (1995): Comparison of management algorithms for the evaluation of women with low-grade cytologic abnormalities. *Obstet Gynecol* 85:202–10.

Young LHY (1996): Therapy for cytomegalovirus retinitis: Still no silver lining. *JAMA* 275:149–50.

# PART II: APPLICATIONS

# Quantification of Gene Expression by Competitive RT-PCR: The hCGβ/LHβ Gene Cluster

Vladimir Lazar, Ivan Bièche, Michel Bahuau, Yves Giovangrandi, Dominique Bellet, and Michel Vidaud

## 1. Introduction

hCG is a dimeric hormone synthesized by trophoblast cells, and it promotes steroidogenesis by the luteal body during early pregnancy. This hormone is composed of two noncovalently bound α- and β-subunits (1). The clinical relevance of the measurement of hCG and its free beta subunit is in the diagnosis of early pregnancy, and in prenatal screening for trisomy 21 (2). In addition, hCG and its free β-subunit are used as biological markers, mainly of trophoblast-derived and gonadal malignancies, but also of neoplasms of nongonadal origin, such as bladder carcinoma (3). The hCGβ subunit can be encoded by four distinct genes, namely hCGβ-7, -8, -5, and -3, highly homologous with regard to their nucleotide sequences. These four genes are clustered together with the luteinizing hormone beta subunit gene (hLHβ) and map to 19q13 (4, 5). The discriminative quantification of the hCGβ genes is achieved using reverse transcription in combination with competitive PCR.

## 2. Reverse Transcription and Competitive PCR

### 2.1. Principle

Competitive PCR is understood as simultaneous amplification, in a single tube, of two templates of equal or similar lengths, using the same primer pair. One of the templates is the sequence of interest, of unknown amount (target), and the other one is the standard template, of known concentration (quantitative standard or QS). PCR products of the target and of the QS must be clearly and easily distinguishable and their amount accurately measured. QS allows the monitoring of the variability in the PCR assays, and measurement of the target is achieved by determining the concentration at which the QS and target PCR amounts are identical. Competitive PCR can be used to quantify cDNA, and therefore mRNAs, after a reverse transcription step. In this case, a normalization procedure must be performed in order to take into account the extent of possible fluctuations in RNA preparations and in reverse transcription yields.

### *2.2. Design of the quantitative standards, titration, and quality control procedures*

Using a QS as similar as possible to the target offers the advantage that PCR can be run either at exponential or plateau phase and therefore offers operational simplicity (6). Nevertheless, when setting up a quantitative protocol, it is necessary to test the impact of the number of PCR cycles (7), by comparing the accumulation of two PCR products generated from a known amount of QS and target, and to check whether or not the initial ratio of target to QS remains constant irrespective of the number of cycles. Different types of homologous competitors can be designed, e.g., with a different length from that of target or harboring a mutation that creates a new restriction site (6, 8). We opted for the first, more straightforward alternative.

The QS designed for the quantification of hCG$\beta$ cDNAs was generated by PCR using the "looped oligo" method (8). The resulting QS is similar in sequence to the target sequence except for a 12-bp insertion that enables discrimination between the two PCR products. The 12-bp insertion does not modify the efficiency of the amplification of each target and its corresponding QS, and it confers an excellent resolution after electrophoresis through 6% acrylamide gels.

The QS is synthesized using a two-step PCR procedure. In the first step, the CG1 and the looped CG3 primers generate a 359-bp fragment (Figure 1). In the second step, CG1 and CG2 primers are used for a larger-scale synthesis.

The QS is purified using Microspin SH-300 columns (Pharmacia) and recovered in TE 10/1 buffer. At this step, an estimation of the QS copy number can be performed by classical procedures such as absorbance spectrophotometry. The QS can be co-amplified with the target (hCG$\beta$ cDNA) using primers CG1 and CG2, resulting in a 347-bp fragment (target) and a 359-bp fragment (QS). The CG2 primer is fluorescein-5′end-labeled (Fluoreprime, Pharmacia) for analysis of the PCR product with a DNA sequencer (ABI-373 Perkin Elmer).

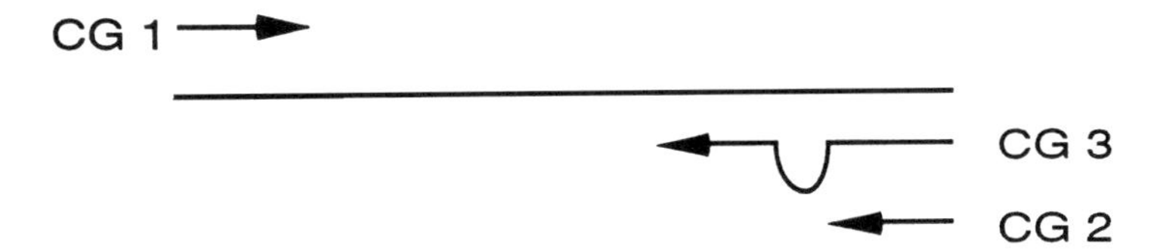


FIGURE 1. Design of the QS with the "looped oligo" method. Primers CG1 and CG3 generate a 359-bp fragment from hCG$\beta$ cDNA. Primer CG3 contains, from 5′ to 3′, a sequence identical to CG2, a trinucleotide repeat $(CTG)_4$, and a 20-bp sequence specific to the target and immediately 3′ to CG2.

The aim of our study is a relative estimate of the targets, and a comparison of different samples for the expression of a given gene. Therefore, target and a fixed amount of QS are co-amplified in the same tube, assuming that any variable affecting amplification has the same effect on both template species. The adequate concentration of the QS to be used in order to obtain equal amounts of PCR products for both target and QS is assessed by testing serial dilutions of the QS. For this purpose, a 1 log dilution of the QS is performed, spanning an appropriate range of concentrations, e.g., from $10^{-4}$ to $10^{-9}$ of the original QS PCR product.

## 2.3. *Template preparation*

Total RNA is prepared using the acid-phenol guanidium method (9). Following quantification by absorbance spectrophotometry at 260 nm, the quality of an aliquot of one microgram of total RNA is determined by electrophoresis through a denaturing agarose gel, and by staining with ethidium bromide and visualization of the 18S and 28S RNA bands under ultraviolet transillumination. One mg total RNA is reverse transcribed using the oligo (dT)16 primer and MuLV reverse transcriptase according to the protocol supplied with the GeneAmp RNA PCR Core kit (Perkin Elmer). RT products are then diluted 1 to 4 in the same RT mix, devoid of enzymes, primers, and dNTPs, and stored at +4°C prior to amplifications.

## 2.4. *Design of the competitive PCR and protocol*

Primers were specifically selected to avoid amplification of LH$\beta$ mRNA or contaminating genomic DNA, and to commonly amplify the $\beta$-7, -8, -5, and -3 genes. For this purpose, primer CG1 was placed at the junction between exon 2 and exon 3 of hCG$\beta$ genes. In addition, a 2-bp mismatch with the LH$\beta$ mRNA (CC versus TG) and a 3-bp mismatch with the genomic sequence of hCG$\beta$ genes from the contaminating DNA (ACC versus GTG), have been placed at the 3′ end of CG 1 primer. The existence of these mismatches at the 3′ end of CG 1 completely aborted the amplification of the undesirable sequences. PCR was carried out in a 20-$\mu$l reaction volume containing 4$\mu$l of the diluted RT product (50ng equivalent cDNA), 6pmol of each primer, 200$\mu$M of each dNTP, 1.5mM $MgCl_2$, 10mM Tris-HCl pH 8.3, 50mM KCl, 1 unit of Taq DNA polymerase, and a known concentration of the QS. Amplification reactions were carried out in 20–30 sequential cycles of 94°C for 30s, 65°C for 30s, and 72°C for 30s in a 9600 DNA thermocycler (Perkin Elmer). Aliquots of the PCR products were added to 2.5$\mu$l of deionized formamide containing 0.5$\mu$l of a molecular size marker (Genescan 500 ROX, Perkin Elmer), denatured at 90°C for 3min, and loaded onto a 6% denaturing polyacrylamide sequencing gel. Gels were electrophoresed for 2hr at 1000V in a 373 DNA sequencer. The resulting gel data were analyzed for peak color, fragment size, and peak area using the Genescan 672 Fragment Analysis software. Measurement of the target was achieved by determining the concentration at which the QS and target signals were identical.

## *2.5. Results*

Measure of the amounts of target and QS PCR products must be as accurate as possible. In our opinion, use of an automated DNA sequencer and detection of fluorescently-labeled PCR products is the option of choice for at least two reasons: i) High sensitivity implies that only a small number of cycles are required for quantitation, and thus quantitation occurs during the exponential phase (7), and ii) Data are directly computed, and results can be expressed in a numeric form (Figure 2).

The steady-state level of hCGβ mRNAs is dramatically different in trophoblast and nontrophoblast tissues (Figure 3).

Indeed, a thousandfold higher, or even greater, amounts of hCGβ QS were necessary to obtain equivalent amplification of target and QS in placental tissue, in comparison to nontrophoblastic tissues such as prostate, testis, or breast.

## *2.6. Normalization of the results*

In order to take into account possible fluctuations in RNA extraction, RNA integrity, and reverse transcription yields, quantification of hCG$\beta$ mRNAs was normalized by using a housekeeping gene rather than by adding an internal RNA standard (6). There is no ideal housekeeping gene for internal control. We chose the TFIID-encoding gene, which is ubiquitously

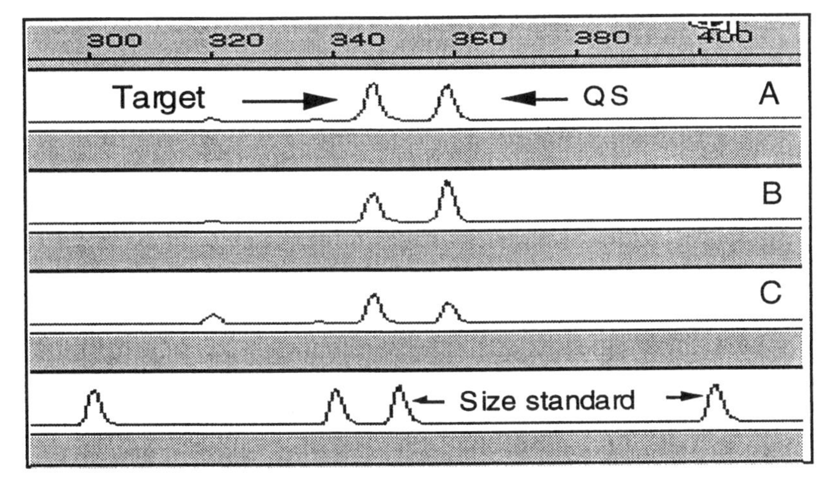


FIGURE 2. Electrophoretogram as recorded by the sequencer. The ratio between the peak area of target (347-bp fragment) and QS (359-bp fragment) is ascertained with various concentrations of QS. (A: target/QS = 1.1; B: target/QS = 0.8; C: target/QS = 1.5)

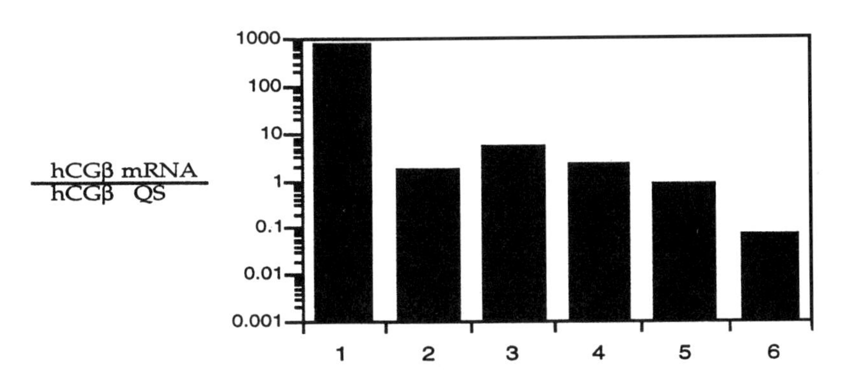


FIGURE 3. hCG$\beta$ mRNA steady-state levels in normal tissue specimens are obtained by calculating the ratios of hCG$\beta$ over QS peak areas. Lane 1: trophoblast, lane 2: prostate, lane 3: testis, lane 4: breast, lane 5: skeletal muscle, and lane 6: uterus.

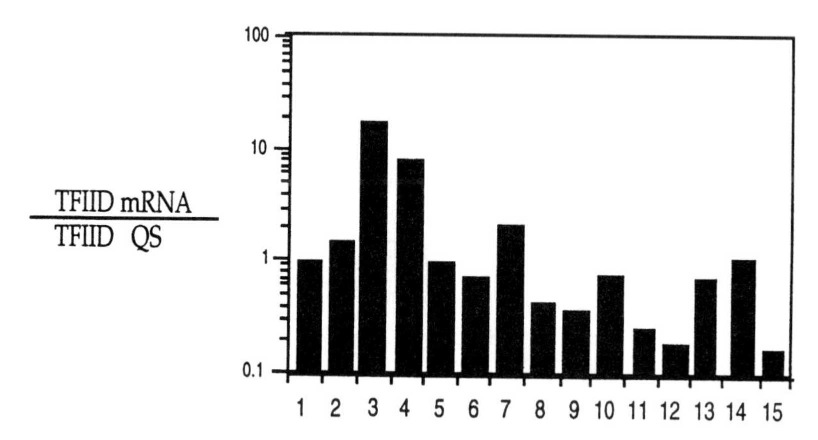


FIGURE 4. TFIID mRNA steady-state levels in normal tissue specimens are obtained by calculating the ratios of TFIID over QS peak areas. Lane 1: trophoblast, lane 2: prostate, lane 3: testis, lane 4: breast, lane 5: skeletal muscle, lane 6: uterus, lane 7: adrenal, lane 8: liver, lane 9: stomach, lane 10: pituitary, lane 11: bladder, lane 12: brain, lane 13: thyroid, lane 14: colon, and lane 15: lung.

expressed. We quantified the TFIID mRNAs, using the same method as described above, and found that the steady-state level of TFIID is similar in trophoblastic and diverse nontrophoblastic tissues (10). Results are presented in Figure 4.

This result indicates that the steady-state level of TFIID transcripts offers a reliable basis for normalization of gene expression quantification. Normalized results are expressed as hCG$\beta$/TFIID ratios (Figure 5).

The hCGβ mRNA quantitation results (Figure 3) are not significantly changed when results are expressed as hCG$\beta$/TFIID ratios (Figure 5), even if corrections do occur (e.g., testis).

## *2.7 Reproducibility and linearity of the competitive RT-PCR method*

The reproducibility and linearity of the quantification can be assessed by serial dilutions of the target with a constant amount of QS. For that purpose, cDNAs from normal prostate were mixed with cDNA prepared from lymphocytes in which hCG$\beta$ genes are not transcribed, using the following proportion: 0%, 25%, 50%, 75%, and 100%. Each dilution point was run in triplicate, in the presence of a constant amount of QS, in two independent assays. Results are presented in Table I.

All CV values were lower than 20%. In addition, the linearity of the relationship between the expected hCG$\beta$/QS ratio values as deduced from

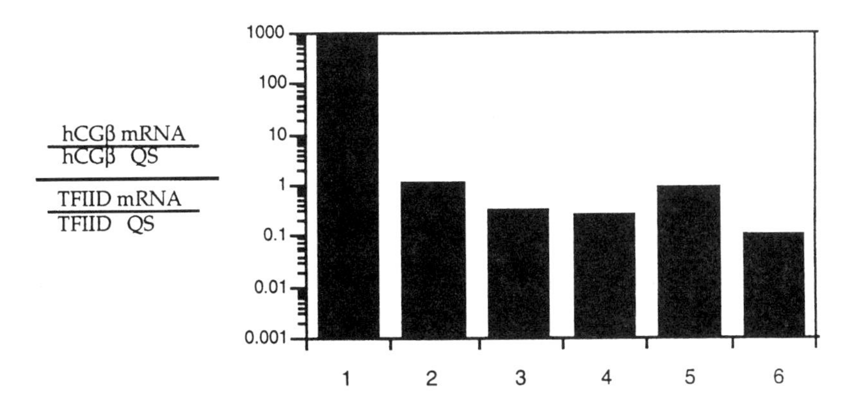


FIGURE 5. hCG$\beta$ mRNA steady-state levels in normal tissue specimens normalized by TFIID mRNA steady-state levels. Lane 1: trophoblast, lane 2: prostate, lane 3: testis, lane 4: breast, lane 5: skeletal muscle, and lane 6: uterus.

TABLE I. Reproducibility and linearity of the competitive RT-PCR method.

| DAY 1 | | | DAY 2 | | |
|---|---|---|---|---|---|
| hCG$\beta$/QS | Mean | CV% | hCG$\beta$/QS | Mean | CV% |
| 0 | | | 0 | | |
| 0 | 0 | **0** | 0 | 0 | **0** |
| 0 | | | 0 | | |
| 0.34 | | | 0.38 | | |
| 0.47 | 0.43 | **19.5** | 0.44 | 0.40 | **8.7** |
| 0.50 | | | 0.38 | | |
| 0.66 | | | 0.74 | | |
| 0.81 | 0.70 | **12.7** | 0.66 | 0.64 | **16.5** |
| 0.65 | | | 0.53 | | |
| 0.99 | | | 0.96 | | |
| 0.99 | 1.02 | **6.2** | 1.30 | 1.15 | **15.2** |
| 1.10 | | | 1.20 | | |
| 1.37 | | | 1.31 | | |
| 1.38 | 1.42 | **5.9** | 1.68 | 1.63 | **14.9** |
| 1.52 | | | 1.9 | | |

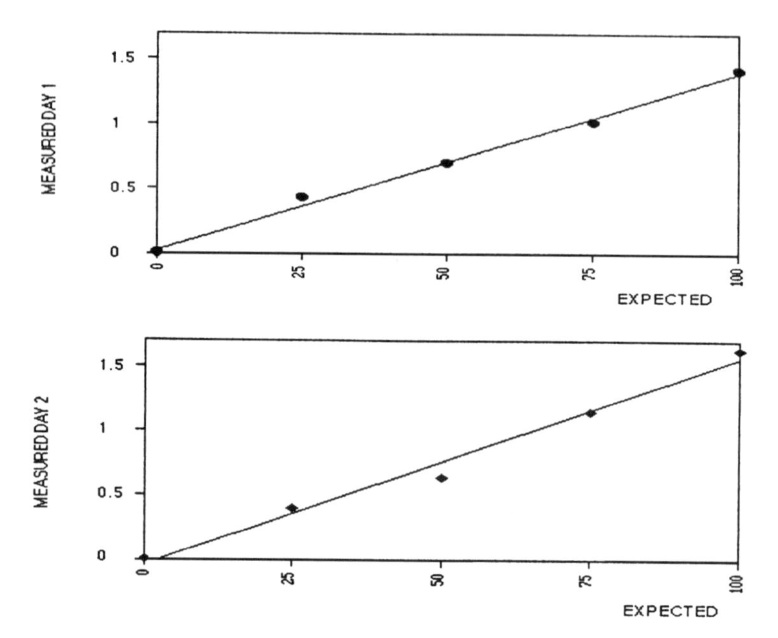


FIGURE 6. Linearity of the relationship between the expected and measured hCG$\beta$/QS ratio values.

serial dilutions and the hCG$\beta$/QS ratio values as actually observed during the assays further demonstrates the quantitation ability of this assay (Figure 6).

## 2.8. *Main pitfalls and limits to the method*

In our experience, the time and effort required to obtain the QS by PCR with the "looped oligo" method is a limit to this approach. One of the most serious problems is long-term conservation of the QS and long-term reproducibility of the method. To address this issue, one might prepare larger amounts of the QS and of serial dilutions, followed by freezing of enough aliquots for separate thawing prior to running the assay. From this point of view, a relative estimate of the number of target molecules appears more reliable than an estimation of their absolute number.

Another limiting factor is adequate preparation of the RNA templates. This depends mainly on the interval between the surgical excision of tissues and their freezing (preferably in liquid nitrogen).

No major analytical pitfall is specific to the competitive PCR protocol. Troubleshooting procedures are common to other PCR protocols (i.e., regarding failure of amplification, contamination, etc.) and they mandate careful monitoring of controls. As a general rule, runs should be performed in duplicate.

To achieve adequate quantitation, it is very important to respect the analytic scale of the DNA sequencer and avoid overloading it with excesses of fluorescently-labeled products.

## 3. Competitive Oligo Priming and Specific Quantification of hCG$\beta$ Genes

### *3.1. Design of the method and protocol*

The hCG$\beta$-7, -8, -5 and -3 genes display up to 96% homology with regards to their nucleotide sequence, making their discrimination by PCR elusive. It is of interest to discriminate the $\beta$-7 gene from the $\beta$-5, -3, and -8 group, particularly when investigating tumor biopsy specimens. Indeed, we have recently analyzed the transcription of the hCG$\beta$ genes from a series of normal and tumoral bladder biopsy specimens, and observed that hCG$\beta$ mRNAs were detected in both normal urothelial and carcinomatous cells (10). However, tumor progression was characterized by different hCG$\beta$ gene transcription patterns; the $\beta$-7 gene was the only gene transcribed in normal urothelia and superficial tumors whereas, in addition to $\beta$-7, the $\beta$-5, -8, and -3 genes were transcribed in invasive tumors (10). Here, we present a simple technique to discriminate between the $\beta$-7 and the $\beta$-5, -3, -8 mRNAs and to quantify them in a specific manner.

The present technique is based on the existence, within the CG1-CG2 PCR fragment, of an A to C nucleotide substitution at codon 117. Presence of nucleotide A characterizes the $\beta$-5, -8, and -3 genes, whereas nucleotide C is specific for the $\beta$-7 gene. We used a competitive oligo primer (COP) technique in which two allele-specific primers, conjugated to a different fluorescent dye, are used in conjunction with a common unlabeled primer. The fluorescence of each dye with respect to its amplified DNA locus, is scored on the 373 DNA sequencer. The two allele-specific primers (CG4 and CG5) were labeled with TET and FAM, respectively. The allele specificity was conferred by the 3′-end nucleotide of each primer: C for CG4 and A for CG5. The specificity of the resulting mismatches C:T and A:G has been shown to be excellent and was enhanced by primer competition and the use of the Stoffel fragment of DNA polymerase (Perkin Elmer).

Briefly, COP was carried out in a 10-$\mu$l reaction volume containing 0.1 $\mu$l of the CG1-CG2 PCR product, 0.1 pmol of each allele-specific primer (CG4, CG5), 0.1 pmol of the common primer CG2, 50 $\mu$M of each dNTP, 3 mM $MgCl_2$, 10 mM Tris-HCl pH 8.3, 10 mM KCl, and 2 units of the Stoffel fragment. Amplifications were processed through five cycles of 30 s at 95°C and 30 s at 65°C. The COP products were analyzed on a 373 DNA sequencer as previously described. The resulting data were analyzed for peak color, fragment size, and peak area with the Genescan 672 software. The ratio between the relative fluorescent units of the two resulting blue and green peak areas of the 119-bp COP products is representative of the expression of type $\beta$-7 and type $\beta$-5, -8, -3 genes, respectively, and enables the determination of the so-called CG117 index, defined as follows:

$$\text{CG117 index } (\%) = \frac{\text{Type } \beta\text{-5, -8, -3mRNA}}{\text{Type } \beta\text{-7mRNA+ Type } \beta\text{-5, -8, -3mRNA}} \times 100$$

### 3.2. *Quality control of the COP technique*

PCR products using primers CG1 and CG2 from trophoblast, in which only the $\beta$-3, -5, -8 genes are expressed, were mixed in increasing proportions (0, 20, 40, 60, and 80%) with PCR products obtained from normal prostate, in which only the $\beta$-7 gene is expressed. Each dilution point was run with the COP technique in triplicate. Results are presented in Figure 7.

All CV values concerning the COP data are lower than 20%. Figure 7 illustrates the quantitation capability of this assay in discriminating between the hCG$\beta$-7 and hCG$\beta$ -5, -8, and -3 transcripts.

## 4. Conclusion

The hCG$\beta$ subunit is trackable in sera of 47% of patients with bladder cancer, and for this reason it is considered a tumor marker. To understand the molecular basis of hCG$\beta$ production by the tumor, we analyzed the

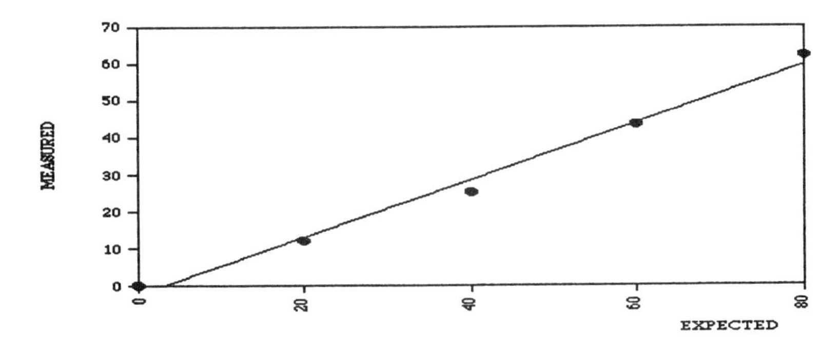


FIGURE 7. Expected values of the CG117 index (abscissa) are plotted against measured values (ordinate) and demonstrate linear regression.

expression of hCG$\beta$-7, -8, -5, and -3 genes in normal versus tumoral urothelia. Using standard RT-PCR procedures, hCG$\beta$ transcripts were detected in both normal and tumoral cells. In order to clarify this issue, the competitive RT-PCR approach was then used to measure the steady-state level of hCG$\beta$ in normal and tumor bladder cells (10). Surprisingly, no statistically significant difference in the overall steady-state level of hCG$\beta$ transcripts was observed between normal and tumoral specimens. Therefore, the competitive RT-PCR and the COP technique were combined in order to quantify the different hCG$\beta$ transcripts.

The COP technique allows researchers to estimate the relative proportion of a given hCG$\beta$ mRNA, and therefore to infer its copy numbers from the total hCG$\beta$ quantitation by competitive RT-PCR. Using this approach, we were able to demonstrate that the expression pattern of various hCG$\beta$ genes differs in normal urothelia and bladder carcinomas. Indeed, normal bladder cells express only the hCG$\beta$-7 gene. In contrast, malignant transformation and disease progression are characterized by expression of the hCG$\beta$-5, -8, and -3 genes, resulting in a transcription pattern similar to that of trophoblast cells.

A broader study including normal tissue specimens and 86 tumor samples from breast, bladder, prostate, and thyroid demonstrated that the expression of the hCG$\beta$-7 gene is common to all above-cited normal tissues, whereas the expression of hCG$\beta$-5, -8, and -3 genes was a hallmark of the corresponding cognate tumors (11).

Finally, these results suggest that discriminative quantitation of the hCG$\beta$ transcripts could find its place in the biological monitoring of a wide array of human tumors.

# REFERENCES


1. Pierce JG and Parsons TF (1981): Glycoprotein hormones: structure and function. *Annu Rev Biochem* 50:465–95.

2. Haddow JE, et al. (1992): Prenatal screening for Down's syndrome with use of maternal serum markers. *N Engl J Med.* 27:588–93.

3. Marcillac I, et al. (1992): Free human chorionic gonadotropin $\beta$ subunit in gonadal and non gonadal neoplasms. *Cancer Res* 52:3901–7.

4. Talmadge K, Vamvakopoulos NC, and Fiddes JC (1984): Evolution of the genes for the $\beta$ subunits of human chorionic gonadotropin and luteinizing hormone. *Nature* 307:37–40.

5. Policastro P, et al. (1983): The $\beta$ subunit of human chorionic gonadotropin is encoded by multiple genes. *J Biol Chem* 258:11492–9.

6. Pannentier C, et al. (1993): Quantitative titration of nucleic acids by enzymatic amplification reactions run to saturation. *Nucleic Acids Res* 21:577–83.

7. Wiesner RJ (1993): Quantitative PCR. *Nature* 366:416.

8. Sarkar G, et al. (1994): The looped oligo method for generating reference molecules for quantitative PCR. *Biotechniques* 17,5:864–6.

9. Chomczynski P and Sacchi N (1987): Single-step method of RNA isolation by acid guanidium thiocyanate-phenol-chloroform extraction. *Anal Biochem* 162:156–9.

10. Lazar V, et al. (1995): Expression of human chorionic gonadotropin $\beta$ subunit genes in superficial and invasive bladder carcinomas. *Cancer Res* 55:3735–8.

11. Bellet D, et al. (1997): Malignant transformation of nontrophoblastic cells is associated with the expression of chorionic gonadotropin $\beta$ genes normally transcribed in trophoblast cells. *Cancer Res* 57:516–23.

# Quantitative Detection of Mycoplasma DNA Using Competitive PCR

Maninder K. Sidhu, Mei-June Liao, and Abbas Rashidbaigi

## Introduction

Mycoplasma are free-living pleomorphic prokaryotes that can pass through regular bacteriological filters. These organisms have a small genome with low G + C content, and lack the rigid peptidoglycan cell wall of eubacteria. There are more than 80 species known to date, the major cell culture contaminants of human, bovine, or porcine origin being: *Mycoplasma orale*, *M. fermentans*, *M. hyorhinus*, *M. arginini*, *M. salivarium*, and *Acholeplasma laidlawii* (McGarrity et al., 1985). More than 15–30% of all the cell lines used are contaminated with mycoplasma species (Mowles et al., 1988; Gignac et al., 1991). Mycoplasma grow in the cell culture supernatant and are attached to the extracellular surface of the cell membranes. The growth of these microorganisms may not necessarily destroy the cells but produces significant changes in the culture medium composition, which leads to alterations in the host cell metabolism, thus rendering infected cells unacceptable for use in any research or diagnostic procedure. Cell culture infections by various mycoplasma species, as well as their effects, are well documented. Some of the effects of mycoplasma on cell cultures include induction or suppression of enzymes and cytokines, alteration of cell surface antigenicity, interference with viral expression, and induction of chromosomal breaks. A few mycoplasma species have also been reported to produce transformation of cells in culture (Macpherson and Russel, 1966).

A number of diseases in humans, plants, and animals are known to be caused by mycoplasma. *Mycoplasma pneumoniae* is the common cause of pneumonia in children and adults (Denny et al., 1971; Foy et al., 1979). *Mycoplasma hominis* and *Ureaplasma urealyticum* have been implicated in acute infections of the urogenital tract and nongonococcal urethritis (Thomsen, 1978; Taylor Robinson et al., 1977). Recently, *M. fermentans*, a known contaminant of cell cultures, and *M. penetrans* were found in AIDS patients (Wright, 1990). Therefore, it is important to develop and improve the diagnostic procedures for mycoplasma infections to better understand their roles in the progression of the disease process and at the same time prevent cell culture problems caused by these organisms.

Many direct and indirect procedures currently exist for the detection of mycoplasma infections (McGarrity et al., 1985; Hay et al., 1989; Uphoff et al., 1992). Direct biological culture method is considered the most sensitive method and is widely accepted. In practice, this assay takes several weeks to perform. Also, due to difficulty in culturing certain strains of mycoplasma, the use of direct culture as a single-detection method is limited. Use of both direct and indirect methods is now recommended by the FDA for the evaluation of finished products and for validation studies. Indirect tests measuring specific markers include: DNA fluorochrome stains, ELISA, DNA probes, polymerase chain reaction (PCR), electron microscopy, and biochemical assays. The PCR technique is simple and sensitive, and has been used and compared to other indirect tests (Blanchard et al., 1993; Hopert et al., 1993; Van Kuppeveld et al., 1994) for the detection of mycoplasma infections in tissue cultures and clinical samples. In a study performed by Blanchard et al. (1993), it was shown that *M. genitalium* was only detected by PCR and not by culture in 11% of patients with urethritis.

Our aim in this study was to develop a quantitative PCR assay for the detection of mycoplasma DNA. We used PCR with a single primer set targeted to the conserved region of 16s rRNA gene for the detection of several species of mycoplasma. In addition, internal controls (mimic DNA) were constructed to serve as positive controls and competitor DNA for the quantification of mycoplasma. The internal control DNA contained an 80 bp addition as compared to the wild-type 16s rRNA gene. Upon amplification using the same primer set, the competitor DNA generated a product, which was distinguished by size from the wild-type. This approach has been successfully used previously for the quantitative detection of various viral RNA and DNA sequences (Piatak et al., 1993; Gilliland et al., 1990). We also developed an internal control that can be used for selective quantification of *M. pneumoniae* DNA.

## Results and Discussion

### *Mycoplasma detection*

A fast and simple method to detect and quantitate mycoplasma infections by PCR was developed. The following mycoplasma species that either are cell culture contaminants or cause infections in humans were investigated: *M. orale*, *M. hyorhinus*, *M. synoviae*, *M. gallisepticum*, and *M. pneumoniae*. Primers for amplification were selected from the conserved region of the 16s rRNA gene of mycoplasma after a comparison among nucleotide sequences of 30 different species. The sequence of the sense primer was 5′ATG RGG RTG CGG CGT ATT AG 3′, and the anti-sense primer was 5′CKG CTG GCA CAT AGT TAG CCRT 3′. The symbol K stands for mixed nucleotides of G and T, and R stands for A and G. Specific detection of *M. pneumoniae* DNA was performed with primers P1 sense (5′GGG GTG CTG GAG GGG GTT CTT 3′) and P2 antisense (5′CCA TTT TAG GGT TCA CAC TTG 3′), which were selected from the attachment protein gene (P1) unique to *M. pneumoniae* (Inamine et al., 1988).

Amplification of DNA was performed with procedures described by Perkin Elmer except for the addition of DMSO (2.5%) in the reactions. Cell lysates from the five mycoplasma species were tested using the primer set for the rRNA gene. Figure 1 shows the result where all the five species could be amplified (301 bp product). The primer set for *M. pneumoniae* could detect 20–50 cfu of *M. pneumoniae* but gave negative results with all the other species tested, thus demonstrating its specificity (data not shown). Bacterial and human DNA did not show any cross-reaction with these primers. Specificity of the amplified products was also confirmed by Southern blot hybridization with an internal probe and restriction enzyme analysis. On the basis of the information of these experiments with purified DNA, the conserved mycoplasma primers were examined on extracts from different cell lines including HeLa-1, HeLa-229, HEp-2a, HEp-2b, Daudi, Namalwa, Human osteosarcoma (HOS), RK13 (rabbit kidney), MDBK (bovine kidney), and U937. Preparation of cell extracts and PCR procedure were as described in Sidhu et al. Out of the 11 cell lines tested, four cultures, namely HeLa-1, RK13, MDBK2, and U937, displayed a distinct 301 bp band on agarose gels, while the others showed no DNA amplification (Figure 2, lanes 2–6, 9, 15, and 16). A faint band visible in lane 14 was nonspecific amplification as determined by restriction enzyme analysis. Electron micrographs of one of the cell lines, RK13 (rabbit kidney), which tested positive by PCR, also showed the presence of mycoplasma (Figure 3) thereby confirming the PCR results. This cell line was also infected with vesicular stomatitis virus for size comparisons between the two. Different dilutions of an infected culture (HeLa-1 cells contaminated with mycoplasma) ranging from 1 to $10^6$ cells were centrifuged to collect the cells, and PCR was performed on these cell extracts to identify the number of cells required for the assay. As few as 10 pelleted cells can be used for mycoplasma detection (Figure 2, Lanes 1–6). Cell concentrations of $>10^5$ from a positive culture did not show a specific amplified product, possibly because

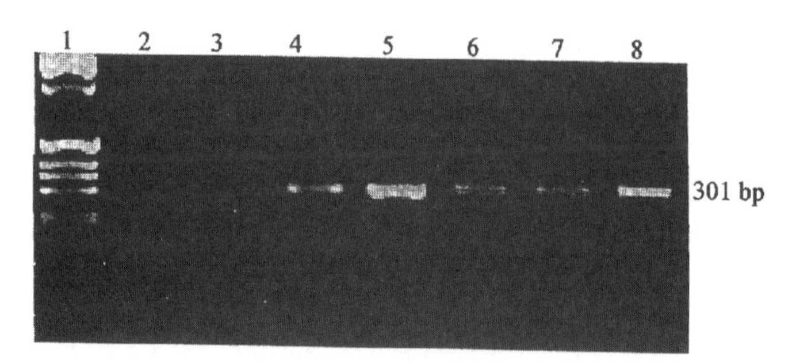


FIGURE 1. Electrophoretic analysis of PCR amplified DNA from mycoplasma species using 16s ribosomal gene-specific primers. Amplified samples were visualized on ethidium-bromide-stained 2% agarose gels. Lane 1: 1 kb ladder, Lane 2: negative control, Lane 3: leukocyte DNA (1pg), Lane 4: *M. orale,* Lane 5: *M. hyorhinus*, Lane 6: *M. synoviae,* Lane 7: *M. pneumoniae,* Lane 8: *M. gallisepticum.* Lanes 4–8 contain 200 cfu of mycoplasma from each species.

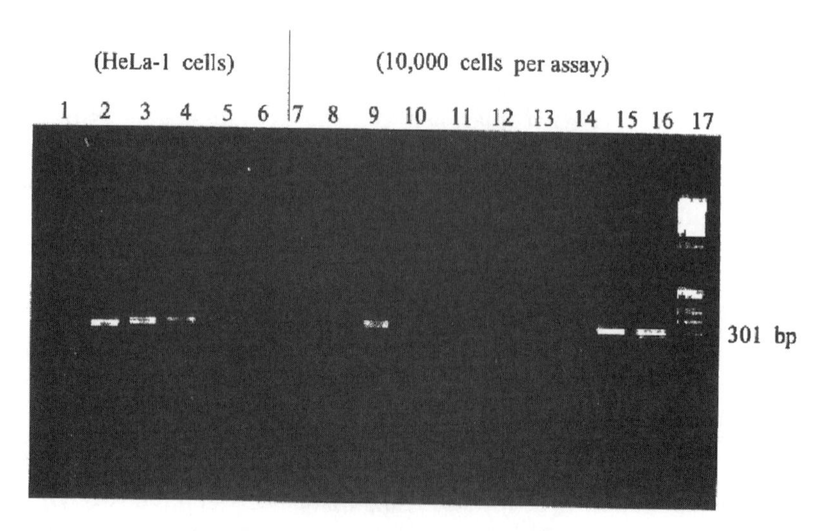


FIGURE 2. Detection of mycoplasma contamination in various cell lines. Lane 1: $10^6$, Lane 2: $10^5$, Lane 3: $10^4$, Lane 4: $10^2$, Lane 5: 10, Lane 6: 1 cells from a contaminated HeLa-1 cell culture. Lanes 7–16: Cells from Daudi, HEp-2a, RK13, Namalwa, MDBK1, HeLa-229, HEp-2b, Human osteosarcoma (HOS), MDBK2, and U937 cultures, respectively. Lane 17: 1 kb ladder. PCR amplified DNA was analyzed on 2% agarose gels.

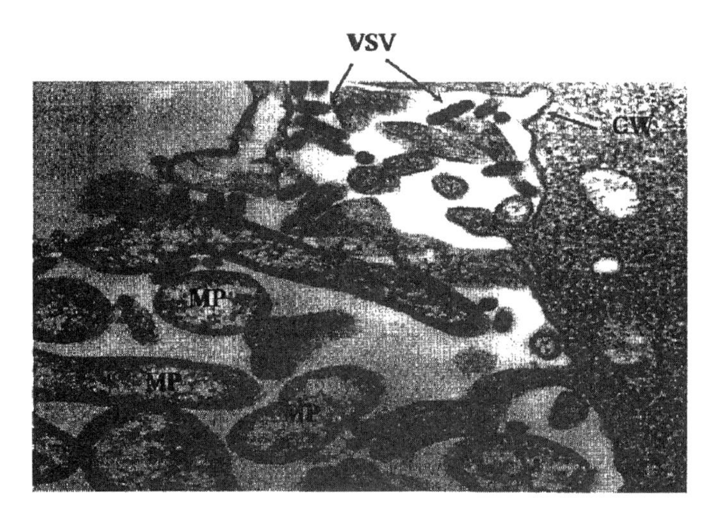


FIGURE 3. Electron micrograph of RK13 (rabbit kidney) cells infected with mycoplasma and vesicular stomatitis virus (VSV). CW: cell wall, MP: Mycoplasma.

of the presence of excessive eukaryotic DNA and the degenerate nature of the primers used. Under these conditions $10^4$ cells per assay are optimal for the detection of mycoplasma using this single-tube method.

## *Decontamination of cell cultures*

Elimination of mycoplasma contamination from cell cultures is possible. Various methods including the use of antibiotics, complement sera, exposure to macrophages, culture in soft agar, and several other techniques have been employed in the past (Lincoln and Lundin, 1990; Kotani et al., 1991).

Antibiotic treatment is considered an effective and simple method with a cure rate of about 75%. We used ciprofloxacin to cure one of our irreplaceable cell lines contaminated with mycoplasma. Ciprofloxacin was used for 14 days at concentrations of 10 and 50 $\mu$g/ml. Following treatment, cells were cultured in antibiotic-free medium for 10 days and tested for the presence of mycoplasma DNA by PCR. A concentration of 50 $\mu$g/ml was found to be effective in removing the mycoplasma contamination (Figure 4).

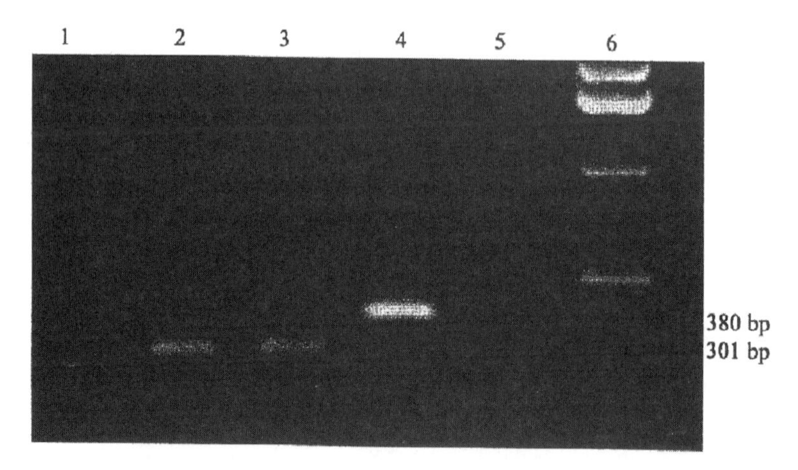


FIGURE 4. Treatment of mycoplasma-infected cell culture with ciprofloxacin. Lane 1: 50 $\mu$g/ml, Lane 2: 10 $\mu$g/ml, Lane 3: untreated, Lane 4: internal standard, Lane 5: negative control, Lane 6: 1 kb ladder. PCR was performed on cell extracts, and the amplified products were analyzed on 2% agarose gels.

### *Mycoplasma quantitation*

Detection of DNA and mRNA by quantitative PCR is simple, rapid, and capable of discriminating as little as twofold differences in either DNA or RNA copy numbers (Pannetier et al., 1993). Several methods described for quantitative PCR require either an unrelated endogenous DNA (e.g., a housekeeping gene or a gene that is structurally related to the target), or some form of an internal standard, to compare the efficiency of amplification in different reactions. When an unrelated endogenous control is used, amplification may vary significantly from one primer to another such that the two templates cannot be amplified in the same tube (Wang et al., 1989). In another method, internal standards are constructed to compete for the same primers to alleviate these problems (Becker-Andre and Hahlbrock, 1989). The PCR products are then distinguished by a second step of either hybridization or restriction enzyme analysis.

We have developed a competition-based quantitative PCR in which homologous internal standards can be distinguished from the wild-type by size differences on an agarose gel, without requiring additional manipulations after PCR. The internal standards are constructed by modifying the sizes of the wild-type PCR products and cloning them into plasmid vectors.

Restriction enzymes are used to introduce an 80 bp insertion in the center of the 301 bp amplified product of the 16s rRNA gene. This 381 bp fragment is then cloned to generate an internal control DNA template that can be used in competitive PCR to quantitate mycoplasma DNA.

We also constructed an internal control template for the quantitative detection of *M. pneumoniae* DNA with PCR primers targeted to the P1 gene. Sequences of this gene have been used previously for the specific detection and discrimination of *M. pneumoniae* from other mycoplasma species (Cadieux et al., 1993). A deletion of 96 bp was introduced in the wild-type amplified product of the P1 gene (424 bp) using suitable restriction enzymes. This fragment (328 bp) was cloned and used as a positive control for *M. pneumoniae* quantitation.

Using PCR primers specific for the 16s rRNA gene and the P1 gene of *M. pneumoniae*, the quantitative titration procedure was tested by adding known concentrations of the internal control DNA to the PCR amplification reactions containing a constant amount of mycoplasma DNA. Following PCR, the amount of products generated by the target and mimic sequences were analyzed by agarose gel electrophoresis. A comparison of the PCR products was then used to determine the quantity of the target gene product. The reciprocal of the dilution of the mimic DNA at which the target and mimic PCR products appear equal determines the amount of target DNA added to the reaction. Upon amplification, both the 16s rRNA and the P1 gene competitor templates were easily distinguished from their wild-type homologues. The point at which target and mimic PCR products were equal was reached between 700 and 7000 copies of ribosomal mimic DNA (Figure 5B). This is in agreement with the amount of target added to the PCR reaction (800 cells equivalent or 1600 copies, since each genome has two copies of the rRNA gene). In the case of *M. pneumoniae* DNA, equivalence was reached between 200 and 1000 copies of mimic DNA, when 500 cells were used in the assay (Figure 5A). These results are based on visual observation of the ethidium-bromide-stained PCR products. However, for an exact quantification, fluorescence data can be obtained from computer-based video imaging of gels, by the use of radioactive labeling during PCR, or by separating relative amplified products on RP-HPLC. The quantitative PCR assay we describe here has a detection limit of 5–25 genome copies per assay for the 16s rRNA gene and 40–60 copies per assay for the P1 gene.

### *Mycoplasma clearance/inactivation studies*

We used this quantitative PCR system in conjunction with the biological assay to evaluate samples from a mycoplasma clearance/inactivation study.

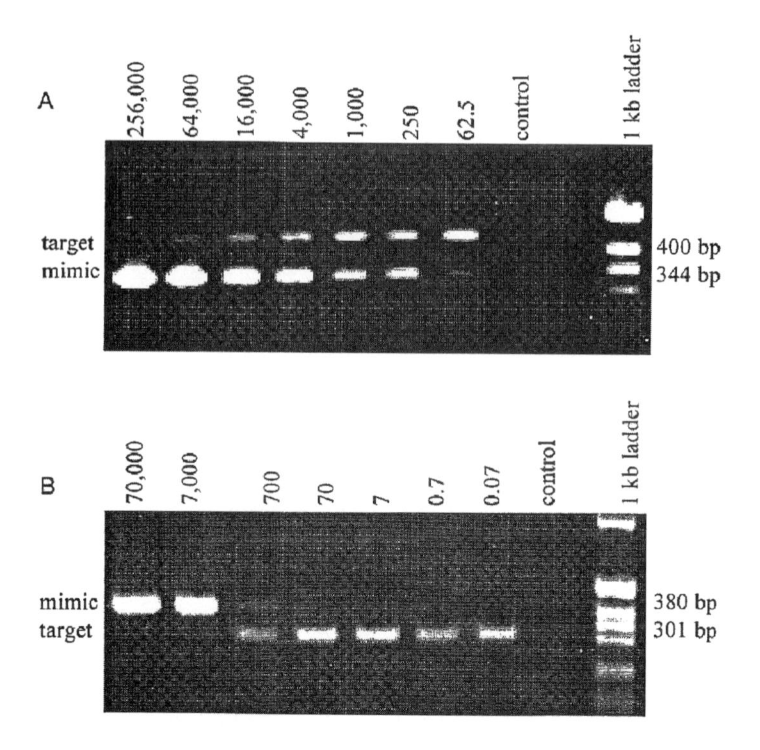


FIGURE 5. (A) Competitive PCR on *M. pneumoniae* DNA. Serial dilutions (fourfold) of P1 gene mimic were added to the reactions containing a constant amount of mycoplasma DNA equivalent to 500 cells. (B) *M. orale* DNA equivalent to 800 cells was added to the PCR reaction containing 16s rRNA gene mimic serially diluted tenfold.

Immunoaffinity purified interferon preparation derived from human peripheral blood leukocytes was spiked with mycoplasma before and after a 5-day treatment at pH2. This treatment is one of the steps in the purification of the interferon for inactivation of adventitious agents. Following pH2 treatment, samples were analyzed by biological assay and quantitative PCR. Part of the results from this study, which have both PCR and biological data available, are presented in Table 1. There was a $>7\log_{10}$ reduction in the growth of *Mycoplasma orale* as observed with the biological assay. However, no change in the DNA copy numbers was noticed, since DNA amplification allows the detection of both active and inactive mycoplasma. Filtration of interferon samples through $0.22\,\mu m$ filters reduced

TABLE 1. Inactivation and clearance of *Mycoplasma orale* spiked into interferon preparation. DNA was prepared from spiked interferon samples, and competetive PCR was performed with 16s rRNA primers. Bioassay was performed by an outside company using standard culture methods. PDC: purified drug concentrate, IFN: interferon.

| Sample | Total CFU (Bioassay) | $Log_{10}$ reduction (Bioassay) | Total gene copies (PCR) | $Log_{10}$ reduction (PCR) |
|---|---|---|---|---|
| Spiked IFN | $1.8 \times 10^7$ | NA | $3.7 \times 10^6$ | NA |
| pH2 treated IFN | $<4.5 \times 10^2$ | >4.61 | $3.7 \times 10^6$ | 0 |
| pH2 treated neutralized IFN | 0 | 7.26 | $3.9 \times 10^6$ | 0 |
| Spiked PDC | $1.7 \times 10^7$ | NA | $1.16 \times 10^8$ | NA |
| PDC filtrate | $5.8 \times 10^5$ | 1.47 | $3.06 \times 10^7$ | 0.58 |

97% (1.47 log) of mycoplasma by direct culture method and 73% (0.58 $log_{10}$) by PCR (Table 1). The discrepancy could be due to the presence of cell-free DNA, which can pass through the 0.22 $\mu$m filter and be detected by PCR. Our results indicate that PCR detection efficiently measures both active and inactive DNA but at the same time does not differentiate between the two. This suggests that for validation of the product during inactivation of spiked infectious agents it is necessary to use biological assay to determine active contamination. However, in a clearance study, PCR is useful for rapid quantification of spiked contaminants.

In summary, the quantitative PCR method is a valuable tool for the detection of mycoplasma contamination of cell cultures and products derived from *in vitro* systems. It also helps in monitoring the progression of mycoplasma in the disease process as well as in basic research involving mycoplasma.

## *Acknowledgments*

We thank Nancy Schocklin for her helpful comments on the manuscript. Some of the figures have been modified from Sidhu et al. (1995): Competitor internal standards for quantitative detection of mycoplasma DNA. *FEMS Microbiol Lett.* 128:207–212. With kind permission from Elsevier Science-NL, Sara Burgerhartstraat 25 1055KV Amsterdam, The Netherlands.

# REFERENCES

Becker-Andre M and Hahlbrock K (1989): Absolute mRNA quantification using the polymerase chain reaction (PCR): A novel approach by a PCR aided transcript titration assay (PATTY). *Nucleic Acids Res* 17:9437–46.

Blanchard A, Hamrick W, Duffy L, Baldus K, and Cassell GH (1991): Use of polymerase chain reaction for detection of *Mycoplasma fermentans* and *Mycoplasma genitalium* in the urinogenital tract and amniotic fluid. *Clin Infect Dis* 17:S272–9.

Cadieux N, Lebel P, and Brousseau R (1993): Use of triplex polymerase chain reaction for the detection and differentiation of *Mycoplasma pneumoniae*. *J Gen Microbiol* 139:2431–7.

Denny FW, Clyde WA, Jr., and Glezen WP (1971): *Mycoplasma pneumoniae* disease: clinical spectrum, pathophysiology, epidemiology, and control. *J Infect Dis* 123:74–94.

Foy HM, Kenny GE, Cooney MK, and Allan ID (1979): Long term epidemiology of infections with *Mycoplasma pneumoniae*. *J Infect Dis* 139:681–7.

Gignac SM, Brauer S, Hane B, Quentmeier H, and Drexler HG (1991): Elimination of mycoplasma from infected leukemia cell lines. *Leukemia* 5:162–5.

Gilliland G, Perrin S, Blanchard K, and Bunn HF (1990): Analysis of cytokine mRNA and DNA: Detection and quantitation by competitive polymerase chain reaction. *Proc Natl Acad Sci* 87:2725–8.

Hay RJ, Macy ML, and Chen TR (1989): Mycoplasma infection of cultured cells. *Nature* 339:487–8.

Hopert A, Uphoff CC, Wirth M, Hauser H, and Drexler HG (1993): Specificity and sensitivity of polymerase chain reaction (PCR) in comparison with other methods for the detection of mycoplasma contamination in cell lines. *J Immunol Meth* 164:91–100.

Inamine JM, Denny TP, Loechel S, Schaper U, Huang C-H, Bott KF, and Hu P-C (1988): Nucleotide sequence of P1 attachment protein gene of *Mycoplasma pneumoniae*. *Gene* 64:217–19.

Kotani H, Butler G, Heggan D, and McGarrity GJ (1991): Elimination of mycoplasmas from cell cultures by a novel soft agar technique. *In Vitro Cell Dev Biol* 27A:509–13.

Lincoln CK and Lundin DJ (1990): Mycoplasma detection and control. *US FCC Newsletter* 20:1–3.

Macpherson I and Russel W (1966): Transformation in hamster cells mediated by mycoplasmas. *Nature* 210:1343–5.

McGarrity GJ, Sarama J, and Vanaman V (1985): Cell culture techniques. *Am Soc Microbiol* 51:170–83.

Mowles JM (1988): The use of ciprofloxacin for the elimination of mycoplasma from naturally infected cell lines. *Cytotechnology* 1:355–8.

Pannetier C, Delassus S, Dorche S, Sancier C, and Kourilspy P (1993): Quantitative titration of nucleic acids by enzymatic amplification reactions run to saturation. *Nucleic Acids Res* 21:577–83.

Piatak G, Luk K-C, Williams B, and Lifson JD (1993): Quantitative competetive polymerase chain reaction for accurate quantitation of HIV-1 DNA and RNA species. *Biotechniques* 14:70–81.

Sidhu MK, Rashidbaigi A, Testa D, and Liao M-J (1995): Competitor internal standards for quantitative detection of mycoplasma DNA. *FEMS Microbiol Lett* 128:207–12.

Taylor-Robinson D, Csonka GW, and Prentice MJ (1977): Human intra urethral inoculation of ureaplasma. *Q J Med* 46:309–26.

Thomsen AC (1978): Occurence of mycoplasmas in urinary tracts of patients with acute pyelonephritis. *J Clin Microbiol* 8:84–8.

Uphoff CC, Brauer S, Grunicke D, Gignac SM, Macleod RAF, Quentmeier H, Steube K, Tummler M, Voges M, Wagner B, and Drexler HG (1992):

Sensitivity and specificity of five different mycoplasma detection assays. *Leukemia* 6:335–41.

Van Kuppeveld FJ, Johansson KE, Galama JM, Kissing J, Bolske G, Van der logt JT, and Melchers WJ (1994): Detection of mycoplasma contamination in cell cultures by a mycoplasma group specific PCR. *Appl Environ Microbiol* 60:149–52.

Wang AM, Doyle MV, and Mark DF (1989): Quantitation of mRNA by the polymerase chain reaction. *Proc Natl Acad Sci* 86:9717–21.

Wright K (1990): Mycoplasmas in the AIDS spotlight. *Science* 248:682–3.

Zhao S and Yamamoto R (1993): Detection of *Mycoplasma meleagridis* by polymerase chain reaction. *Vet Microbiol* 36:91–7.

# The Detection and Quantification of *bcr-abl* in Chronic Myeloid Leukemia Following Marrow Transplantation

Jerald Radich

## Introduction

The definition of remission is the central issue in the therapy of cancer. Remission is defined in hematologic malignancies by gross pathological examination of the bone marrow following therapy. The absence of obvious leukemic cells in a fully regenerated marrow defines remission. Unfortunately, leukemic blasts appear similar to normal early hematopoetic progenitors, which may be seen at a frequency of up to 5% of cells in normal marrow. Thus, the sensitivity of the detection of residual leukemia by morphology is roughly 5%. Relapse rates in patients who have achieved remission demonstrate that this conventional definition of remission based on morphologic evaluation is too insensitive to detect clinically significant residual disease. For example in adult acute myeloid and lymphoblastic leukemia (AML and ALL, respectively), approximately 70% of patients will achieve remission with induction chemotherapy, yet most will eventually relapse. In some leukemias, the malignant clone can be specifically identified by a clonal chromosomal abnormality, and this marker can be searched for by cytogenetic methods in remission marrows. However, since most cytogenetic evaluations typically evaluate only 20 metaphases, the lower limit of detecting leukemia is one cytogenetic abnormality in 20 metaphases, or 5%.

At each stage of leukemia therapy, important decisions relating to the efficacy of treatment, future therapeutic options, and long-term prognosis hinge upon the remission status. Thus, the relative insensitivity of routine procedures for the evaluation of remission hinder the ability to accurately judge the effectiveness of therapy, and cripple our attempts to tailor the intensity of therapy to the burden, or residual disease. Could more sensitive techniques for detecting minimal residual disease (MRD) discriminate between those who are truly leukemia-free, and need no further therapy, and those patients who have MRD and need more aggressive therapy to be optimally treated? The goal of the study of

MRD can be distilled to the redefinition of what remission truly is, and to providing a fulcrum for optimal clinical decisions.

*The detection of MRD*

The advent of the polymerase chain reaction (PCR) has allowed for the amplification of molecular sequences (either DNA or RNA) that are specific to the patient's leukemic clone. These molecular "fingerprints" can arise from the juxtaposition of two genes resulting from a translocation, such as the BCR gene on chromosome 22 being translocated to the 5′ region of the ABL gene on chromosome 9 in the Philadelphia chromosome (Heistercamp et al., 1983; Groffen et al., 1984; Bernards et al., 1987), or from rearrangements that occur in cell lineage differentiation, such as gene rearrangement in the immunoglobulin heavy genes and the T cell receptor genes (Hanssen-Hagge et al., 1989; Yamada et al., 1990). Thus PCR can amplify these leukemia-specific nucleic acid sequences without competition from normal sequences, allowing the detection of one leukemic cell in a background of $10^3$–$10^6$ normal cells.

Opinion as to the clinical utility of MRD is still in evolution. It has become clear that the detection of MRD by the PCR amplification of the molecular "fingerprint" of a leukemia does not always foretell relapse; rather, careful consideration must take place within the context of the type of disease, the molecular lesion itself, and the type of therapy employed. Thus, while the detection of MRD in ALL or in the *t*(15;17) subtype of AML has a high correlation with relapse (Miller et al., 1993), detection of MRD in *t*(8;21) AML has little correlation with relapse (Maruyama et al., 1994). In some diseases (e.g., CML), the clinical relevance of MRD detection may be strengthened by quantification, although this in itself adds another layer of complexity and potential pitfalls. Nevertheless, we are moving toward the day when the molecular definition of disease status will guide therapy.

## MRD Detection in Leukemia

The molecular targets for MRD detection obviously differ depending on the type of leukemia (i.e., ALL vs. AML). Moreover, even in a given type of leukemia (AML), different translocations may be used as markers of MRD. Surprisingly, however, the study of MRD has revealed that the detection of MRD may have a different clinical significance not only across different types of leukemias, but even within the same type of leukemia, given different chromosomal targets, e.g., *t*(8;21) vs. *t*(15;17) AML.

### *The significance of MRD in ALL*

In ALL the detection of MRD is based on the PCR amplification and nucleotide sequencing of clonogeneic rearrangements of the immunoglobulin heavy chain variable, diversity, and junctional genes (V-D-J) or of the T-cell receptor (TCR) genes. This allows for the construction of leukemia-specific oligonucleotide primers or probes that can then amplify and detect the leukemic V-D-J or TCR "fingerprint" in "remission" samples. The clinical efficacy of the detection of MRD after conventional chemotherapy has been suggested in large-scale pediatric studies. These studies suggest that more than 50% of patients with MRD present at "remission" from chemotherapy will relapse, compared to a less than 10% relapse rate for patients who are negative for MRD (Brisco et al., 1994). There are no such comprehensive studies of patients with adult ALL who have been treated with conventional chemotherapy. We have published a study of the use of V-D-J detection in children and adults following BMT, reporting our findings that the detection of MRD post-BMT was associated with a sixfold risk of relapse, with a Kaplen-Meier estimation of relapse of 100% in PCR-positive patients compared to 14% in patients who were persistently PCR-negative (Radich et al., 1995).

### *The curious case of MRD in AML*

In AML there are no universal chromosomal rearrangements that facilitate MRD detection. Rather, in AML the emphasis in MRD detection revolves around the molecular detection of specific translocations. Recent work has established PCR protocols for the amplification of the chimeric mRNAs found in the three most frequent cytogenetic syndromes in AML: the PML/RAR-alpha fusion mRNA found in *t*(15;17) AML (Miller et al., 1993), the AML1/ETO fusion mRNA product found in *t*(8;21) AML (Maruyama et al., 1994; Kusec et al., 1994), and for the CBFB/MYH11 chimeric mRNA found in inv. (16) typically found in AML-M4Eo (Claxton et al., 1994; Hebert et al., 1994). The cumulative frequency of cytogenetic lesions is approximately 30% in *de novo* AML cases. The literature suggests that the significance of MRD detection in AML following conventional chemotherapy depends on the actual molecular lesion. In *t*(15;17) AML, the detection of the PML/RAR-alpha fusion transcript clearly places the patient at high risk of relapse, because patients who have PCR evidence of PML/RAR while in hematological and cytogenetic remission have a greater than 50% risk of subsequent relapse, compared to a risk of less than 10% for patients who are persistently PCR negative (Miller et al., 1993; Fukutani et al., 1995). The data from *t*(8;21) AML is dramatically different: over 25 long-term remission patients (up to 8 years) have been tested for MRD by PCR, and *all* have evidence of the AML1/ETO fusion mRNA (Maruyama et al., 1994; Kusec et al., 1994). The molecular breakpoints that occur in inv. (16) AMLs have been described, but the significance of their detection after therapy is unknown.

These studies of MRD detection in AML patients following conventional chemotherapy may offer insight into the biology of these heterogeneous diseases, and run contrary to the expected results that MRD in acute leukemias inevitably portends relapse. The fact that some *t*(8;21) bearing cells persist long into remission suggests that this translocation may be a very early event in leukemogenesis, and its persistence at apparently low levels suggests that the lesion may reside in a slowly dividing compartment of the myeloid cell lineage. In contrast, flow cytometry studies with *t*(15;17) leukemia show that the translocation occurs later in myeloid development, because early myeloid cells (defined by CD34+ and CD33– antigens) do not show evidence of the *t*(15;17) by PCR or fluorescence *in situ* hybridization studies (Turhan et al., 1995). What does this tell us about the biology of these subtypes of AML, and what are the implications for MRD detection?

### *The detection and significance of MRD in CML*

CML is a hematopoetic stem cell malignancy with a distinct molecular and clinical profile. The molecular hallmark of CML is the Philadelphia chromosome (Ph), which occurs in over 90% of cases. Molecular analyses have established that the Ph translocation results in the joining of 3′ sequences of the tyrosine kinase c-ABL proto-oncogene on chromosome 9 to the 5′ sequences of the BCR gene on chromosome 22. In CML this specific translocation generates an aberrant 8.5 kb *bcr-abl* chimeric messenger RNA (mRNA) that is expressed as a 210-kilodalton (kd) fusion protein. The precise region of breakage on BCR and c-ABL varies, but in all cases 5′ sequences of BCR are joined to 3′ sequences of c-ABL. In CML, the majority of Ph breakpoints occur within a 5.8 kilobase (kb) region on chromosome 22, known as the major breakpoint cluster region (M-bcr) of the BCR gene (Figure 1).

BMT is the only curative modality for CML, but relapse occurs in 20% of patients transplanted in chronic phase (Thomas et al., 1986) and in over 50% of patients transplanted in accelerated or blast crisis (Clift et al., 1994). PCR can detect CML cells through the amplification of the unique *bcr-abl* fusion mRNA transcript (the "p210" *bcr-abl*) with a sensitivity equivalent to detecting a single Ph+ cell in a background of $10^5$–$10^6$ normal cells. Several studies have examined the association of *bcr-abl* PCR-positivity post-BMT in relatively small patient cohorts (the largest with $N = 64$), but results led to contradictory conclusions, with some studies suggesting that the

**BCR gene, Ch. 22**

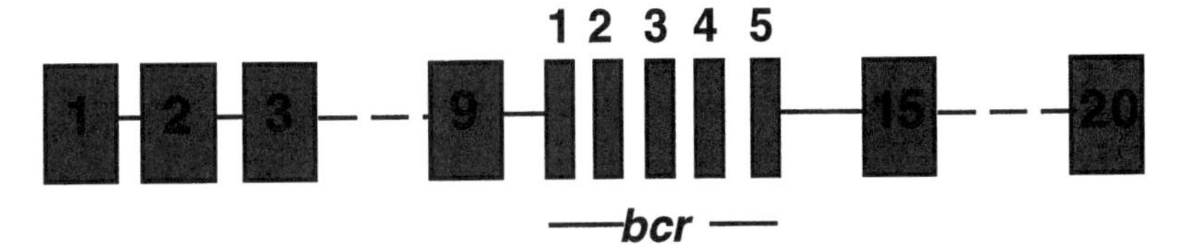


**ABL gene, Ch. 9**

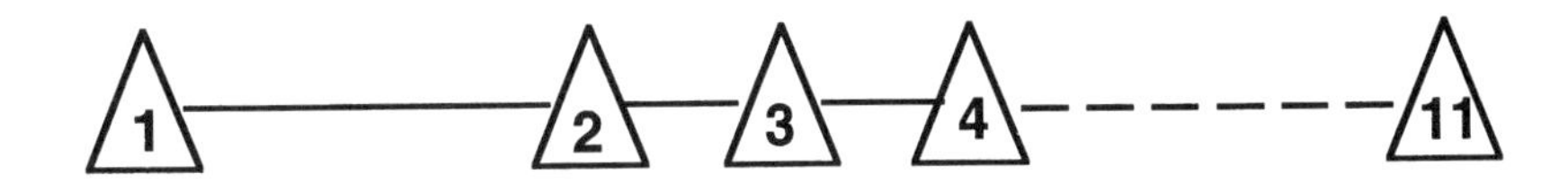


**Chimeric *bcr-abl* mRNA**

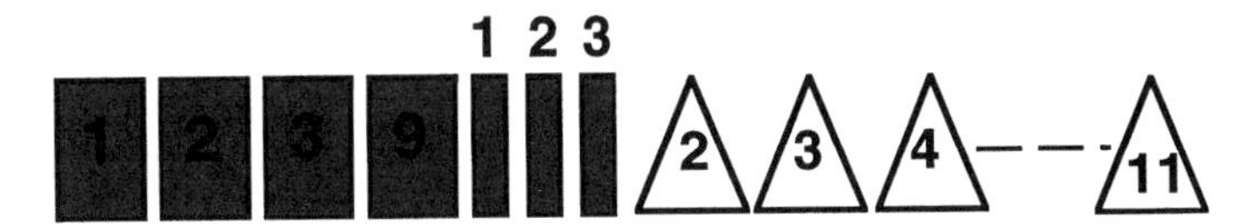


FIGURE 1. The chimeric *bcr-abl* transcript in CML. The recriprocal translocation in the Philadelphia chromosome causes the 5′ domains of BCR from chromosome 22 to become fused with the 3′ domains of ABL from chromosome 9. The chimeric *bcr-abl* mRNA is thus the leukemia-specific target for PCR detection of residual disease.

persistence of *bcr-abl* post-BMT heralded relapse, while others showed no relationship (Roth et al., 1992; Miramuya et al., 1993). Moreover, the relatively small size of these studies precluded control for other clinical variables that are known to influence relapse rates, such as phase of disease and the presence or absence of graft-versus-host disease (GVHD).

We have recently completed and published the largest study to date describing the results and implications of a positive PCR for *bcr-abl* in 346 CML patients post-BMT (Radich et al., 1995). A total of 636 samples of bone marrow and/or peripheral blood was obtained for PCR analysis between 3 and 192 months after BMT. A positive PCR test at 3 months post-BMT was not statistically significantly associated with an increased risk of relapse, compared to PCR-negative patients. However, a positive PCR assay at 6 months and beyond was highly associated with subsequent relapse. At 6–12 months, the PCR test is quite powerful, because 42% of PCR-positive patients relapse, compared to 3% of PCR-negative patients ($p < 0.0001$). After 12 months, the predictive power of a positive PCR assay decreases steadily (Figure 2). Multivariable analysis, including the PCR result and other variables potentially associated with relapse such as type of BMT donor (allogeneic matched donor versus mismatched or unrelated), phase of disease, and presence or absence of GVHD, indicated that only a PCR-positive result at 6 or 12 months post-BMT and the donor status were independent risk factors for subsequent relapse. Surprisingly, the outcome of long-term patients (>36 months post-BMT) who tested PCR-positive was much better. While 14 of 58 (24%) of such patients tested positive for *bcr-abl*, only one such patient relapsed. While this study has documented that *bcr-abl* persistence after BMT is an *independent* risk factor for relapse, the results logically lead to new questions to be explored, such as: 1) Can we improve the PCR assay to better discriminate which patients who are positive 3 months post-BMT will relapse, so that the window of therapeutic opportunity preceding relapse is wider? 2) Why don't all patients who are positive for *bcr-abl* relapse? Is this a question of burden of cells expressing *bcr-abl*, or, alternately, in some patients do rare host lymphoid cells persisting after transplant express *bcr-abl*? 3) Why do patients who become PCR-positive late after transplant fail to relapse? Is the kinetics of growth of the malignant clone slower than in patients who become *bcr-abl* positive earlier? These are just some of the questions about the clinical biology of CML that can be addressed by a quantitative *bcr-abl* assay.

## CML as a Model for Quantitative PCR (Q-PCR)

We have found that approximately 30% of patients testing positive at 3 months post-BMT will eventually relapse, compared to 23% of PCR-negative patients. This pattern of poor discrimination by the PCR assay early after BMT has been noted by other investigators, but the reasons underlying the phenomenon are not understood. In contrast, a positive PCR assay 6–12 months post-BMT strongly predicts relapse, as the Kaplan-Meier estimate of relapse was 42% in PCR-positive patients compared to 3% of PCR-negative patients. The fact that 60–70% of patients who test positive *fail* to relapse suggests that the qualitative PCR assay poorly reflects the biology of relapse. Why? Does the risk of relapse correlate with the burden of CML

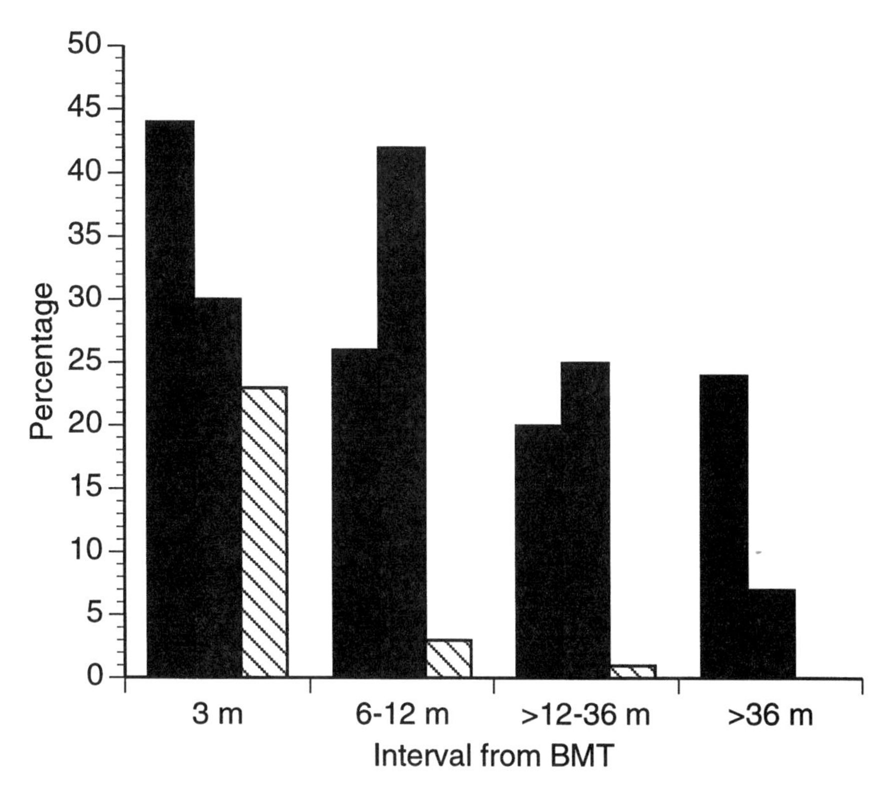


FIGURE 2. The percentage of patients testing PCR-positive and the Kaplan-Meier (K-M) estimate of relapse associated with a PCR-positive and PCR-negative test, defined by time from BMT. The x-axis represents the time interval from BMT, while the y-axis is percentage. The black bars are the percentage of patients testing PCR-positive at each time interval. The gray bars represent the K-M estimate of relapse for PCR-positive patients tested in the designated time interval from BMT. The hatched bars represent the estimate of relapse associated with a negative PCR-test in the time interval. The *p*-values compare the risk of relapse for patients PCR-positive (gray bars) to PCR-negative (hatched bars). Reprinted with permission from *Blood* 85:2632, 1995.

after transplant? If so, the *quantitation* of *bcr-abl* may better delineate patients destined to relapse, compared to patients who will eventually become PCR-negative and remain in remission. Alternately, since CML is known to be a stem cell leukemia, and thus the Ph chromosome can be found in B and T cell lineages (Martin et al., 1986; Skowron et al., 1989), perhaps some patients express *bcr-abl* in host lymphoid cells that have persisted post-BMT. These patients would test PCR+, but would not be expected to relapse. Thus, the development of a quantitative PCR (Q-PCR) for *bcr-abl* may allow us to clarify the significance of residual disease, as well as shed light on the biology of relapse.

### *A competitive Q-PCR for bcr-abl*

Several investigators have utilized a competitive PCR to quantitate mRNA, including *bcr-abl* transcripts in CML (Cross et al., 1993). The process usually calls for the construction of a competitor molecule that has a similar sequence and identical primer binding sites as the target RNA template. This competitor is generally inserted into a plasmid for convenience. The competitive PCR then requires multiple reactions using the same amount of target in different reactions that contain varying amounts of competitor. The PCR reaction is performed, and the reaction products run on a gel. Given similar reaction efficiencies for the target and the competitor, the PCR reaction in which the target product equals the competitor product indicates the starting amount of target.

The fundamental limitation with this procedure is that multiple reactions must be carried out for each target determination. For instance, a single sample may need titration over a 5 log range (hence, perhaps five or more separate PCR reactions), and if duplicate runs are needed, this can quickly add up to a prohibitive number of reactions for a single sample determination. In order to simplify the competitive PCR approach, we have turned the problem on its head, and have established a reaction using the Ph-chromosome-bearing cell line K562 as a standard against which to measure target amounts, using a single level of competitor in each reaction.

### *A competitive Q-PCR using a K562 standard curve*

We have developed a Q-PCR assay for *bcr-abl* that allows quantification over a broad range of *bcr-abl* levels without the need for multiple amplifications. The initial goal was to find a single level of competitor that will compete with the target *bcr-abl* over many logs of *bcr-abl* copy numbers. We first used PCR to construct the synthetic competitor (designated here as bcr-abl), which has been cloned into the pT7 Blue plasmid vector (Figure 3). This synthetic bcr-abl was designed to be 168 base pairs (bp) in size, with a novel *NotI* restriction enzyme site that allows for digestion of any potential "carryover" contamination from the bcr-abl PCR amplimer before a new Q-PCR reaction proceeds. The kinetics of PCR amplification of the 305 or 230 bp *bcr-abl* (the amplimer size depending on the presence or absence of BCR exon 3 in the chimeric mRNA) and the 168 bp bcr-abl

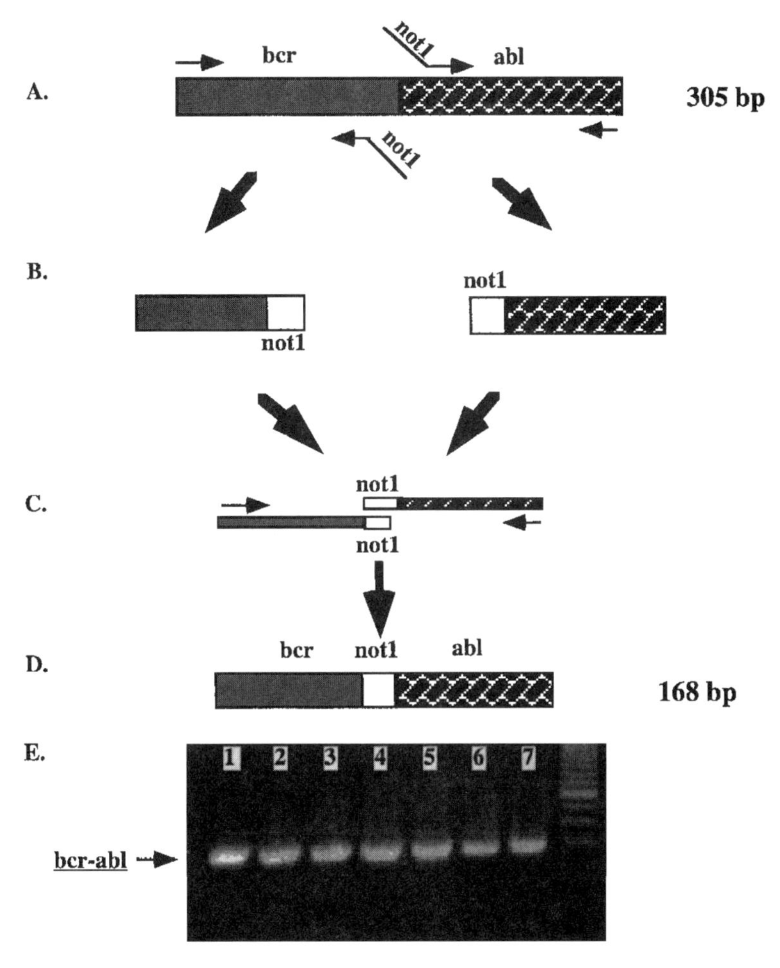


FIGURE 3. PCR synthesis of the bcr-abl competitor derived from the 305 bp *bcr-abl* PCR. (3A) PCR primers were constructed to the 3′ region of *bcr* and the 5′ region of *abl* to include a complementary sequence including the *Not1* restriction enzyme site. Two PCR reactions were performed (3B), yielding *bcr* and *abl* fragments, both of which included the new homologous *Not1* sequence. The PCR products were combined, boiled, and allowed to reanneal (3C). In the presence of *bcr* and *abl* primers, PCR amplification occurred and produced double-stranded bcr-abl competitior (3D), which was then cloned into the pT7 Blue plasmid vector. 3E shows multiple PCR amplifications of the bcr-abl competitor from the engineered plasmid.

competitor was log-linear for both templates over 25 cycles of amplification. Next, we performed PCR amplification of dilutions of RNA from the K562 cell line (which harbors the Ph) into normal bone marrow, with varying amounts of bcr-abl added to each dilution series. We found that 100 copies of bcr-abl would allow for linear amplification of the target *bcr-abl* over 4 logs of *bcr-abl* target copy number. The PCR products were run on an agarose gel and stained with ethidium bromide, and the quantities of the signals estimated by Imagequant™ software (Figure 4). We have found that this assay allows for a log-linear relationship from approximately $1–10^4$ K562 cells. Thus, we have developed a Q-PCR that appears to be rapid and reproducible, and should allow for log-linear quantification over a 3–4 log range. Moreover, the assay is fairly inexpensive, and utilizes technology available in most labs. Armed with this Q-PCR assay, we are now addressing the questions posed previously, in order to understand the clinical significance of *bcr-abl* positivity and the biology of relapse in CML.

While the system described above for quantitating both the target *bcr-abl* and the bcr-abl competitor is relatively quick and reproducible, it still requires a fair amount of "hands on" work. We therefore explored the use of high performance liquid chromatography (HPLC) as a way to separate and quantitate target *bcr-abl* and bcr-abl competitor. The obvious advantages of HPLC are its accuracy and high throughput, since scores of PCR products can be quantitated with a minimum of handling. The 168 bp bcr-abl competitor was designed so that its size could be easily separable on HPLC from the target *bcr-abl* size of 305/230 bp (depending on the presence of *bcr* exon 3). We have found that HPLC can quantitate over 4 logs of an input *bcr-abl* to a level of less than 10 picograms of *bcr-abl* template (approximately <1 K562 cell). This represents 1 log greater sensitivity than ethidium bromide dosimetry. In addition, the reproducibility of the HPLC quantification is excellent. The standard deviations of repeat HPLC measurements of K562 and plasmid PCR products ranged from approximately 3–15% of the mean value. Figure 5 shows the results of multiple HPLC quantitations of identical K562 dilutions. This data demonstrates that HPLC may be adaptable to rapid and reproducible Q-PCR.

An example of Q-PCR using HPLC is shown in Figure 6, which shows the increase of *bcr-abl* over time post-BMT. This patient was *bcr-abl* positive 8 and 12 months post-BMT, but was cytogenetically normal. At 24 months his *bcr-abl* load increased 2 logs, but his cytogenetic exam still showed 0/20 Philadelphia chromosomes. After another log increase in *bcr-abl* load, the Philadelphia chromosome was detected in 2/20 metaphases, signaling a cytogenetic relapse. In the future the evaluation of such *bcr-abl* "kinetics" may be able to target patients for early intervention before relapse.

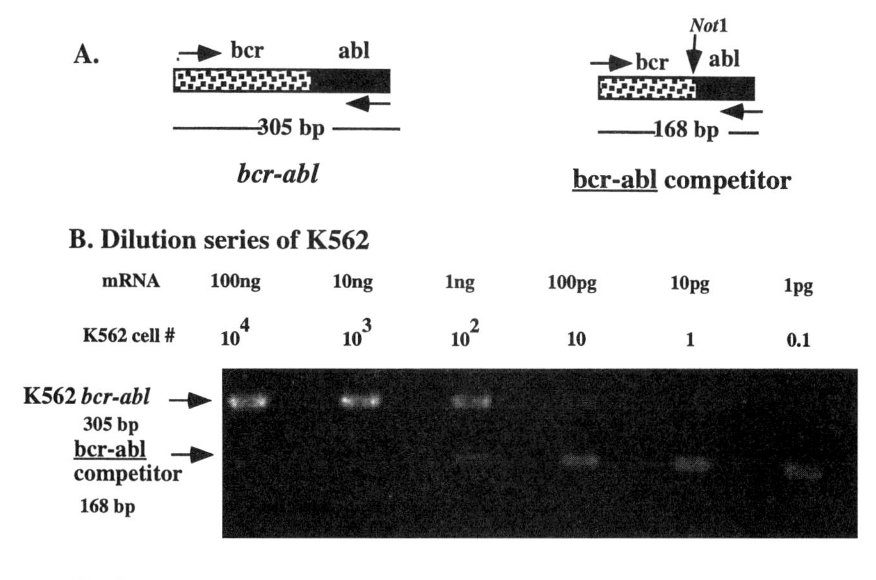


C. Graph of Q-PCR densitometry

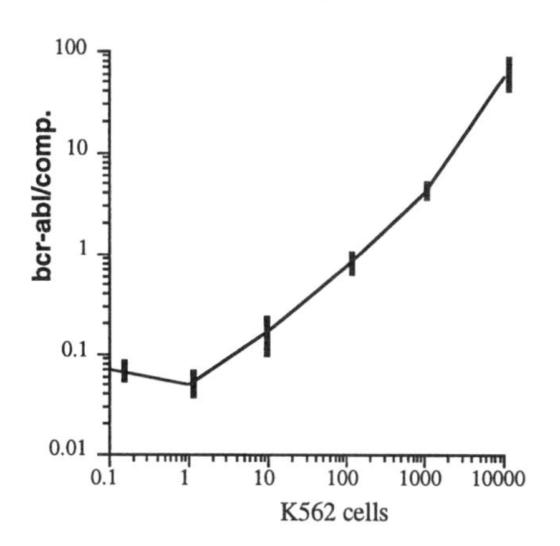


FIGURE 4. The Q-PCR for *bcr-abl*. 4A is a cartoon of the amplification of the 305 bp K562 *bcr-abl* mRNA (left), and the 168 bp bcr-abl competitor (right). 4B shows the results of a PCR of the dilution of K562 RNA into 1 $\mu$g of negative control HL60 RNA. Above each lane is the amount of input K562 RNA and the corresponding number of K562 cells. In each reaction 100 copies of the bcr-abl competitor was co-amplified. Fifteen $\mu$l of the 100 $\mu$l PCR reaction was run on a 2% agarose gel, stained with ethidium bromide, and photographed. The image was scanned and the band intensities quantitated. 4C shows the mean results of five duplicate Q-PCR reaction (with 1 standard deviation) bars.

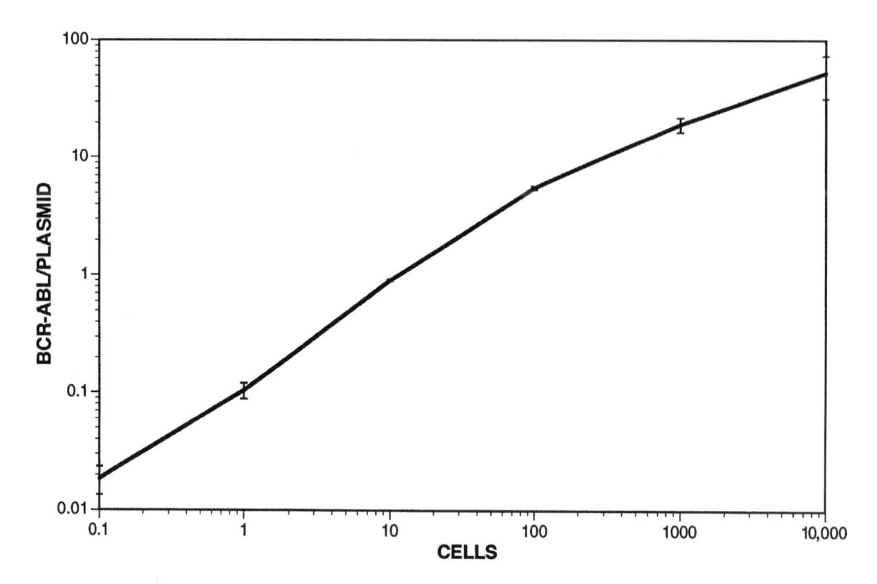


FIGURE 5. HPLC quantification of a competitive Q-PCR. HPLC separation and quantification was performed using a Perkin-Elmer DEAE-NPR nonporous ion exchange column. Buffer A is 25 mM Tris-HCl (pH 9.0), and Buffer B is 25 mM Tris-HCl, 1 M NaCl (pH 9.0). The separation gradient runs at a rate of 1 mL/min, and proceeds with 20–40% B over 0.5 min; 45–65% B over 4.5 min; 65–100% B over 0.5 min; 100% B for 0.5 min; and 100–30% B over 0.5 min. The total injection-to-injection time is 9 min. The y-axis represents the ratio of the K562 *bcr-abl* signal to the plasmid competitor signal; the x-axis is the number of K562 cells.

## The Future of MRD Detection

CML has been a paradigm for molecular oncology. The Philadelphia chromosome was the first tumor-specific cytogenetic lesion discovered (Nowell, 1960; Rowley, 1973), and led to the first demonstration of a chimeric gene in human disease (Stam et al., 1985). Moreover, *bcr-abl* was the first chimeric mRNA amplified by PCR (Kawasaki et al., 1988), and this led to CML being the first model for the study of MRD. As described above, CML is now the model for the intervention at molecular relapse; as such, it again is a paradigm for how we might approach most malignancies in the future. The development of a quantitative assay for *bcr-abl* will further allow investigators to explore issues of tumor burden and kinetics that will only refine our understanding of MRD and relapse, and allow us to further tailor the early treatment of molecular relapse. Thus, the paradigm continues.

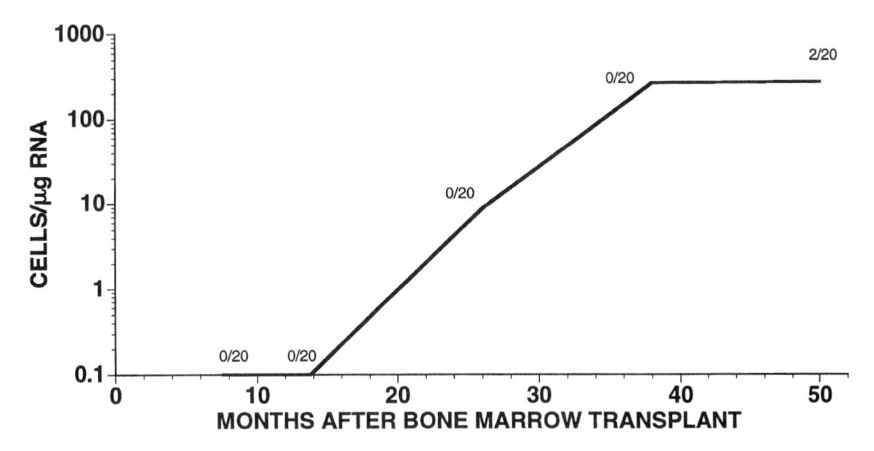


FIGURE 6. Q-PCR in a CML patient post-BMT. The fractions next to the data points represent the number of metaphases on cytogenetic exam showing the Philadelphia chromosome (numerator) out of the total number of metaphases counted (denominator). The y-axis represents the estimated equivalent of K562 cells contained in the 1 $\mu$g of sample RNA.

The detection and significance of MRD in other leukemias and solid tumors is complicated by both the heterogeneous nature of the malignancies themselves, and the various types of treatments employed (i.e., conventional transplantation vs. marrow transplant). For example, after conventional chemotherapy, nearly all patients with follicular cell lymphoma remain PCR-positive for the hallmark bcl2-IgH translocation while in remission (Price et al., 1991); however, the presence of PCR-positivity in the context of bone marrow transplant has been found by some investigators to be associated with a high risk of relapse (Gribben et al., 1993). Furthermore, for a given disease and treatment, the actual molecular defect may influence the interpretation of MRD. For example in acute myeloid leukemia (AML), in the *t*(15;17) subtype, the detection of MRD by PCR following induction chemotherapy heralds relapse; however, in the *t*(8;21) AML variant, *all* patients studied in continued remission up to 8 years posttherapy still are PCR-positive. Therefore, the significance of the detection of MRD, and the usefulness of quantification, must be taken in context with the disease, the treatment, and the specific genetic lesion.

Nonetheless, there are three areas where MRD detection in general, and quantification in specific, hold promise:

1. In guiding optimal therapy by more closely defining response to therapy than conventional morphology. For example, patients treated for acute leukemias who still have MRD after conventional induction therapy

might be good candidates for more intensive therapy (i.e., BMT) earlier rather than later.

2. In evaluating the efficacy of "purging" protocols to rid autologous bone marrow or peripheral blood stem cells of leukemia. Currently, autologous stem cell products are often treated with antibody or chemo therapeutic agents to eliminate residual malignant cells, for fear that the reinfusion of tumor cells might precipitate relapse. However, since these autologous collections are only performed on "remission" patients, tumors cannot be detected before, or after, purging. Thus, it is impossible to determine if purging works, or if malignant cells actually are infused back into the patient. The sensitive quantitation of residual malignant cells would both allow investigators to test the efficacy of purging, and determine the relationship between tumor contamination and relapse.

3. In testing the dose-response of new drugs or new drug combinations. Currently, remission induction is a fairly insensitive endpoint for comparing the efficacy of new therapeutic regimens, since the level of sensitivity of remission judged by conventional pathology is so low (5%). Thus, Agent A might eliminate the malignancy to a level of 1 tumor cell/100 normal cells, while Agent B kills to a level of 1/1000. By conventional pathology, these two agents would be judged equal, since both eliminated cells below the level of detection. The sensitive quantitation of malignant cells would offer a far more informative evaluation of these agents' activity. Moreover, if quantification could be shown to be tightly associated with subsequent relapse, perhaps dose escalation studies could be performed using Q-PCR as an endpoint, rather than waiting for gross hematologic relapse.

The determination of the clinical significance of MRD will depend on large, well-designed coordinated studies utilizing a unified treatment approach. Such studies will depend on standardized approaches that can maximize the "absolute quantification" of leukemia-specific targets (such as the absolute number of *bcr-abl* mRNAs in a particular sample), so that data from one lab can easily be compared to that of another. To that end, we have estimated that each K562 cell contains 100 copies of the *bcr-abl* chimeric mRNA. Further studies of residual disease hold great promise for redefining the concept of "remission" and ushering in an era when therapy can be more precisely tailored to the patient's need.

## REFERENCES

Bernards A, Bubin CM, Westbrook CA, et al. (1987): The first intron in the human c-abl gene is at least 200 kilobases long and is the target for translocations in chronic myelogenous leukemia. *Mol Cell Biol* 7:3231–6.

Brisco MJ, Condon J, Hughes E, et al. (1994): Outcome prediction in childhood acute lymphoblastic leukaemia by molecular quantification of residual disease at the end of induction. *Lancet* 343:196–200.

Claxton DF, Liu P, Hsu HB, et al. (1994): Detection of fusion transcripts generated by the inversion 16 chromosome in acute myelogenous leukemia. *Blood* 83:1750–6.

Clift RA, Buckner CD, Thomas ED, et al. (1994): Marrow transplantation for patients in accelerated phase of chronic myeloid leukemia. *Blood* 84:4368–73.

Cross NCP, Feng L, Chase A, et al. (1993): Competitive polymerase chain reaction to estimate the number of bcr-abl transcripts in chronic myeloid leukemia patients after bone marrow transplant. *Blood* 82:1929–36.

Fukutani H, Naoe T, Ohno R, et al. (1995): Prognostic significance of the RT-PCR assay of PML-RAR-a transcripts in acute promyelocytic leukemia. *Leukemia* 9:588–93.

Gribben JG, Neuberg D, Freedman AS, et al. (1993): Detection by polymerase chain reaction of residual cells with the BCL-2 translocation is associated with increased risk of relapse after autologous bone marrow transplantation for B-cell lymphoma. *Blood* 81:3449–57.

Groffen J, Stephenson JR, Heistercamp J, et al. (1984): Philadelphia chromosome breakpoints are clustered within a region, bcr, on chromosome 22. *Cell* 36:93–9.

Hanssen-Hagge TE, Yokota S, Bartram CR (1989): Detection of minimal residual disease in acute lymphoblastic leukemia by in vitro amplification of rearranged T-cell receptor delta chain sequences. *Blood* 74:1762–7.

Hebert J, Cayuela JM, Daniel MT, et al. (1994): Detection of minimal residual disease in acute myelomonocytic leukemia with abnormal marrow eosinophils by nested polymerase chain reaction with allele specific amplification. *Blood* 84:2291–6.

Heistercamp N, Stephenson JR, Groffen J, et al. (1983): Localization of the c-abl oncogene adjacent to a translocation breakpoint in chronic myelocytic leukemia. *Nature* 306:239–42.

Kawasaki ES, Clark SS, Coyne MY, et al. (1988): Diagnosis of chronic myeloid and acute lymphocytic leukemias by detection of leukemia-specific mRNA sequences amplified in vitro. *PNAS* 85:5698–702.

Kusec R, Laczika K, Knobl P, et al. (1994): AML1/ETO fusion mRNA can be detected in remission blood samples of all patients with *t*(8;21) acute myeloid leukemia after chemotherapy or autologous bone marrow transplantation. *Leukemia* 8:735–9.

Maruyama F, Stass SA, Estey EH, et al. (1994): Detection of AML1/ETO fusion transcript as a tool for diagnosing *t*(8;21) positive acute myelogenous leukemia. *Leukemia* 8:40–5.

Miller WH, Levine K, DeBlasio A, et al. (1993): Detection of minimal residual disease in acute promyelocytic leukemia by a reverse transcription polymerase chain reaction assay for the PML/RAR-a fusion mRNA. *Blood* 82:1687–93.

Miyamura K, Tahara T, Tanimoto M, et al. (1993): Long persistent bcr-abl positive transcript detected by polymerase chain reaction after marrow transplant for chronic myelogenous leukemia without clinical relapse: A study of 64 patients. *Blood* 81:1089–93.

Nowell PC and Hungerford DA (1960): A minute chromosome in human chronic granulocytic leukemias. *Science* 14:97.

Price CGA, Meerabaux J, Murtagh S, et al. (1991): The significance of circulating cells carrying *t*(14;18) in long remission from follicular lymphoma. *J Clin Oncol* 11:1668–73.

Radich J, Bryant E, Collins S, et al. (1995): PCR detection of the *bcr-abl* fusion transcript after allogeneic marrow transplantation for chronic myeloid leukemia: results and implications in 346 patients. *Blood* 85:2632–8.

Radich J, Ladne P, Schoch G, and Gooley T (1995): PCR-based detection of unique immunoglobulin V-D-J gene rearrangements in acute lymphoblastic leukemia predicts relapse after allogeneic bone marrow transplantation. *Biol Blood Marrow Trans* 1:24–31.

Roth MS, Antin JH, Ash R, et al. (1992): Prognostic significance of Philadelphia chromosome-positive cells detected by the polymerase chain reaction after allogeneic bone marrow transplant for chronic myeloid leukemia. *Blood* 79:276–82.

Rowley JD (1973): A new consistent chromosomal abnormality in chronic myelogenous leukaemia identified by quinacrine fluorescence and Giemsa staining. *Nature* 243:290–3.

Stam K, Heisterkamp N, Crosveld G, et al. (1985): Evidence of a new chimeric bcr/c-abl mRNA in patients with chronic myelocytic leukemia and the Philadelphia chromosome. *NEJM* 313:1429–33.

Thomas ED, Clift A, Fefer A, et al. (1986): Marrow transplantation for the treatment of chronic myelogenous leukemia. *Ann Intern Med* 104:155–63.

Turhan AG, Lemoine FM, Debert C, et al. (1995): Highly purified primitive hematopoietic stem cells are PML-RAR-a negative and generate nonclonal progenitors in acute promyelocytic leukemia. *Blood* 85:2154–61.

Yamada M, Wasserman R, Lange B, et al. (1990): Minimal residual disease in childhood B-lineage lymphoblastic leukemia. *NEJM* 323:448–55.

# Competitive RT-PCR Analysis of Brain Gene Expression During Inflammation and Disease

Douglas L. Feinstein and Elena Galea

## Introduction

The developmental, functional, and pathological status of the central nervous system (CNS) requires regulation of several thousands of genes whose levels of expression vary from relatively abundant (>2%) to rare (<.001%). Although in most instances analysis of the expression patterns can be achieved by conventional RNA blot analysis, in many cases low levels of expression or limited amount of material require more sensitive methods, such as RT-PCR. In addition, in many instances the partially degraded nature of the materials precludes analyses dependent upon intact mRNA molecules. The advantages of RT-PCR methods and the use of specific internal competitive standards have allowed us to accurately measure multiple mRNA levels in samples derived from long-term glial cultures, isolated brain regions, and human autopsy material.

Our studies have focused on two genes that are upregulated during inflammation and injury in the brain. The first encodes the glial fibrillary acidic protein (GFAP), a member of the intermediate filament family of proteins, one that is specifically expressed in astrocytes in the central nervous system (CNS) (Eng, 1980). The function of GFAP is not completely understood, but it is known to participate in the establishment of astrocyte morphology, formation and extension of astrocyte processes, and formation of glial scars that surround sites of injury. Upregulation of GFAP expression may reflect a protective reaction of glial cells to prevent further damage. The second gene studied encodes an isoform of the enzyme nitric oxide synthase (iNOS). This enzyme, which catalyzes the synthesis of NO from L-arginine, is normally not expressed in brain cells, but during pathologies involving inflammatory reactions it appears in glial and endothelial cells (Koprowski et al., 1993; Wallace and Bisland, 1994; Dorheim et al., 1994). The iNOS enzyme synthesizes sufficiently large quantities of the free radical gas NO to promote cell damage and death (Schmidt and Walter, 1994). The quantitative analysis of expression of these two genes thereby provides information concerning the ability to simultaneously regulate a protective and a damaging gene within the same tissue, and in the case of astrocytes, within the same cell.

## Methodology

For studies of astrocyte gene expression, we use primary cultures of rat brain astrocytes, prepared from the cerebral cortices of postnatal day 1 rat pups (Feinstein et al., 1992). These cultures consist of >95% astrocytes as determined using cell-specific antibodies. The cells are maintained in Dulbecco's modified Eagle's media (DMEM) containing 10% FCS and antibiotics, and they can be maintained for up to 3 weeks in culture. Cells are typically grown in 6-well (35 $cm^2$ per well) plates, which allows differential treatment of essentially identical sister cultures, and provides sufficient RNA per well for PCR analyses.

Our procedures for RNA isolation follow the phenol:chloroform extraction procedure, in which an initial step is to remove cell debris and intact nuclei by low-speed centrifugation (Sambrook et al., 1989). Although this method is more time-consuming than others, for example those using guanidine isothiocyanate, we have found that this procedure reproducibly yields DNA-free RNA as assessed by PCR analysis of cDNA made in the absence of reverse transcriptase. For cell culture samples, the cells are washed in PBS, collected, lysed in 0.1% NP-40, digested with 100 $\mu$g/ml proteinase K, and RNA-purified by multiple phenol:chloroform extractions and isopropanol precipitation. We typically obtain 10–20 $\mu$g of total cytoplasmic RNA from one 75 $cm^2$ flask of confluent cells, and between 2–5 $\mu$g from one 35 $cm^2$ well. For freshly dissected rat brain samples, the same method is used except that the cell membranes are lysed in hypotonic media rather than detergent. In both cases the resulting RNA is of high quality, as evidenced by 260 nm to 280 nm absorbtion ratios greater than 1.8, and the presence of intact 18S and 28S rRNA bands on nondenaturing agarose gels. The RNA is kept frozen at –80°C in Tris-EDTA, although –20°C is adequate for short term (e.g., a few weeks) storage. RNA was isolated from human brain autopsy samples by a guanidinium thiocyanate method (Chomczynski et al., 1987) (provided by Dr. Brian Apatoff, Department of Neurology, Cornell University Medical College).

Conversion of RNA to cDNA is carried out with random hexamers and murine-MLV reverse transcriptase (RT) or RNase-reduced Superscript RT (GIBCO, BRL). Random hexamers are used rather than specific reverse primers to allow analysis of multiple mRNAs in the same sample. Primers

(400 ng) plus RNA (0.5 to 1 μg) are annealed in 10 μl volume at 70°C for 10 min, cooled gradually, then placed on ice. A solution containing 20 units RT, dNTPS, DTT, and RNasin is added (in 10 μl) to the cold samples, and the samples are incubated at 39°C for 1 hr. The reaction is halted by heating at 95°C for 5 min, the samples are cooled, and the final volume is brought to 100 μl with water. The cDNA is stored at −80°C. For PCRs, 5–10 μl of the resulting cDNA (corresponding to 25–50 ng of starting RNA) are amplified.

For quantitative RT-PCR analysis we have prepared competitive internal standards derived from the full-length PCR products. We have used two methods for the synthesis of deletion standards: initially a restriction-enzyme-based technique and subsequently the more convenient linker-primer procedure (Chuang-Fang et al., 1994). For GFAP, a 167 bp internal standard (ΔGFA167) was produced by restriction enzyme digestion of a 296 bp PCR product followed by ligation of the digested fragments and reamplification of ligated products lacking an internal 130 bp fragment (Galea and Feinstein, 1992). For iNOS, as well as most other genes we have studied, we used the linker-primer method, which does not require the presence of particular restriction enzyme sites. In this method, the 15 bases at the 3′ end of the linker primer are homologous to a region 50–100 bp downstream of the forward primer, while the 15 bases at the 5′ end are identical to the forward primer. After synthesis of an intermediate product using the linker and the reverse primer, initially at a low annealing temperature (in our case we use 37°C for 3 cycles, followed by 60°C for 32 cycles), t he competitive internal standard is synthesized using the forward and reverse primer combination. The final product can then be reamplified at high annealing temperature.

We have used the linker-primer approach to synthesize internal standards for several mRNAs, including GFAP, NOS, G3PDH, BDNF, TNF-α, CD14, and the NFkB subunits p65 and MAD-3. In most cases, we have found that approximately 2–3 days are required to obtain purified standards that can be used for competitive PCR. In all cases, we have verified construct identity by DNA sequencing, and confirmed that amplification of the competitive standard requires both the forward and reverse PCR primers. Occasionally we have found that the standard could be amplified with only a single primer. In one case this was due to hybridization of the forward primer to an area near the 3′ end of the molecule. A second time amplification of the intermediate linker product had occurred with only the reverse primer. In both these situations alternative primers were designed and used successfully.

Primers are selected with assistance of the PRIME program, developed by the University of Wisconsin Genetics Core Group. We routinely select primers having A or T in the 3′ position, and a theoretical annealing temperature, using the formula Tann = 2*(A + T) + 4*(G + C), of at least 60°C. Primers are also checked against current DNA databases to exclude homology to other known genes. PCRs are carried out in a Hybaid Thermoreactor controlled by tube temperature. The reaction components in 45 $\mu$l, except enzyme, are brought to 88°C, and the reaction is started by addition of 0.5 units of Taq polymerase (Promega, Madison, WI) in 5 $\mu$l of 1 × Taq buffer. Typical PCR conditions are 35 cycles of: denaturation at 93°C for 30s, annealing at 60–63°C for 45s, and extension at 72°C for 45s, followed by one 10 min extension at 72°C.

For quantitative analysis, PCRs are carried out in the presence of 100,000 cpm per 50 $\mu$l reaction of $^{32}$P-dATP or $^{33}$P-dATP (both >3000 Ci/mmole). After electrophoresis, the bands are separated through 2 or 2.5% agarose gels containing ethidium bromide, the cDNA and competitive internal standard products are excised from the gels, then placed into vials with scintillation fluor, and incorporated radioactivity is determined by liquid scintillation counting. Background radionucleotide levels are determined by loading an aliquot of a sample that contains all components of the PCR, but which was not subjected to temperature cycling, and then excising and counting the corresponding areas of the gel after electrophoresis.

Initial levels of template cDNA are calculated as described (Gilliland et al., 1990) from linear regression analysis of log–log plots of the ratio of cDNA product to construct product versus the amount of construct initially added. The amount of construct corresponding to the point at which log [ratio] = 0 is equal to the amount of starting template. Corresponding mRNA levels are obtained by multiplying the calculated amount of cDNA by the ratio of mRNA to PCR product sizes. For final calculations of mRNA levels, we have used efficiency values as measured by the enzyme supplier (28% for M-MLV and 40% for Superscript RT), since we did not ourselves directly measure cDNA synthesis efficiencies. For this reason, mass values presented for mRNA content are not absolute. However, since cDNAs are prepared with the same reagents, these values allow us to discuss and compare our results in terms of mRNA levels.

In some cases, only single-point competitive PCRs (Personett and McKinney, 1996) were carried out either because a limited amount of sample was available (the human MS samples) or because a large number of samples had to be analyzed (the rat brain developmental studies). In these cases the competitive PCRs, done in duplicate or triplicate, were done in the presence of a constant amount of competitive standard that was predetermined to be within tenfold of the amount of template initially present. In addition, single-point competitive PCRs were done for G3PDH mRNA.

## Glial Fibrillary Acidic Protein

GFAP is a 50 kDa protein originally isolated from human multiple sclerosis plaques (Eng et al., 1971), and determined to be the major component of astroglial intermediate filaments (IF). The functions of IF proteins are not well established, but it is clear that they are constituents of the cytoskeleton of the cell and play a role in maintaining and modifying the cell's overall morphology. Increases in GFAP levels occur in various neuropathological situations, and are considered diagnostic for astrocytic activation, which may correlate with other inflammatory responses and/or formation of glial scars. While mainly present in the central nervous system (CNS), GFAP is also expressed in other tissues, including the glia (Schwann cells) of the peripheral nervous system (PNS) (Dahl et al., 1982).

We have been interested in examining the changes in GFAP mRNA levels that occur in cell culture and isolated rat brain slices in response to inflammatory stimuli. Our initial studies (Feinstein et al., 1992) led to isolation of a Schwann cell cDNA clone derived from a form of the GFAP mRNA (GFAP-$\beta$) that differed from the previously isolated (Lewis et al., 1984) CNS GFAP clone (GFAP-$\alpha$). The GFAP-$\beta$ is produced by transcription initiation occurring at an alternate upstream start site, and results in an mRNA having a 5′-untranslated region of 183 bases versus 13 bases for GFAP-$\alpha$. The presence of the longer 5′-UTR could influence GFAP mRNA stability and/or translation, and potentially gives rise to a truncated GFA protein by directing translational initiation to a downstream start codon.

Since the relatively similar lengths of the two GFAP mRNAs (GFAP-$\alpha$, 2692 bases; GFAP-$\beta$, 2861 bases) precluded facile analysis by RNA blots, we developed a PCR-based method to allow distinction of the two mRNAs as well as quantitation of their respective levels (Figure 1A). For this purpose, we designed PCR forward primers that would amplify either total ($\alpha$ + $\beta$) GFAP mRNA, or specifically the GFAP-$\beta$. For all PCRs, oligonucleotide GFA+284R was used as the reverse primer. Oligonucleotide GFA-12F, corresponding to bases −12 to +7 of GFAP-$\beta$, yields a 296 bp product when GFAP-$\beta$ cDNA is amplified. However, this primer shares only seven bases with the beginning of the GFAP-$\alpha$ mRNA, and therefore does not serve as a primer for amplification of GFAP-$\alpha$ derived cDNA. Oligonucleotide +1F is complementary to bases 1–21 present in both mRNAs, and therefore efficiently amplifies cDNA derived from both GFAP-$\alpha$ and -$\beta$ to yield a 284 bp product.

For quantitative determinations, we constructed a competitive GFAP standard that consisted of the 296 bp GFAP PCR product obtained by amplification using primers GFA-12F and +284R, from which an internal

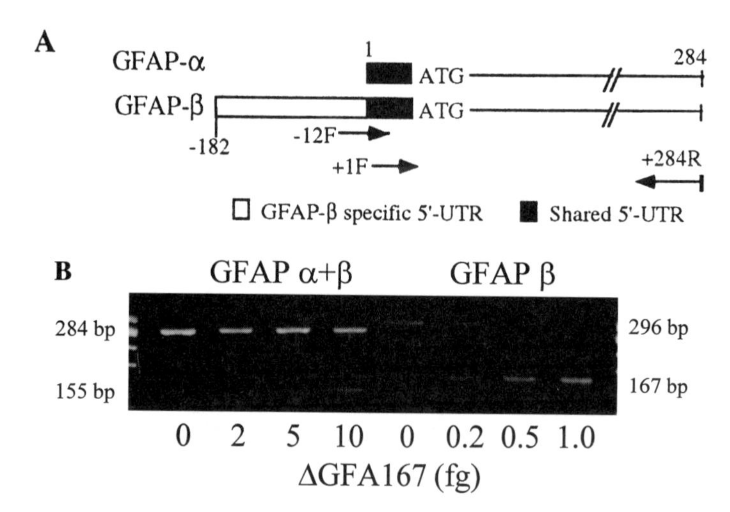


FIGURE 1. (**A**) Schematic illustration of GFAP mRNA isotypes and location of PCR primers. GFAP-$\alpha$ mRNA contains a 5′-UTR of 13 bases (filled box), while the GFAP-$\beta$ mRNA contains an extended 5′-UTR of 183 bases (open plus filled boxes). The beginning of GFAP-$\alpha$ 5′UTR is designated is +1. The initial ATG codon (at +14) is indicated. (**B**) Representative gel showing competitive PCR of total GFAP ($\alpha + \beta$) and GFAP-$\beta$. Brain total RNA (1 $\mu$g) was converted to cDNA, and equal aliquots (corresponding to 50 ng of original RNA) amplified with forward primer GFA+1F for total GFAP, or GFA–12F for GFAP-$\beta$, together with GFA+284R, in the presence of indicated amounts of ΔGFA167. The cDNA and ΔGFA167 PCR product sizes are shown. Adapted from Galea et al., 1995.

130 bp HaeIII fragment was removed (Galea and Feinstein, 1992). The resulting 167 bp fragment (ΔGFA167) competed with GFAP cDNA when included in the same PCR using primer combination +1F and +284R to amplify total GFAP, or –12F and +284R to amplify GFAP-$\beta$ specifically (Figure 1B). The efficiency of amplification of ΔGFA167 was identical to that of the full-length cDNA, since the ratio of cDNA to ΔGFA167 PCR products was constant when analyzed after 22, 25, or 32 amplification cycles.

## *GFAP expression in astrocyte cultures*

The levels of GFAP-$\alpha$ and -$\beta$ were determined in cDNA samples prepared from either cultured astrocytes or peripheral Schwann cell RNA. Aliquots of the resulting cDNAs were amplified in the presence of increasing amounts of construct ΔGFA167 with primer combinations +1F/+284R or –12F/+284R. Total GFAP mRNA levels in peripheral Schwann cells were 4.3 pg per $\mu$g RNA, and in these cells the GFAP-$\beta$ constituted over 75% of the GFAP mRNA (3.4pg per $\mu$g RNA). In astrocytes, total GFAP was higher (15.6pg per $\mu$g RNA), and GFAP-$\beta$ was 0.9pg per $\mu$g RNA (approximately 5% of total GFAP mRNA). Amplification of GFAP-$\beta$ in astrocyte cDNA was not observed if the reverse transcriptase was omitted from the cDNA reaction. Assuming that mRNA constitutes roughly 2% of total cytoplasmic RNA, we can estimate that in astrocytes GFAP mRNA is present at 0.08% relative abundancy, and 0.02% in the Schwann cells. These results demonstrated that GFAP-$\beta$, originally isolated from a Schwann cell cDNA library, is also expressed at significant levels in the CNS.

## *GFAP isoform expression during inflammation*

Levels of GFAP mRNA expression are influenced by exposure to cytokines such as tumor necrosis factor-$\alpha$, and interleukin 1-$\beta$ (Selmaj et al., 1991; Oh et al., 1993). However, little was known concerning inflammatory effects of interferon-$\gamma$ (IFN) on GFAP expression, or of cytokine effects of GFAP isoform expression. We examined the effects of IFN upon GFAP mRNA levels in freshly dissected rat brain cortical slices and in astrocyte cultures.

In brain slices, a 4hr perfusion with saline did not effect total GFAP mRNA levels (Figure 2). However, after 4 hr perfusion with 25 units/ml IFN, total GFAP mRNA levels were decreased over tenfold (from $43 \pm 7$ to $1.8 \pm 1.5$; $n = 3$; $p < 0.05$; fg cDNA/$\mu$g total RNA). At the same time, IFN had no significant effect upon GFAP-$\beta$ mRNA levels (not shown). Owing to the large decrease in total GFAP mRNA levels, the percentage of GFAP-$\beta$ was increased from roughly 5% in control slices to over 50% in IFN-treated samples (Figure 2B, right side). In primary astrocyte cultures, incubation with 25 units/ml IFN caused a 1.5–2.0-fold decrease in total GFAP levels and at the same time a modest increase (approximately twofold versus DMEM) in GFAP-$\beta$ mRNA levels. Overall, in cultures, IFN resulted in a final GFAP-$\beta$ content 3–4 times control values ($18 \pm 8$% versus $5 \pm 1$%; $n = 2$; $p < 0.05$; treated versus control). Thus, in both cortical slices and primary cultures, IFN-dependent decreases in total GFAP mRNA levels were accompanied by either no change or increased GFAP-$\beta$mRNA levels.

The above results demonstrate the utility of the RT-PCR method in being able to 1) measure levels of two alternate transcripts of similar size and 2) provide quantitative measurements of the two isoforms, although one is expressed at much reduced levels. It is clear from our observations that changes in total GFAP mRNA can be accompanied by either the same or the opposite changes in the GFAP-$\beta$ form. This suggests that certain

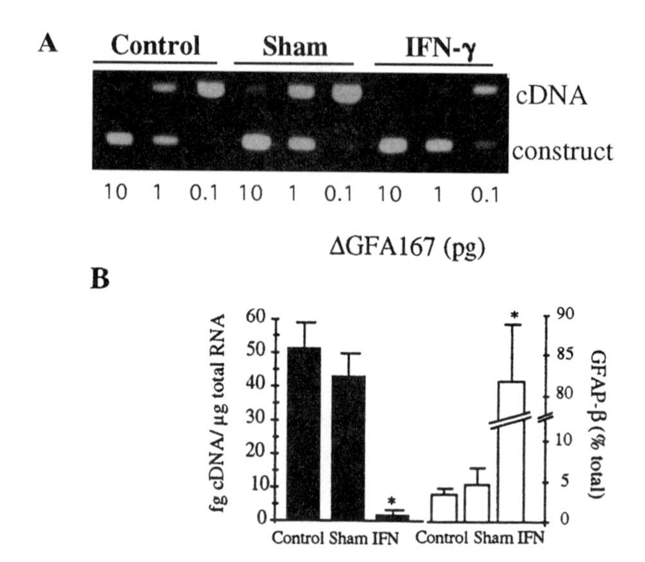


FIGURE 2. Effect of IFN treatment of GFAP mRNA isoforms in brain slices. (**A**) Representative gel showing results for total GFAP levels. Total cytoplasmic RNA was prepared from freshly dissected rat cortical slices (control) or from slices after 4 hr perfusion with 0 (sham) or 25 units/ml IFN. PCR was carried out on equal aliquots of cDNA (corresponding to 50 ng of initial RNA), in the presence of the indicated amount of internal standard with primers +1F and +284R to determine total GFAP $\alpha + \beta$ levels. (**B**) Calculation of GFAP levels. The data is presented as fg of GFAP cDNA per $\mu$g of total RNA (left side), and as amount of GFAP-$\beta$ (as % of total GFAP) on the right side. Data is mean ± s.e.m. of three separate experiments. *, $p < 0.05$ versus sham. Adapted from Galea et al., 1995.

types of astrocyte activation, perhaps those associated with inflammatory responses, would not be revealed by measurements of total GFAP content.

## Inducible Nitric Oxide Synthase

NO is synthesized from L-arginine by three principal isoforms distinguishable by dependence upon calcium for catalytic activity, tissue distribution, and whether they are expressed constitutively or transiently. While two NOS isoforms are modulated by changes in intracellular calcium, the third isoform, iNOS, is permanently active even when calcium produces large amounts of NO for sustained periods of time (Schmidt and Walter, 1994). In contrast to the calcium-dependent NOSs, which are constitutively present in neurons and endothelial cells, iNOS is not present in normal adult brain. However, iNOS appears in microglia, astrocytes or blood vessels during demyelinating disorders (Koprowski et al., 1993; Bo et al., 1994), ischaemia (Iadecola et al., 1995; Wallace and Bisland, 1994), Alzheimer's disease (Dorheim et al., 1994), and viral infections (Zheng et al., 1993). In some cases, the use of specific iNOS inhibitors decreased neural damage, suggesting a direct role of NO in the genesis of the pathology (Cross et al., 1995; Iadecola et al., 1995).

Here we describe the characterization of iNOS expression in astrocytes, using cells in culture, and in whole brain during embryonic and postnatal development. We developed a competitive RT-PCR using primers based upon the sequence of the cloned rat astrocyte iNOS (Galea et al., 1994). PCR primers were designed against a region of the 3′-UTR to avoid possible cross-hybridization to other NOS isoforms, and were designed to amplify a 395 bp fragment extending from base 3191 to 3585. The linker-

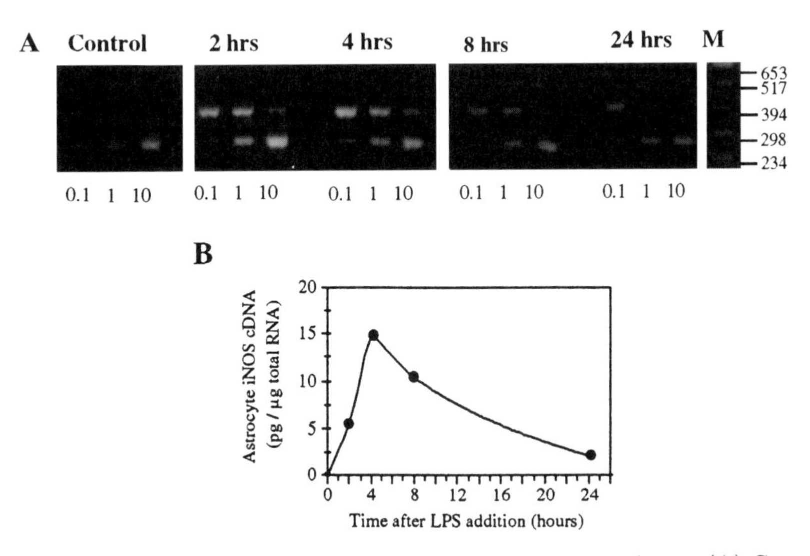


FIGURE 3. Time course of iNOS expression in astrocyte cultures. (**A**) Competition between 395 bp iNOS cDNA and 283 bp iNOS construct. The indicated amounts (pg) of construct were co-amplified with cDNA obtained from 100 ng total cytoplasmic astrocyte RNA. RNA was isolated from cells incubated with 1 µg/ml LPS for the indicated lengths of time. Control is cells in media only, for 4 hr. (**B**) Levels of iNOS cDNA 395 bp product were determined from log–log plots of incorporated $^{32}$P-dATP. Adapted from Galea et al., 1994.

primer method was used to synthesize a 283 bp competitive standard.

*iNOS expression in astrocyte cultures*

We had previously shown that in astrocyte cultures, incubation with LPS caused the induction of iNOS activity, which appeared between 4–6 hr after LPS addition, produced NO at a constant rate, and was still present when assayed after 48 hr of incubation (Galea et al., 1992). To establish the relationship between iNOS activity and presence of iNOS mRNA, we measured iNOS mRNA levels at different times after LPS addition (Figure 3).

Astrocyte iNOS cDNA was detectable after 2 hr exposure to LPS, peaked at 4 hr, and declined over the next 20 hr, with the value at 24 hr being 15% of the maximal observed. Incubation in fresh media alone (Control) did not produce detectable iNOS expression. Hence, LPS induced the transient appearance of astrocyte iNOS mRNA, beginning approximately 2 hr prior to iNOS catalytic activity but decreasing during times when iNOS activity was still high. These findings have led to current interest in defining *cis*-acting elements in the iNOS 3′-UTR that confer instability onto the iNOS mRNA.

*iNOS expression in developing brain*

Since pathological expression of some genes often reproduces a developmental expression pattern, it was important to determine whether or not the iNOS mRNA occurred at any point in normal brain development, and if so, what was the regional, cellular, and temporal pattern of that expression. Expression was analyzed in cortex, diencephalon, cerebellum, hippocampus, and striatum, from the last third of embryonic development to the adult age. The large number of samples to analyze, together with the small amount of RNA obtained from embryonic brain regions, prompted us to use single-point competitive PCR for iNOS as well as for G3PDH. This approach, although not allowing the measurement of absolute cDNA levels, provided quantitative and rapid comparison of cDNA amounts amongst different samples.

As expected, iNOS mRNA was not present in adult brains. In contrast, expression was detected perinatally in all brain areas in two phases, most clearly detected in cortex and diencephalon (Figure 4). The first phase occurred before birth, and the second, more robust phase (reaching 2.5- to 10-fold embryonic levels) arose during the first postnatal week to peak at or near day 7. In embryos, the highest levels of expression were in diencephalon, while in postnatal animals cerebellum showed maximal expression.

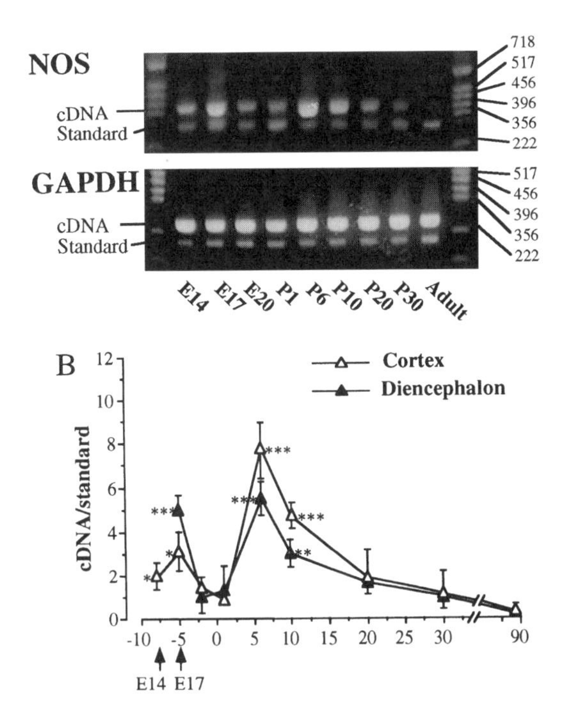


FIGURE 4. Quantitative RT-PCR analysis of iNOS mRNA expression in developing brain areas. The cDNAs were amplified for 40 cycles in the presence of 0.05 fg of iNOS internal standard. (**A**) Representative PCR results for cortex showing the evolution of iNOS and G3PDH cDNA levels over time. (**B**) Ratio of cDNA/standard for cortex and diencephalon. Values are the means ± s.e.m. of three to five pools of animals from different litters. Pools contained 6–10 animals (prenatal age) or 1–3 (postnatal age). (*) $p < 0.001$ and (***) $p < 0.0001$ versus adult (postnatal-day 90), ANOVA followed by Fisher's least significant differences *post hoc* analysis. Relative iNOS cDNA levels from adult were either undetectable or less than 0.5 (ratio cDNA/standard). Adapted from Galea et al., 1995.

After postnatal-day 7, expression diminished in all regions to become negligible in the adult.

To rule out the possibility that variations in iNOS mRNA levels were nonspecific, and thus reflected a general developmental pattern of gene expression, we also measured mRNA levels of G3DPH (Figure 4A). In contrast to the biphasic pattern obtained for iNOS, relatively constant levels of G3DPH mRNA were detected in all brain areas over the time examined (embryonic-day 17 to adult), except for cerebellum, in which expression increased around the second postnatal week, to reach a plateau around postnatal-day 20.

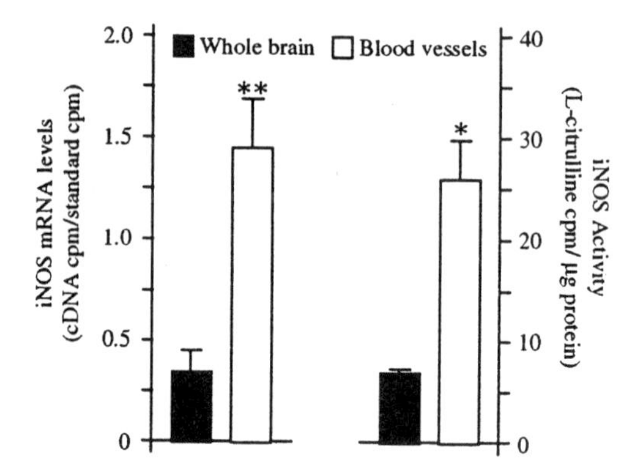


FIGURE 5. iNOS mRNA and enzyme activity in postnatal-day 7 whole brain parenchyma and isolated vessels. Values are the means ± s.e.m. of 4–5 animals. iNOS mRNA levels were determined by quantitative RT-PCR, using 1 fg of internal standard. NOS activity was determined by the production of L-citrulline assays in tissue homogenates. *, $p < 0.05$ and **, $p < 0.01$ versus whole brain, unpaired Student *t*-test. Adapted from Galea et al., 1995.

Immunohistochemistry with specific iNOS antibodies revealed that iNOS mRNA was translated into protein and that the cellular source was not the expected one, i.e., the astrocytes. Instead, iNOS protein was localized to the blood microvasculature, most probably to the endothelial cells. To confirm this localization, we carried out direct measurements of iNOS mRNA levels and enzyme activity in parenchymal blood vessels isolated from postnatal-day 7 and adult brains. Microvessels isolated from adult rats ($n = 3$) failed to express iNOS mRNA or enzyme activity. However, postnatal-day 7 vessels expressed both the iNOS mRNA and activity (Figure 5), and both values were significantly enriched compared to postnatal-day 7 whole brain parenchyma (enrichment was 4.7-fold for iNOS mRNA levels, and 3.4-fold for iNOS activity). These results confirm that microvessels are a cellular source of iNOS mRNA and protein in perinatal brain. Further, the comparable degree of enrichment in mRNA and enzyme activity validates the use of the RT-PCR approach to quantitate iNOS mRNA. Although the functional significance of iNOS in developing brain is unknown, upregulation of vascular iNOS in brain diseases suggests similar mechanisms of regulation during adult pathology and development.

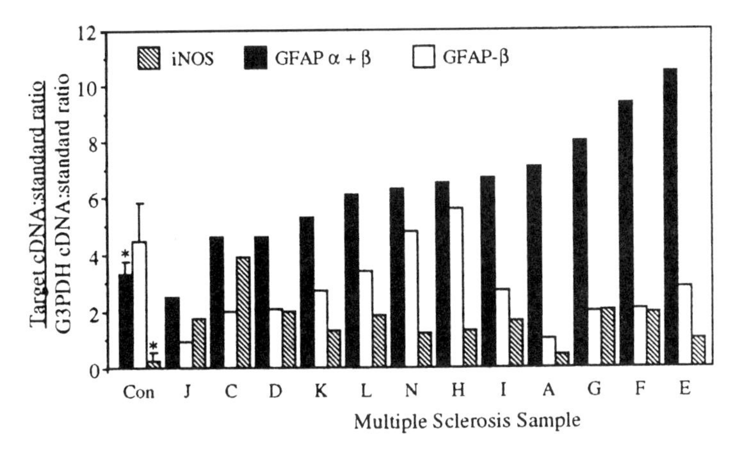


FIGURE 6. Single-point competitive PCR analysis of MS and control samples. Individual results for 12 MS samples (labeled A, C–L, N) and mean ± s.e.m. for three control samples (Con) are shown. *, $p < 0.05$ versus mean of MS samples, Fisher's *post hoc* analysis. MS samples are presented in order of increasing total GFAP content.

## Gene Expression in Multiple Sclerosis

Since inflammatory reactions play a role in the pathogenesis of multiple sclerosis (MS), we sought to determine whether GFAP and iNOS genes were upregulated in this disease. With this aim we analyzed RNA isolated from brain samples of normal and MS patients. In contrast to cell culture and freshly dissected brain samples, RNA isolated from autopsy material presents two major difficulties: 1) the general integrity of the RNA (degree of breakdown as observed in agarose gels) is worse, and 2) there is large variability in RNA integrity dependent upon individual patients, methods of isolation, and length and temperature of storage. These problems, along with the limited amount of material, preclude the use of conventional RNA blot analysis and make PCR a superior method for quantitation of gene expression in autopsy material. First, the use of random hexamers for cDNA synthesis combined with amplification of small target regions (e.g., 300–400 bp) allows efficient synthesis of small regions to occur even with highly degraded termplate RNAs. Second, normalization for the quality of the RNA can be obtained by carrying out quantitative determination of a housekeeping gene such as G3PDH in aliquots of cDNA from the same sample.

RNA samples were isolated using a guanidinium isothiocyanate method from sclerotic plaque material dissected from 12 MS patients (labeled A, C–

TABLE 1. Single-point competitive PCR was carried out with equal aliquots of cDNA prepared from control and MS tissues for indicated genes, in the presence of specific primers and internal standard. The amount of internal standard used was: GFAP $\alpha$+$\beta$, 100 fg; GFAP-$\beta$, 1 fg; iNOS, 1 fg; G3PDH, 10 fg. Data is shown as means ± s.e.m. for pooled control and MS samples.

| Ratio of cDNA to Internal Standard PCR Product | | |
|---|---|---|
| Gene | Control ($n = 3$) | MS ($n = 12$) |
| GFAP $\alpha$+$\beta$ | 2.2 ± 0.5 | 3.9 ± 0.9 |
| GFAP-$\beta$ | 1.9 ± 0.8 | 1.5 ± 0.2 |
| iNOS | 0.1 ± 0.1 | 1.1 ± 0.2* |
| GDH | 7.3 ± 2 | 6.4 ± 1 |
| Normalized to G3PDH | | |
| GFAP $\alpha$+$\beta$ | 0.33 ± 0.1 | 0.62 ± 0.06* |
| GFAP-$\beta$ | 0.41 ± 0.2 | 0.26 ± 0.04 |
| iNOS | 0.33 ± 0.3 | 1.70 ± 0.2** |

*, $p < 0.05$ versus control; **, $p < 0.02$ versus control.

L, N) and from normal cortical tissue from three healthy brains. After cDNA synthesis, single-point competitive PCR was carried out for total GFAP, GFAP-$\beta$, and iNOS mRNAs, as well as for levels of G3PDH. For each gene, a single concentration of the internal standard, based upon preliminary experiments, was included in the PCR. Individual values for three mRNAs, normalized to G3PDH values, are shown in Figure 6, and mean values are summarized in Table 1. The levels of total GFAP were approximately twofold higher in the MS versus control samples (Table 1), ranging from 1.5-fold (C) to over threefold (E). Normalization to the individually measured G3PDH levels did not substantially modify the overall fold-difference, but decreased the intersample variation and thus made the difference between control and MS samples statistically significant. At the same time, GFAP-$\beta$ levels were lower in the MS samples than in the controls; however, this difference was not statistically significant even after normalization to G3PDH values.

These results demonstrated that astrocytic activation occurring in MS results in overall increased GFAP mRNA levels, while the GFAP-$\beta$ content is unaffected or slightly decreased. The iNOS mRNA was detected in only one out of three control samples, but was present in all 12 MS samples

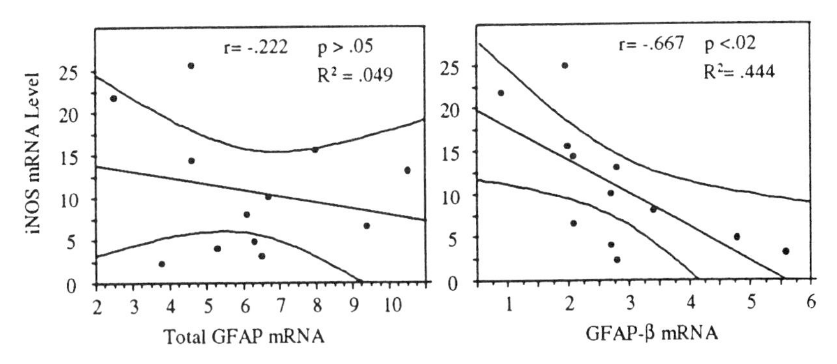


FIGURE 7. Correlative analysis of GFAP to iNOS mRNA levels. The individual values obtained for total GFAP mRNA and GFAP-$\beta$ were compared to iNOS levels measured in the same samples.

(Figure 6), and was significantly different from the controls regardless of normalization to G3PDH values (Table 1).

Further analysis, achieved by comparison of individual iNOS mRNA values to the total GFAP or GFAP-$\beta$ levels measured in the same samples was used to detect possible relationships not identified by comparison of population means (Figure 7). Whereas no significant correlation existed between total GFAP and iNOS levels, a negative correlation was present between GFAP-$\beta$ and iNOS levels ($p < 0.02$). Why a correlation exists between GFAP-$\beta$ and iNOS levels, but not between total GFAP mRNA and iNOS levels, is as yet unknown.

## Summary

We have described the use of quantitative PT-PCR to study expression of GFAP and iNOS genes *in vitro* and *in vivo*. This method has allowed us to: 1) analyze independently two mRNA isoforms, GFAP $\alpha$ and $\beta$, whose similarity in size does not allow easy separation by RNA blot analysis, 2) analyze iNOS mRNA levels and regional and cellular distribution, during brain development, despite the limited amount of starting material available in embryonic brains and isolated brain microvessels, 3) compare large numbers of samples from developing brain in which iNOS expression ranges from low or absent (in the adult) to high (at postnatal-day 7), and 4) analyze iNOS and GFAP gene expression in brain tissue from MS patients, using autopsy material that provides low yields of RNA and in addition, presents variable integrity among samples.

## REFERENCES

Bo L, Dawson TM, Wesselingh S, Mork S, Choi S, Kong, Hanley D, and Trapp BD (1994): Induction of nitric oxide synthase in demyelinating regions of multiple sclerosis brain. *Ann Neurol* 36:778.

Chomczynski P and Sacchi N (1987): Single step method of RNA isolation by acid guanidinium thiocyanate phenol chloroform extraction. *Anal Biochem* 162:156–9.

Chuan-Fang J, Mata M, and Fink DJ (1994): Rapid construction of deleted DNA fragments for use as internal standards in competitive PCR. *PCR Meth Appl* 3:252–5.

Cross A, Misko TP, Lin RF, Hickey, Trotter JL, and Tilton RG (1994): Aminoguanidine, an inhibitor of inducible nitric oxide synthase, ameliorates experimental autoimmune encephalomyelitis in SJL mice. *J Clin Invest* 93:2684.

Dahl D, Chi NH, Miles LE, and Nguyen BT, and Bignami A (1982): Glial fibrillary acidic protein in Schwann cells—fact or artefact. *J Histochem Cytochem* 30:912–18.

Dorheim MA, Tracey WR, Pollock JS, and Grammas P (1994): Nitric oxide synthase activity is elevated in brain microvessels in Alzheimer disease. *Biochem Biophys Res Commun* 205:659–65.

Eng LF, Vanderhaeghen JJ, Bignami A, and Gerste B (1971): An acidic protein isolated from fibrous astrocytes. *Brain Res* 28:351–4.

Eng LF (1980): The glial fibrillary acidic protein. In: *Proteins of the Nervous System*, Vol. 2 (Bradshaw and Schneider, eds.), pp. 85–117.

Feinstein DL, Weinmaster G, and Milner RJ (1992): Isolation of cDNA clones encoding rat glial fibrillary acid protein: Expression in astrocytes and in Schwann cells. *J Neurosci Res* 32:1–14.

Galea E and Feinstein DL (1992): Rapid synthesis of DNA deletion constructs for mRNA quantitation: Analysis of astrocyte mRNAs. *PCR Meth Appl* 2:66–9.

Galea E, Feinstein DL, and Reis DJ (1992): Induction of calcium-independent nitric oxide synthase activity in primary rat glial cultures. *Proc Natl Acad USA* 89:10945–9.

Galea E, Reis DJ, and Feinstein DL (1994): Cloning and expression of astroglial inducible nitric oxide synthase. *J Neurosci Res* 37:406–11.

Galea E, Reis DJ, and Feinstein DL (1995): Transient expression of calcium-independent nitric oxide synthase in blood vessels during brain development. *FASEB J* 9:1632–7.

Galea E, Dupouey P, and Feinstein DL (1995): Glial fibrillary acidic protein mRNA isotypes: Expression in vitro and in vivo. *J Neurosci Res* 41:452–61.

Gilliland G, Perrin S, and Bunn H (1989): Competitive PCR for quantitation of mRNA. In: *PCR Protocols* (Innis et al., eds.), Academic Press, pp. 60–8.

Iadecola C, Zhange F, and Xu X (1995): Inhibition of nitric oxide synthase ameliorates cerebral ischemic damage. *Am J Physiol* 268:R286–92.

Iadecola C, Xu X, Zhang F, Casey R, and Ross ME (1995): Inducible nitric oxide synthase gene expression in brain following cerebral ischemia. *J Cereb Blood Flow Metab* 15:378–84.

Koprowski H, Zheng Y, Heber-Katz E, Fraser E, Rorke L, Fang F, Hanlon C, and Dietzshold B (1993): In vivo expression of inducible nitric oxide synthase in experimentally induced neurological diseases. *Proc Natl Acad Sci USA* 90:3024–6.

Lewis SA, Balcarek JM, Krek V, Shelanski M, and Cowan NJ (1984): Sequence of a cDNA clone encoding mouse glial fibrillary acidic protein: Structural conservation of intermediate filaments. *Proc Natl Acad Sci USA* 81:2743–6.

Oh YJ, Merkelonis GJ, and Oh TH (1993): Effects of interleukin 1-$\beta$ and tumor necrosis factor-$\alpha$ on the expresson of glial fibrillary acidic protein and transferrin in cultured astrocytes *GLIA* 8:77–86.

Personett DA, Chovinard M, Sugaya K, and McKinney M (1996): Simplified RT/PCR quantitation of gene transcripts in cultured neuroblastoma (SN49) and microglial (BV-2) cells using capillary electrophoresis and laser-induced fluorescence. *Mol Brain Res* (in press).

Sambrook J, Fritsch EF, and Maniatis T (1989): Molecular Cloning. *A Laboratory Manual*, 2nd Ed., Cold Spring Harbor, New York, pp. 192–4.

Schmidt HHW and Walter U (1994): NO at work. *Cell* 78:919–25.

Selmaj K, Shafit-Zagardo B, Aquino DA, Farooq M, Raine CS, Norton WT, and Brosnan CF (1991): Tumor necrosis factor-a induced proliferation of astrocytes from mature brain is associated with down regulation of glial fibrillary acidic protein. *J Neurochem* 57:823–30.

Wallace CV and Bisland SK (1994): NADPH-diaphorase activity in activated astrocytes represents inducible nitric oxide synthase. *Neuroscience* 59:905–19.

Zheng YM, Schafer MK, Weihe E, Sheng H, Corisdeo S, Fu ZF, Koprowsky H, and Dietzschold B (1993): Severity of neurological signs and degree of inflammatory lesions in the brains of rats with Borna Disease correlate with the induction of nitric oxide synthase. *J Virol* 67:5786–91.

# Development and Application of Real-Time Quantitative PCR

P. Mickey Williams, Todd Giles, Ayly Tucker, Jane Winer, and Chris Heid

Since the original description of the polymerase chain reaction (PCR), novel applications of this technique have been reported at an exponential rate. The power of the method lies in the ability to analyze minute amounts of samples (e.g., dried blood, microliters of liquid, several cells) or even to detect a single molecule. Virtually every area of biological science has been impacted by PCR, including forensics, genetics, clinical diagnostics, drug discovery, and environmental biology.

The topic of this book, gene quantitation, has had a central role in understanding and monitoring of various biological responses. Original techniques, such as Southern blots (DNA analysis) and Northern blots (RNA analysis) led to many related hybridization and quantitation techniques (e.g., RNase Protection, dot blots, etc.). Generally these original techniques share several features that limit their application: the amount of time necessary to obtain results, often days, and the requirement for a considerable amount of starting material, often a minimum of 10 $\mu$g of nucleic acid. PCR quantitation offers the advantages of speed and sensitivity when compared to the original approaches. Using PCR, one can begin with nanograms of sample and obtain results in a day or less. Two major PCR methods have been applied to quantitation of DNA and RNA; in the first method quantitation is performed during the exponential phase of amplification, and in the second method a competitor nucleic acid is co-amplified with the target—this method is called quantitative competitive PCR (QC-PCR) (Becker-Andre, 1989; Becker-Andre, 1991; Becker-Andre, 1993; Ferré, 1992; Kellogg et al., 1990; Pang et al., 1990; Piatak et al., 1993; Raeymaekers, 1995).

The use of PCR for nucleic acid quantitation requires the experimenter to overcome two hurdles: the problems of quantitation in the plateau region of the reaction and the issues of reaction inhibitors. The amplification of a target by PCR usually proceeds with exponential accumulation of the product (each cycle should result in the doubling of the product). The exponential phase of the reaction lasts for many cycles, during which huge amounts of product are generated. Eventually the reaction becomes less efficient and is no longer in an exponential product accumulation phase. This may occur for several reasons; late in the PCR cycles, primer concentrations drop, and the concentration of product is so dramatically increased that the product reassociates faster than primers can efficiently anneal, and at the same time critical reagents can become limiting (enzyme, nucleotides etc.). At this stage, analysis of PCR product alone cannot be predictive of starting amounts of target molecules. Indeed, if the reaction reaches plateau because the concentration of product is driving product reassociation, different samples containing different starting amounts of target will achieve plateau at the same product concentration and at that time will contain identical amounts of product. The other hurdle for PCR nucleic acid quantitation is sample inhibitors. Sample preparation reagents and biological sample components may contaminate nucleic acid preparations. Many such compounds inhibit PCR enzymes (heme, salts, phenol, alcohol, etc.). If these contaminants are found in samples, they will affect the efficiency and thus the amount of product generated during PCR. Hence, sample contaminants need to be controlled in order to achieve quantitation using PCR.

Of these two methods of PCR quantitation, analysis during exponential product accumulation and quantitative competitive PCR (QC-PCR), QC-PCR seems better suited to overcome both aforementioned hurdles that may stand in the path of successful PCR quantitation. Therefore, QC-PCR has proven to be the method of choice for establishing quantitative nucleic acid assays.

Several recent advances have made quantitative PCR even more user-friendly. Higuchi and co-workers described the simultaneous amplification and detection of specific DNA sequences (Higuchi et al., 1992). Several other major advances have been made in PCR chemistry, instrumentation, and software (see accompanying chapters by Wittwer and McBride). One such advance in PCR chemistry was the description by Holland and co-workers of the 5′ nucleolytic activity of Taq polymerase (Holland et al., 1991). Another advance was the development of fluorescent energy transfer hybridization probes (Bassler et al., 1995; Lee et al., 1993; Livak et al., 1995a; Livak et al., 1995b). Instrumentation advances have been made by several companies. ABI-Perkin-Elmer has developed the Model 7700 Sequence Detector, which monitors fluorescent emission in "real-time" from the 96 tubes in a thermal cycler. Idaho Technologies has recently developed the "Light Cycler," which also monitors fluorescence in real-time in capillary tubes during thermal cycling. Finally, AcuGen has developed a thermal cycler with fluorescent detection. All three companies have also developed software for data analysis and quantitative measurements on their respective machines.

The ABI system utilizes TaqMan fluorescent chemistry, which is described in detail in the chapter by McBride. Briefly, a fluorescent hybridization probe is designed to recognize a sequence within an amplicon (the region of DNA that is amplified during PCR). This probe contains two fluorescent molecules; one is the donor, and the other serves as an acceptor for fluorescent energy transfer. When the hybridization probe is intact, excitation of the donor dye by 488 nm laser light results in minimal fluorescent emission in the donor dye spectrum, since most of the energy is directly transferred to the acceptor dye molecule. During the extension step of a PCR cycle, the polymerase moves down the template, synthesizing nascent product strands. During this extension, if a hybridization probe is encountered, the polymerase begins to degrade the probe by virtue of its previously mentioned 5′ nucleolytic activity. Once the probe is hydrolyzed, the fluorescent donor molecule is no longer in close proximity to the acceptor dye, and therefore no transfer of donor dye energy occurs. This results in the release of the donor dye energy as fluorescent emission. The ABI Model 7700 Sequence Detector monitors donor fluorescent emission in "real-time" during PCR cycles. As PCR products increase exponentially, so does the amount of probe degradation and the amount of donor fluorescent emission. The donor fluorescent emission intensity is monitored and plotted as a function of time (or cycle numbers). This plot is known as an amplification plot (Figure 1). The Sequence Detector software analyzes the amplification plot and determines a point at which the donor fluorescence reaches an arbitrary threshold (usually ten standard deviations above the average reaction background determined by the early cycles of PCR). The point at which the donor fluorescent emission reaches this threshold is called the $C_T$. This $C_T$ value is a quantitative value. The greater the quantity of starting target, the less the amount of time (in cycles) required to degrade sufficient probe to reach the threshold, resulting in a lower $C_T$ value (Figure 2a). The data can also be represented as shown in Figure 2b, in which the $C_T$ values are plotted against the quantity of input target. A sample containing a known amount of target can be used to make a dilution curve. The standard curve generated from the known sample is used to calculate amounts of target in an unknown sample. This "real-time" approach to quantitation differs from other previously described quantitative PCR approaches, such as exponential analysis or QC-PCR.

"Real-time" PCR monitors and determines relative quantity concomitantly with PCR amplification. Therefore, no post-PCR steps are necessary, saving time and preventing potential laboratory contamination. Quantitative data are collected while reactions are in the exponential phase of product accumulation, preventing problems associated with plateau phase.

"Real-time" PCR allows for very easy validation of methods of DNA and RNA sample preparation. Sample preparation methods can be examined for

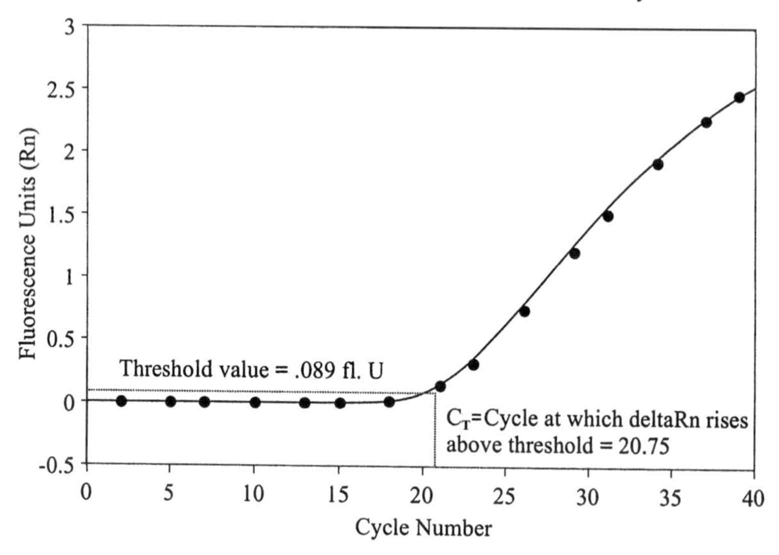


Figure 1. Amplification plot. Reporter (FAM) fluorescent emission is normalized to a reference dye (ROX or TAMRA) and plotted as a function of time (in PCR cycle numbers). Model 7700 Sequence Detector software uses an algorithm to determine baseline fluorescence and calculate the $C_T$ value of the sample. Each data point represents the average of the last three data points during the extension phase of the PCR cycle.

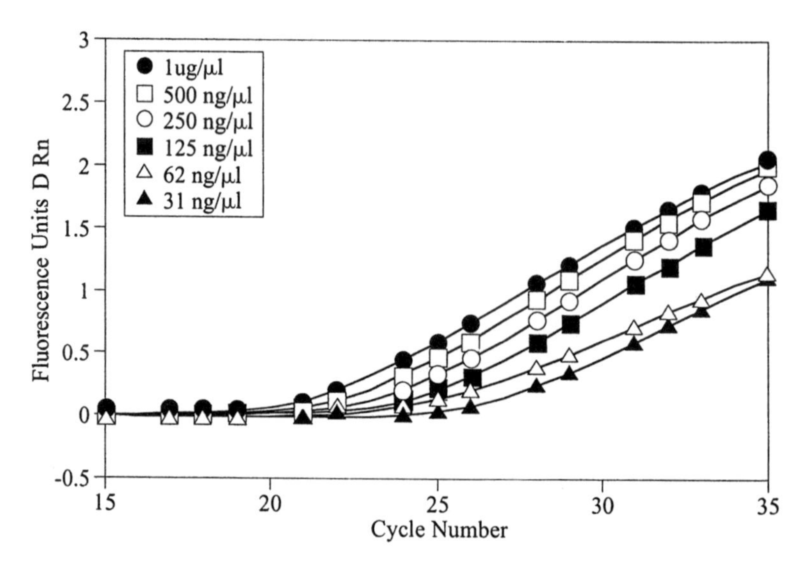


Figure 2a. Amplification plots of a serial dilution of RNA. Six 1:2 serial dilutions were made from rat pancreatic total RNA and analyzed by "real-time" RT-PCR with primers and a probe specific for the rat flute-2 gene.

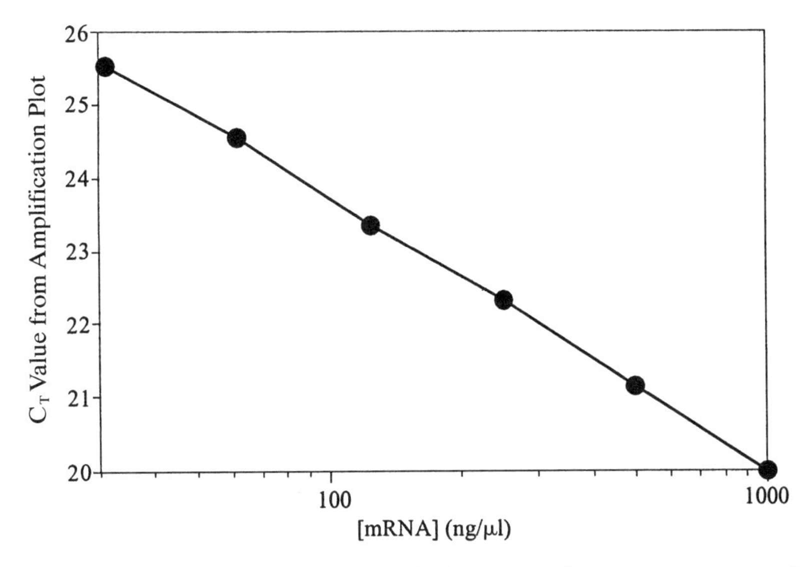


Figure 2b. Plot of $C_T$ values from serial dilution. A standard curve is generated for a gene of interest by analyzing serial dilutions of a known amount of RNA. The standard curve can be used to calculate the amount of the gene's mRNA that is present in an unknown sample.

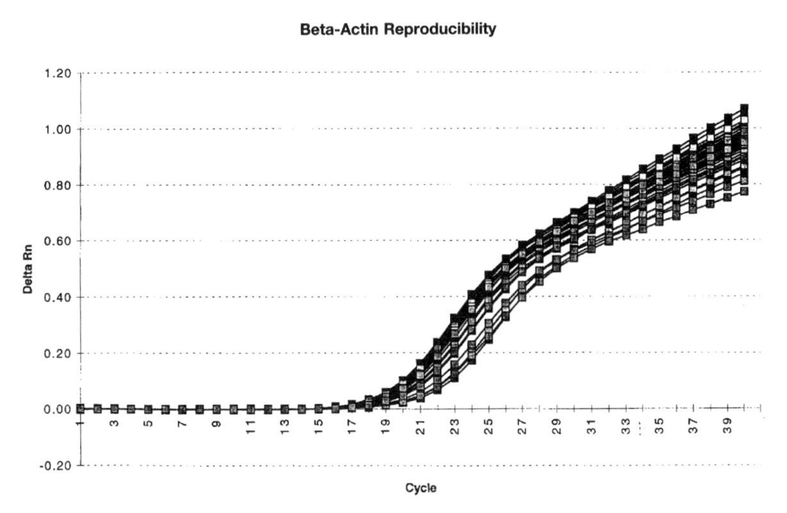


Figure 3. Equal numbers of A549 cells were plated in each well of a 96 well tissue culture plate. When cells were confluent, various amounts of interferon-$\alpha$ were added to the cells, ranging from 500 units/ml to none. Cells were harvested for cytoplasmic RNA using Rnaeasy 96 (Qiagen) as described in the protocol. Equal aliquots of RNA from each of 40 wells were analyzed for beta-actin.

reproducible nucleic acid recovery and resultant amplification efficiency of samples. A large number of identically prepared samples can be analyzed simultaneously by monitoring an invariant gene or mRNA. Sample preparation methods that are highly reproducible should have nearly identical $C_T$ values, and the rates of $C_T$ change with sample serial dilution should also be identical. Further, "housekeeping" genes or other nonfluctuating sequences useful for internal controls can be validated. Housekeeping genes can be used to normalize RNA load differences and to control for amplification inhibitors for each sample in quantitative RT-PCR. Figure 3 depicts the beta-actin amplification plots of 40 "real-time" RT-PCR analyses. Each reaction is a different RNA sample from a plate of A549 cells. These samples represent RNA, some of which is from cells that have been treated with interferon-$\alpha$, and some of which is from untreated cells. As is seen, the plots and hence the $C_T$ values are very similar. This experiment demonstrates that the RNA preparation method is very reproducible and that beta-actin mRNA will serve as a good "housekeeping" gene for this biological system.

"Real-time" PCR allows for a large number of samples to be analyzed using minimal amounts of starting material. "Real-time" PCR has found applications in screening experiments of genetic analysis, and monitoring gene expression (Gibson et al., 1996; Luoh et al., 1997), and monitoring plasmid levels in cells after transfection (Heid et al., 1996); further, it is currently being developed for monitoring adventitious agents of tissue culture.

Genotyping has become an extremely useful tool for the diagnosis of known genetic disease, for the discovery of DNA sequences that may be associated with disease states or disease stage (e.g., obesity, cancer, etc.), for monitoring genotypes of infectious agents (e.g., HIV), and for the analysis of genetic models. A multitude of techniques have been developed for genetic analysis. The appropriate design of fluorescent hybridization probes permits single base discrimination, and thus "real-time" PCR can be used for single base allotyping (Livak et al., 1995a). Additionally, we have developed a high-throughput method for genotyping knock-out mice.

Knock-out mice are routinely used as models for the study of gene function. A knock-out mouse has a particular gene of interest genetically removed so that the function of the gene can be determined. The development of homozygous knock-out mice requires a considerable genotyping effort and many resultant matings. PCR primers and matching fluorescent hybridization probe can be designed to recognize and amplify a region of the wild-type gene under investigation (i.e., IL-8 receptor, see Figure 4a). The wild-type gene primers are selected so that this sequence is present in unaltered wild-type alleles but removed in a knocked-out allele. During the process of knocking out this gene, a neomycin resistance gene is inserted in place of the wild-type sequence. PCR primers

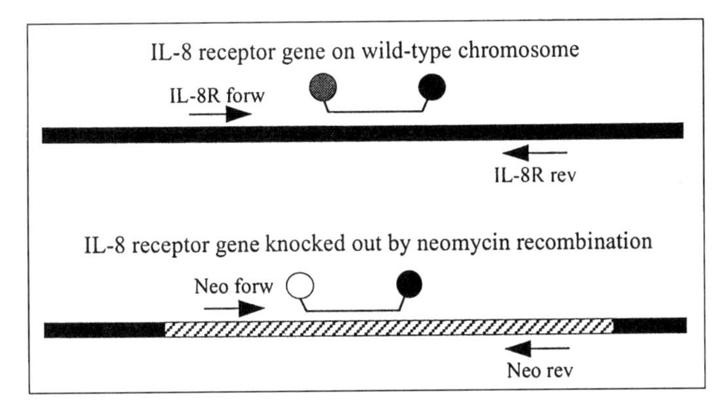


Figure 4a. Dual-color probes for knock-out mouse analysis. Primer and probe combinations are designed to detect the wild-type allele IL-8r and the neomycin resistance gene. A FAM label is used for the reporter dye on the IL-8r gene probe. A JOE label is used for the neomycin resistance gene probe. TAMRA is used for the quenching dye on both probes.

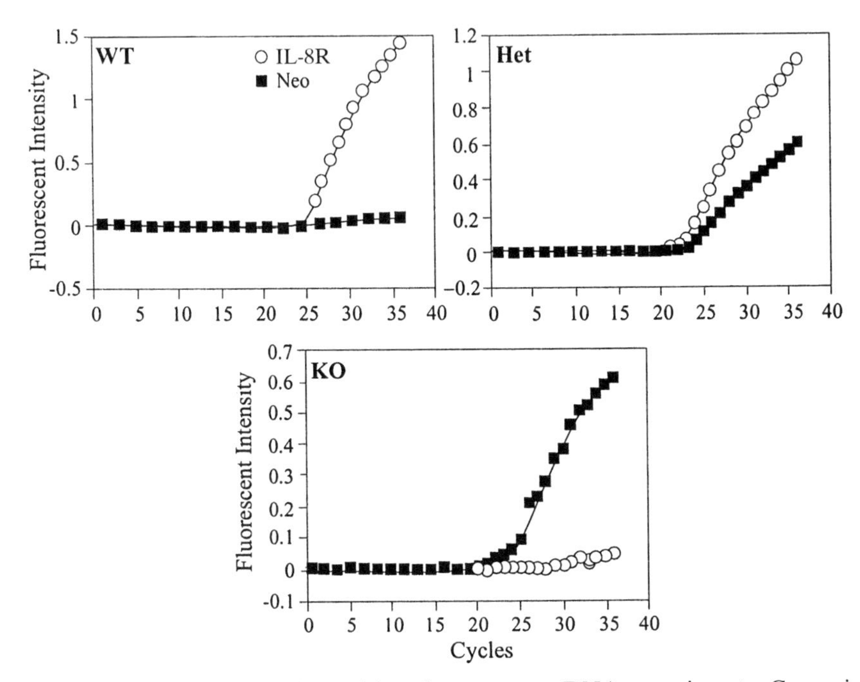


Figure 4b. Amplification plots of knock-out mouse DNA experiments. Genomic DNA is analyzed for the presence of either the IL-8r wild-type gene or the neomycin resistance gene. Three different experimental samples are shown: homozygous wild-type IL-8r mouse, keterozygous mouse and a IL-8r knock-out mouse.

and hybridization probe are also designed to recognize the inserted neomycin resistance gene sequence. Genomic DNA is prepared from mouse tail tissue and analyzed in PCR amplifications using the wild-type primers /probe and neomycin resistance primers/probe in the same reaction tube. In these experiments, the donor dye used for labeling the wild-type probe is FAM (6-carboxy-fluorescein), and the donor dye used for the neomycin resistance probe is JOE (2, 7-dimethoxy- 4, 5 dichloro-6-carboxy fluorescein). The FAM and JOE dyes have slightly different emission spectra and can be distinguished, monitored, and quantitated within the same tube (FAM emission optima is approximately 518 nm and JOE emission optima is approximately 556 nm). However, the two dyes do have some overlap in their spectral emissions. This overlap interferes with detection sensitivity but is not problematic in genotyping experiments, in which at most 50/50 mixtures must be analyzed. (The overlap is undesirable in gene quantitation experiments, in which the two different signals from different genes can be present at any level.) A PCR on DNA from a homozygous wild-type mouse (both alleles contain wild-type gene sequence) exhibits amplification plots showing product for the wild-type gene, but no detectable amplification of the neomycin resistance gene (Figure 4b). A "real-time" PCR on DNA from a heterozygous mouse, containing one allele of wild-type sequence and one allele of neomycin resistance (knocking out the wild-type gene), results in amplification of both target sequences. Analysis of homozygous knock-out mouse DNA shows no detectable amplification of the wild-type gene sequence, but does show amplification of the neomycin resistance sequence. Automation is being developed that will speed up the process and increase the throughput of preparation of the mouse genomic DNA.

As reported by Heid (Heid et al., 1996), "real-time" PCR has been developed to monitor plasmid quantity after transfection of tissue culture cells. This method will be useful for application in gene therapy experiments. Many strategies are being developed to increase the efficiency of delivery of therapeutic genes to diseased target organs, and to increase expression levels and duration of the therapeutic gene. Quantitative PCR and RT-PCR are ideally suited for this purpose since they require very small sample size. "Real-time" PCR can quickly be developed for many different targets and permits the screening of large numbers of samples.

Finally, a major application of "real-time" quantitative RT-PCR is the analysis of gene expression. For example, quantitation of gene expression can be useful in monitoring the response of cells or tissues to external stimuli. Data from these types of experiments should guide researchers in understanding what biological effects a particular stimulus has. Does the stimulus lead to cell proliferation, differentiation, or programmed cell death? We have used "real-time" quantitative RT-PCR extensively in a variety of research projects. Gibson and co-workers describe the use of "real-time" RT-PCR with an internal control to monitor mRNA expression from a therapeutic CFTR gene vector (Gibson et al., 1996). In this system, a synthetic internal control was designed that used identical primers and was similar in product size and G&C content to the target amplicon. Experimental data verified that the target and internal control amplicons exhibited very similar amplification efficiencies. Therefore, a known quantity of the internal control can be spiked into an unknown sample and used as an internal quantitative reference. In another project, the hypertrophy of rat neonatal cardiac myocytes was investigated. This is a well-studied system that can be characterized through a number of different criteria, such as protein content, cell surface area, and visual inspection (King et al., 1996). We have designed a series of primers and probes to a variety of mRNAs that are known to change their level of expression during hypertrophy. We have shown that glyceraldehyde phosphate dehydrogenase mRNA (i.e., GAPDH) expression is constant in this system, which makes it a valid housekeeping gene. Indeed, equal amounts of mRNA derived from equal numbers of nonhypertrophied and hypertrophied cells have identical amounts of GAPDH mRNA. Atrial nitruretic factor (ANF) mRNA levels have been shown to increase dramatically during the course of hypertrophy. Figure 5a shows the ANF mRNA amplification plots of quadruplicate samples derived from cells that have undergone complete hypertropy (7), intermediate hypertrophy (5), or no hypertrophy (control cells) (3). As shown in Figure 5a, the amplification plots shift to the left (smaller $C_T$ values) with increased amounts of ANF mRNA. These data also demonstrate the reproducibility of the biological system, of the instrument, and of the assay. Figure 5b depicts the same mRNA samples analyzed with the GAPDH "housekeeping" gene. The amplification plots are very similar. This demonstrates that equal amounts of mRNA were present in each sample and that all the samples had similar amplification efficiencies. One plot does appear to have a different efficiency at later cycles, and it is important to point out that this has minimal impact on the earlier cycles from which the $C_T$ values are derived. This point also demonstrates the advantage of "real-time" analysis as compared to endpoint analysis.

Another example of a gene expression analysis using "real-time" quantitative RT-PCR is the analysis of leptin receptor splice variants. Shortly after the cloning of the leptin receptor was reported, an assay was developed to monitor the relative amounts of two different spliced forms of the receptor in human tissues (Luoh et al., 1997). This assay relied on designing primers that would differentiate the two different forms of the receptor. Dilution experiments using both forms of the receptor demonstrated similar relationships between the amount of targets and the $C_T$ values. Thus the two different spliced leptin receptor targets are amplified with equal efficiency, and $C_T$ values can be compared directly. The $C_T$ values of both forms were compared to determine the relative amounts of each form in a particular

a)

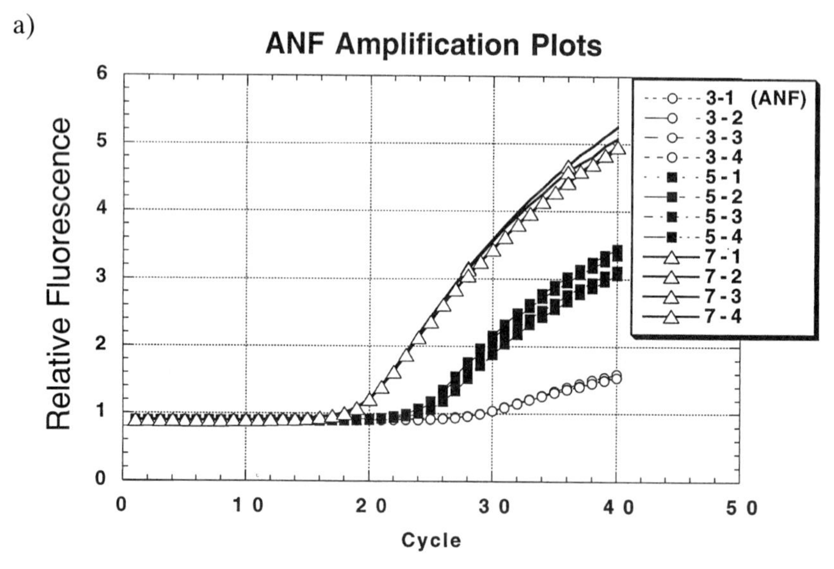


b)

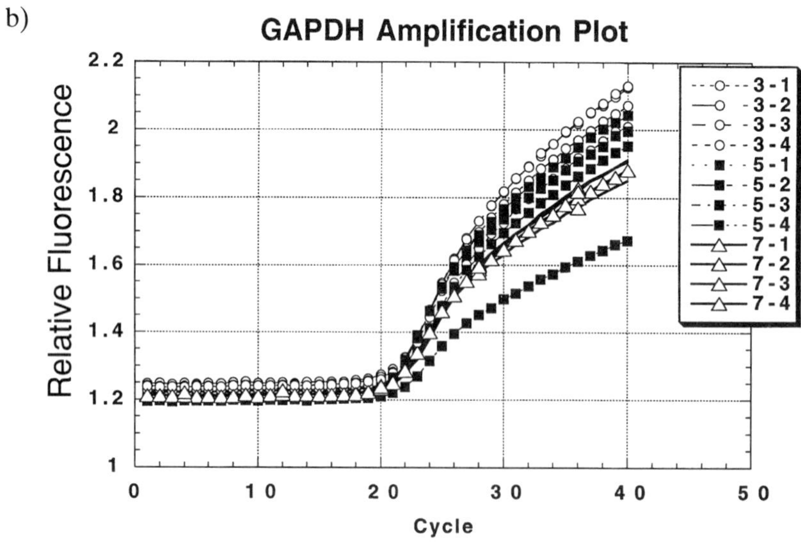


Figure 5. Rat neonatal cardiac myocytes were harvested and seeded into 96-well tissue culture plates (King et al., 1996). Cells were treated with different doses of phenylepherine, and mRNA was harvested (Poly A Tract; Promega as described in the kit) from four wells of cells that were visually scored as complete hypertrophy (7), four wells of intermediate hypertrophy (5), and four wells of control cells (3). Equal aliquots of mRNA were analyzed for ANF(a) and GAPDH(b) mRNA levels.

tissue. "Real-time" quantitative RT-PCR assays are now routinely designed to monitor expression of other genes of interest.

## Conclusion

The use of "real-time" PCR and RT-PCR will greatly facilitate the application of gene quantification to many areas of biological rescarch. The future holds other technological advances such as readily available DNA chips containing hundreds, if not thousands, of different interrogating probes for multiple gene analysis. These chips will have the tremendous advantage of simultaneous quantitation of multiple genes. Finally, the potential for small hand-held instrumentation suited for field study has already begun to be fulfilled.

## REFERENCES

Bassler HA, Flood SJ, Livak KJ, Marmaro J, Knorr R, and Batt CA (1995): Use of a fluorogenic probe in a PCR-based assay for the detection of Listeria monocytogenes. *Appl Environ Microbiol* 61:3724–8.

Becker-Andre M (1989): Absolute mRNA quantification using the polymerase chain reaction (PCR). A novel approach by a PCR aided transcript titration assay (PATTY). *Nucleic Acids Res* 17:9437–46.

Becker-Andre M (1991): Quantitative evaluation of mRNA levels. *Meth Mol Cell Biol* 2:189–201.

Becker-Andre M (1993): Absolute levels of mRNA by polymerase chain reaction-aided transcript titration assay. *Meth Enzymol* 218:420–45.

Ferré F (1992): Quantitative or semi-quantitative PCR: reality versus myth [Review]. *PCR Meth Appl* 2:1–9.

Gibson UEM, Heid C, and Williams PM (1996): A novel method for real time quantitative RT-PCR. *Genome Res* 6:995–1001.

Heid C, Stevens J, Livak K, and Williams PM (1996): Real time quantitative PCR. *Genome Res* 6:986–94.

Higuchi R, Dollinger G, Walsh PS, and Griffith R (1992): Simultaneous amplification and detection of specific DNA sequences. *Biotechnology* 10:413–17.

Holland PM, Abramson RD, Watson R, and Gelfand DH (1991): Detection of specific polymerase chain reaction product by utilizing the 5′–3′ exonuclease activity of Thermus aquaticus DNA polymerase. *Proc Natl Acad Sci USA* 88:7276–80.

Kellogg DE, Sninsky JJ, and Kwok S (1990): Quantitation of HIV-1 proviral DNA relative to cellular DNA by the polymerase chain reaction. *Anal Biochem* 189:202–8.

King KL, Jadine L, Winer J, Luis E, Yen R, Hooley J, Williams PM, and Mather J (1996): Cardiac fibroblasts produce leukemia inhibitory factor and endothelin, which combine to induce cardiac myocyte hyperthrophy in vitro. *Endocrine* 5:85–93.

Lee LG, Connell CR, and Bloch W (1993): Allelic discrimination by nick-translation PCR with fluorogenic probes. *Nucleic Acids Res* 21: 3761–6.

Livak KJ, Flood SJ, Marmaro J, Giusti W, and Deetz K (1995a): Oligonucleotides with fluorescent dyes at opposite ends provide a quenched probe system useful for detecting PCR product and nucleic acid hybridization. *PCR Meth Appl* 4:357–62.

Livak KJ, Marmaro J, and Todd JA (1995b): Towards fully automated genome-wide polymorphism screening [letter]. *Nat Genet* 9:341–2.

Luoh S-M, Di Marco F, Levin N, Armanini M, Xie M-H, Armanini M, Xie M-H, Nelson C, Bennett GL, Williams PM, Spencer SA, Gurney A, and de Sauvage FJ (1997): Cloning and characterization of a human leptin receptor using a biologically active leptin immunoadhesin. *J Mol Endocrinol* 18:77–85.

Pang S, Koyanagi Y, Miles S, Wiley C, Vinters HV, and Chen IS (1990): High levels of unintegrated HIV-1 DNA in brain tissue of AIDS dementia patients. *Nature* 343:85–9.

Piatak MJ, Luk KC, Williams B, and Lifson JD (1993): Quantitative competitive polymerase chain reaction for accurate quantitation of HIV DNA and RNA species. *Biotechniques* 14:70–81.

Raeymaekers L (1995): A commentary on the practical applications of competitive PCR. *Genome Res* 5:91–4.

# Branched DNA (bDNA) Technology for Direct Quantification of Nucleic Acids: Research and Clinical Applications

Janice A. Kolberg, Douglas N. Ludtke, Lu-Ping Shen, Will Cao, Darrah O'Conner, Mickey S. Urdea, Linda J. Wuestehube, and Marcia E. Lewis

## Introduction

Over the past decade a number of new technologies, including the polymerase chain reaction (PCR) (Mullis and Faloona, 1987), have emerged for the detection and quantification of nucleic acid molecules. Distinct among these is the branched DNA (bDNA) assay which, unlike its target amplification counterparts, uses a signal amplification scheme to enhance detection of physiologic concentrations of target nucleic acids. The basis of the bDNA assay involves the specific hybridization of bDNA and enzyme-labeled oligonucleotide probes to target nucleic acids; this is described in detail in the accompanying chapter in this book, entitled "Branched DNA Technology for Direct Quantification of Nucleic Acids: Design and Performance." The bDNA assay is inherently quantitative and nonradioactive, and has proven to be a reproducible and accurate means of quantifying nucleic acid molecules. The bDNA assay offers several advantages for research and clinical applications; it:

- provides direct quantification of nucleic acid molecules at physiological levels. Since bDNA assays do not require highly purified nucleic acid preparations, inhibitors of enzyme-dependent amplification techniques are of no concern.
- allows a wide diversity of specimen types to be used.
- can be used with reference standards to ensure accuracy for applications requiring absolute quantification, or without standards for applications requiring only assessment of relative changes in nucleic acid concentrations.
- exhibits a high level of sensitivity (moderately expressed genes can be detected using as few as 50–100 cells). The level of sensitivity can be

modulated by altering the number of target-specific extenders to accommodate specific applications.
- is thoroughly tested for within-lot and between-lot precision to ensure reproducibility (2.2- to 3-fold changes can be discerned as statistically significant).

One of the earliest applications of the bDNA assay was the quantification of viral nucleic acids as a measure of viral load — the amount of virus in the patient. In the management of patients with viral diseases, viral nucleic acid measurement serves to complement other serological, biochemical, and histological measures of disease. For some viral diseases, viral nucleic acid levels have been correlated with disease progression and clinical outcome. Moreover, changes in viral nucleic acid levels in response to therapy may be clinically relevant. Thus, the use of the bDNA assay to measure viral nucleic acids may facilitate a more rational approach to therapy by providing clinicians with information needed to follow viral load throughout the course of disease, select and adjust treatment protocols, and evaluate the efficacy of therapeutic regimens.

The potential prognostic and therapeutic value of measuring viral load using the bDNA assay is only just beginning to be realized. For example, studies of hepatitis C virus (HCV) infected patients undergoing interferon therapy have shown that lower pretreatment serum HCV RNA levels measured with the bDNA assay are predictive of response (Davis et al., 1994; Lau et al., 1993; Yamada et al., 1995). Also, the rate at which serum HCV RNA levels decrease during interferon (IFN) therapy as measured with the bDNA assay has been shown to be informative (Orito et al., 1995; Terrault et al., 1994). A recent study found that the bDNA assay distinguished a more significant difference in serum HCV RNA concentration between complete and incomplete responders than did competitive RT-PCR (Toyoda et al., 1996), indicating that serum HCV RNA concentrations when measured with the bDNA assay are a better predictor of clinical response than when measured by competitive RT-PCR. Similar studies using the bDNA assay to measure hepatitis B virus (HBV) DNA have shown that different patterns of change in serum HBV DNA levels may be useful in distinguishing response to therapy and in modifying the therapeutic regimen of patients with hepatitis B (Hendricks et al., 1995; Hosotsubo et al., 1994; Watanabe et al., 1993). As was the case for comparative studies of HCV RNA assays, comparative studies of HBV DNA quantification assays showed that the bDNA assay exhibits greater sensitivity and precision in measuring HBV DNA in clinical specimens than do other assays (Kapke et al., in press; Zaaijer et al., 1994).

The bDNA assay also is rapidly becoming the method of choice for monitoring disease progression in patients infected with HIV-1. One of the most striking demonstrations of the prognostic value of measuring HIV-1 RNA with the bDNA assay was described in a recent report by Mellors et al. (Mellors et al., 1996). They showed that a single plasma HIV-1 RNA measurement provided important prognostic information concerning the time to development of AIDS and death, occurring as long as 10 years later. These and other studies have emphasized the importance of treatment strategies aimed at reducing viral load at all stages of disease.

Considered one of the most reproducible HIV-1 RNA quantification assays ready for use in clinical trials (Lin et al., 1994), the bDNA assay for quantification of HIV-1 RNA has facilitated the clinical evaluation of licensed and investigational antiretroviral drugs used to treat HIV disease. Relatively rapid drops in HIV-1 RNA levels, occurring within 1–2 weeks of therapy, have been discerned using the bDNA assay (Ho et al., 1995). Hence, it is possible to quickly assess the antiretroviral effect of drugs in vivo and thereby expedite the screening of potential antiretroviral agents. With newer, more potent drugs, even greater changes in HIV-1 RNA levels as measured with an enhance sensitivity bDNA assay (Kern et al., 1996; Todd et al., 1995) have been observed. These findings have revolutionized current thought on the dynamics of HIV-1 production (Ho et al., 1995; Perelson et al., 1996; Wei et al., 1995). The bDNA assay also was used to show that plasma HIV-1 RNA levels decrease upon treatment of subjects with zidovudine, and subsequently increase upon selection and proliferation of resistant virus or upon removal of drug therapy (Cao et al., 1995; Eastman et al., 1995). Similarly, the antiviral effect and emergence of viral resistance to drugs used in treatment of CMV disease, a frequent cause of morbidity and mortality in patients with HIV infection, can be effectively monitored by measuring changes in CMV DNA levels with the bDNA assay (Chernoff et al., submitted; Flood et al., submitted; Lalezari et al., 1995).

Given its ease of use, sensitivity, and reproducibility, the bDNA assay can readily be used for a broad range of clinical and research applications. In addition to viral load measurement, the bDNA assay recently has been applied to the measurement of cellular mRNA molecules. Since oligonucleotide probes can be designed for any target, bDNA assays can be configured for a variety of uses. bDNA technology has been used to measure cellular mRNA levels in a wide diversity of situations — from monitoring changes in cytokine mRNA levels in healthy and immunocompromised patient populations, to evaluating stress gene induction for molecular toxicology applications. This chapter describes the results from some of these recent studies using the bDNA assay for measurement of cellular mRNAs.

## Cytokine Messenger RNA Quantification

Cytokines play a central role in immunomodulation and thus may be useful as indicators of the presence and severity of disease. Recently, investigators have used the bDNA assay for direct quantification of multiple human cytokine mRNAs, including tumor necrosis factor$\alpha$ (TNF$\alpha$), interleukin-2 (IL-2), IL-4, IL-6, IL-10, and interferon-$\gamma$ (IFN-$\gamma$) (Shen et al., in preparation). An example of cytokine mRNA measurement using the bDNA assay is shown in Figure 1. In this experiment, IL-4 mRNA and protein expression were measured in a T cell lymphoma line, HUT-78, with and without phorbal myristate acetate (PMA) stimulation. When HUT-78 cells were stimulated with PMA, IL-4 mRNA levels peaked at 8 hr. By comparison, in unstimulated HUT-78 cells, IL-4 mRNA levels remained relatively stable. As expected, IL-4 protein levels peaked considerably later, after 25 hr of stimulation, and no IL-4 protein was detected in HUT-78 cells without stimulation. This experiment illustrates several features of the bDNA assay for quantification of mRNA. First, the bDNA assay can readily distinguish differences in mRNA levels between stimulated and unstimulated cells. Second, changes in the level of cytokine gene expression can be detected with the bDNA assay in a relatively short period of time. Third, mRNA levels measured with the bDNA assay follow an expected time course, in that the mRNA peak precedes the protein peak.

The bDNA assay for mRNA quantification offers the unique ability to differentiate between various spliced forms of mRNA. By designing oligo-

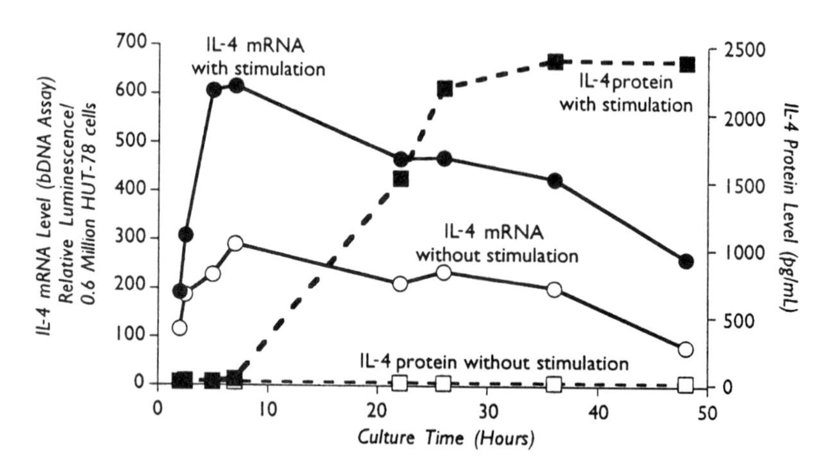


FIGURE 1. IL-4 mRNA and protein levels in HUT-78 cells with and without PMA stimulation.

nucleotide target probes to hybridize to specific regions of processed and unprocessed mRNA, it is possible to measure specifically the amount of precursor versus fully processed mRNA. This approach has been applied to characterize the splicing and processing of insulin mRNA (Wang et al., 1997). The processing of TNF$\alpha$ mRNA also has been investigated using oligonucleotide target probes specific for pre-TNF$\alpha$ mRNA and target probes that recognize both pre-TNF$\alpha$ as well as processed TNF$\alpha$ mRNA. The experiment illustrated in Figure 2 shows the distribution of pre-TNF$\alpha$ versus processed TNF$\alpha$ mRNA in nuclear and cytoplasmic fractions of THP-1 cells (human monocytic leukemia cells). Measurement of the DNA in the nuclear fractions versus total cellular DNA provided a means to assess the integrity of the nuclei during isolation. This experiment showed that whereas over 70% of pre-TNF$\alpha$ mRNA was localized to the nuclear fraction, only ~30% of processed TNF$\alpha$ mRNA was present in the nucleus. By contrast, the cytoplasmic fraction contained ~70% of processed TNF$\alpha$ mRNA and <30% of pre-TNF$\alpha$ mRNA. This intracellular distribution is consistent with what would be expected for processed versus unprocessed forms of TNF$\alpha$ mRNA. Thus, given the ability to select the placement of target probes along the transcript, it is possible to measure various spliced forms of mRNA and thereby gain understanding of the sequence of events in mRNA processing.

One of the potential clinical applications of the bDNA assay for measurement of cytokine mRNA is the evaluation of cytokine mRNA levels in clinical specimens. Figure 3 illustrates an experiment in which cytokine

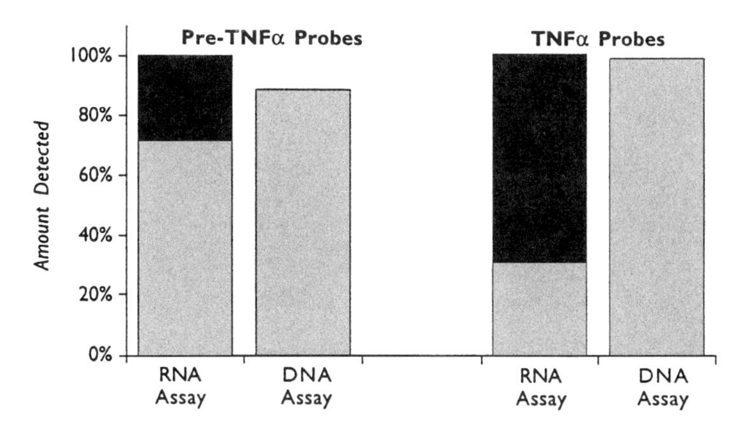


FIGURE 2. Pre-TNF$\alpha$ and processed TNF$\alpha$ mRNA in cytoplasmic (black bars) and nuclear (gray bars) fractions of THP-1 cells. As a control, total cellular DNA also was measured.

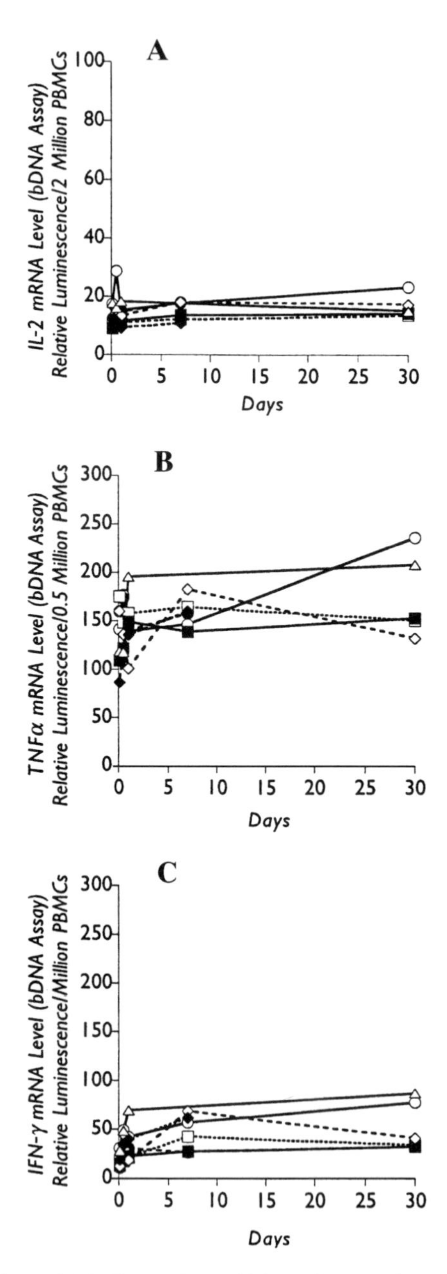


FIGURE 3. IL-2 (panel A), TNF$\alpha$ (panel B), and IFN-$\gamma$ (panel C) mRNA levels in PBMCs from healthy blood donors.

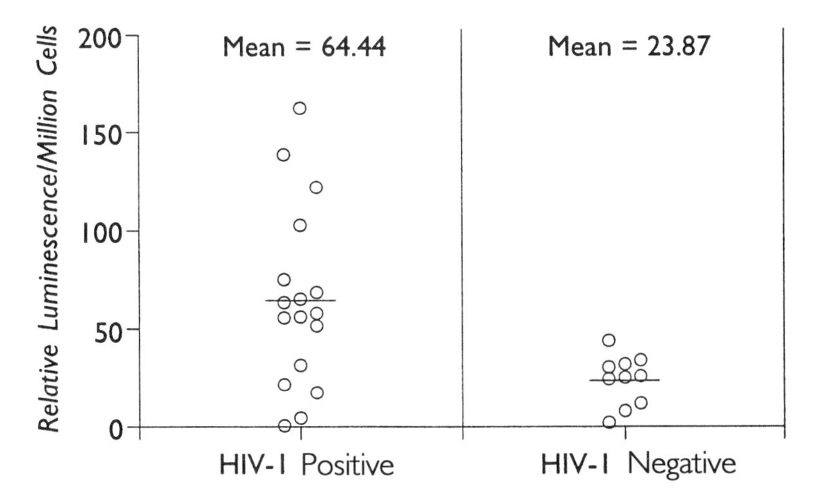


FIGURE 4. IFN-$\gamma$mRNA levels in PBMCs from HIV-1 positive ($n = 17$) and HIV-1 negative ($n = 10$) individuals. The bars indicate mean IFN-$\gamma$mRNA levels.

expression was monitored in peripheral blood mononuclear cells (PBMCs) from healthy blood donors. The results showed no significant changes in IL-2 mRNA levels, and less than two-fold changes in IFN-$\gamma$and TNF$\alpha$ mRNA levels in each individual over the course of 1 month, suggesting that cytokine mRNA levels remain relatively stable in healthy individuals. Hence, changes in normal cytokine mRNA levels could be measured that may indicate changes in the immune status of the individual. Quantification of cytokine mRNA levels also may provide valuable information for monitoring HIV-1-infected individuals. For example, mean levels of IFN-$\gamma$ mRNA in PBMCs from HIV-1 positive individuals were significantly ($p = 0.003$) higher than those of HIV-1 negative individuals (Figure 4). These results reflect the altered status of the immune system in HIV-1-infected individuals.

Given that gene expression may be controlled either at the translational or transcriptional level, different information can be obtained by measurement of mRNA as compared to protein. This difference in information is illustrated by an experiment, shown in Figure 5, in which PBMCs from HIV-1-infected individuals with CD4+ T-cell counts below or above 200 cells/$\mu$L were evaluated for mRNA and protein levels of four cytokines upon stimulation with *Phaseolus vulgarus* lectin (PHA) in vitro (Cao et al., in preparation). Whereas the relative gene expression of all four cytokines was suppressed in both groups, protein levels of these cytokines were either enhanced or suppressed upon stimulation with PHA, depending upon the

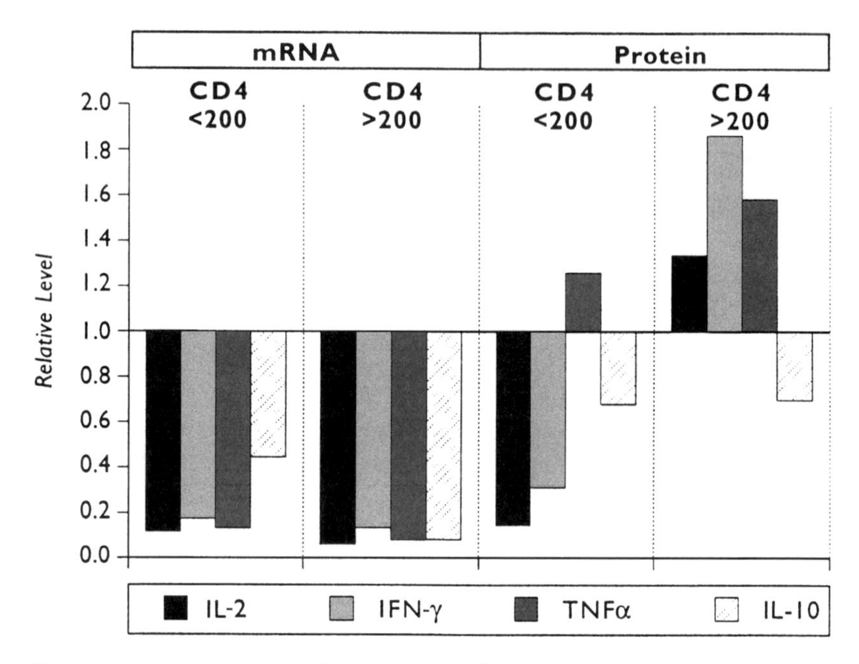


FIGURE 5. Relative levels of cytokine mRNA and protein in PBMCs from HIV-1-infected individuals with CD4+ T-cell counts below and above 200 cells/$\mu$L.

CD4+ T-cell counts of the HIV-1-infected individuals. IL-2 and IFN-$\gamma$ protein levels were suppressed in HIV-1-infected individuals with CD4+ T-cell counts below 200 cells/$\mu$L, but enhanced in those with CD4+ T-cell counts above 200 cells/$\mu$L. By comparison, TNF$\alpha$ protein levels were enhanced while IL-10 protein levels were suppressed in both groups. Thus, quantification of cytokine mRNA levels may provide useful information, unique from that provided by measurement of cytokine protein, for understanding HIV-1 disease progression.

## Quantification of Stress Gene Induction

Researchers also have investigated the potential utility of the bDNA assay in the emerging field of molecular toxicology in which markers of cellular damage are used as a means to evaluate toxicity. Among the array of cellular responses that may be monitored in molecular toxicology applications is the induction of "stress genes," or genes that respond in a characteristic manner to toxic stimuli (MacGregor et al., 1995). Several advantages of using the bDNA assay for measurement of stress gene induction include:

- Gene expression can be measured as a function of transcription under the control of endogenous promoters and/or response elements. No recombinant or fusion constructs are needed.
- Gene expression can be measured from either animal tissues or cell lines, and any animal tissue can be used.
- Transcription of stress genes can be measured at various time points, enabling the generation of a kinetic profile.
- Transcription of stress genes can be compared and correlated to organ damage and/or overall response of the organism.
- Assay components can be modified to deliver sensitivities appropriate to the expression levels of given target genes.

The stress genes for which oligonucleotide target probes have been designed for use in the bDNA assay include *c-fos*, *hsp-70*, and *GADD153*. *c-fos* is a transcription factor that is activated by several forms of stress, including DNA damage, oxidative stress, and exposure to phorbol ester tumor promoters (Deschamps et al., 1985; Hollander and Fornance, 1989). *hsp-70* is the 70-kDa heat shock protein that functions as a protein chaperone to refold or sequester damaged proteins and is induced by a wide range of physiological and chemical insults (Fischbach et al., 1993). *GADD153* is the 153-kDa Growth Arrest and DNA Damage protein, which is induced by a wide variety of DNA-damaging agents, although not by phorbol esters or x-irradiation (Bartlett et al., 1992; Crawford et al., 1996). Oligonucleotide probes also have been designed for control genes, including the ubiquitous cytoskeletal protein, b-actin, and a key enzyme in the glycolytic pathway, glyceraldehyde-3-phosphate dehydrogenase (GAPDH). These control genes were selected since they are moderately and stably expressed in response to most toxic agents; their levels of expression are relatively similar whether in the absence or presence of inducing agents (Biragyn et al., 1994; Zhao et al., 1995). More than one control gene was selected since there may be rare conditions that affect the regulation of one or the other gene.

Figure 6 illustrates the use of the bDNA assay in measuring the upregulation of *c-fos* in HepG2 cells following exposure to the DNA-damaging agent, cis-diamminedichloroplatinum, commonly known as cisplatin. In this experiment, both the optimal dosage of cisplatin and the time of exposure were evaluated in a single bDNA assay run. *c-fos* levels were measured in both induced and uninduced cells, and results were expressed as fold-induction of *c-fos*. Results were normalized to the expression of GADPH to control for any increases in cellular mRNA levels that may be caused by nonspecific effects, or by cell cycling. In addition, normalization corrects for any decreases in cellular mRNA levels due to cell death

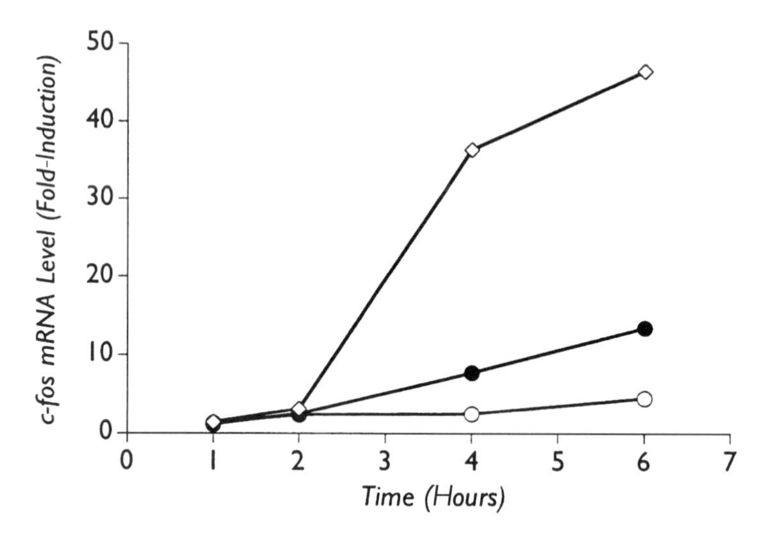


FIGURE 6. Fold-induction of *c-fos* mRNA in HepG2 cells following exposure to cisplatin at concentrations of 25 (○), 50 (●), and 100 (◇) micromolar.

caused by the toxic effects of cisplatin. The results of this experiment indicate that *c-fos* induction in response to cisplatin was dose-dependent and varied with the time of exposure. Thus, in a single bDNA assay run it is possible to optimize experimental conditions, such as the times at which the measurement is taken as well as the appropriate dosages of the inducing agent, to ensure a maximal induction response.

The second experiment, illustrated in Figure 7, compares the upregulation of all three stress genes — *c-fos*, *hsp-70*, and *GADD153* — in HepG2 cells after various times of induction by cisplatin. Again, the expression of stress genes was measured in both induced and uninduced cells, and results were normalized to GADPH expression. The results of this experiment show that, whereas *c-fos* is induced by cisplatin in a time-dependent manner, *hsp-70* and *GADD153* expression remains relatively stable over the 6 hr treatment period. Thus, in a single bDNA assay run, it is possible to distinguish differences in fold-induction by a toxic agent of several genes at various time points.

These examples illustrate the utility of bDNA technology in measuring the expression of *c-fos*, *hsp-70*, and *GADD153*, important markers of the toxic response, following exposure to cisplatin. As bDNA assays are developed for additional target genes, it will be possible to generate a profile of gene induction for any chemical of interest. Gene induction by a given compound, as measured by bDNA methods, can be confirmed by reference

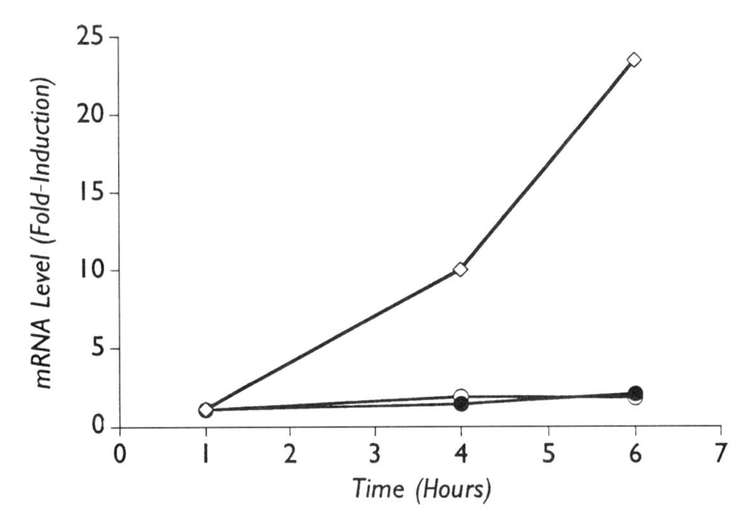


FIGURE 7. Fold-induction of *GADD153* mRNA (○), *hsp-70* mRNA (●), and *c-fos* mRNA (◇) in HepG2 cells following exposure to 50 micromolar cisplatin.

methods such as Northern blot analysis or ribonuclease protection assays. Such comparisons will reinforce the reliability, sensitivity, and ease of use of bDNA assays.

The bDNA assay format is particularly well suited for the measurement of stress gene induction. As opposed to other methods of molecular toxicity measurement, which use recombinant constructs, bDNA methods quantify expression from genes in their natural configuration, i.e., off endogenous promoters. A microtiter format of cell growth, exposure to agents, in situ lysis, and nucleic acid quantification by bDNA analysis provide a very sensitive and specific means for measuring expression of genes involved in the "stress response." The ultimate utility of in vitro cell-based assays will be proven when correlations among cell-based assays, tissue-specific markers of toxicity, and markers of organismal response, such as LD50, can be made for a broad range of toxic compounds. The bDNA assay can be applied toward the screening of a set of related compounds or combinatorial libraries. If a molecular profile of toxicity can be defined in terms of changes in expression among a set of genes, then compounds can be screened and chosen according to a desired response or profile. Indeed, bDNA technology can be applied toward any gene of interest in a number of cellular pathways that are impacted by exposure to toxic chemicals. These include genes involved in response to DNA damage, in redox, protein, ionic, or energy stress, in stress to cell surface receptor mediated processes, and genes that are transcriptional modulators, or activators or repressors of apoptosis.

## Concluding Remarks

The bDNA assay is an inherently quantitative and nonradioactive method that uses signal amplification for sensitive and reproducible quantification of nucleic acid molecules. As illustrated by the examples described in this chapter, the bDNA assay can be applied to a wide variety of applications. A key advantage of bDNA technology is that it can be adapted to in vitro or in vivo applications in multiple species. As long as sequence information is available for a given gene, specific target probes for capture and labeling can be designed. The versatility of the bDNA assay format, as well as its ease of use and potential for high-throughput automation, ensures that the bDNA assay will be an ascendant technology, not only for biomedical and toxicology applications, but for any application in which the quantification of gene expression is paramount.

## REFERENCES

Bartlett JD, Luethy JD, Carlson SG, Sollott SJ, and Holbrook NJ (1992): Calcium ionophore A23187 induces expression of the growth arrest and DNA damage inducible CCAAT/enhancer-binding protein (C/EBP)-related gene, gadd153. Ca2+ increases transcriptional activity and mRNA stability. *J Biol Chem* 267:20465–70.

Biragyn A, Arkins S, and Kelley KW (1994): Riboprobe expression cassettes for measuring IGF-I, beta-actin and glyceraldehyde 3-phosphate dehydrogenase transcripts. *J Immunol Meth* 168:235–44.

Cao Y, Ho DD, Todd J, Kokka R, Urdea M, Lifson JD, Piatak M, Jr., Chen S, Hahn BH, Saag MS, and Shaw GM (1995): Clinical evaluation of branched DNA (bDNA) signal amplification for quantifying HIV-1 in human plasma. *AIDS Res Hum Retroviruses* 11:353–61.

Cao WW, Landay AL, Shen L-P, Siegel J, DePaz N, and Kolberg J: Quantitative analysis of cytokine expression in peripheral blood mononuclear cells from HIV-infected individuals (in preparation).

Chernoff DN, Hoo BS, Shen L-P, Kelso RJ, Kolberg JA, Drew WL, Miner RC, Lalezari JP, and Urdea MS: Quantification of CMV DNA in peripheral blood leukocytes using a branched DNA (bDNA) signal amplification assay (submitted).

Crawford DR, Schools GP, and Davies KJ (1996): Oxidant-inducible adapt 15 RNA is associated with growth arrest- and DNA damage-inducible gadd153 and gadd45. *Arch Biochem Biophys* 329:137–44.

Davis GL, Lau JY, Urdea MS, Neuwald PD, Wilber JC, Lindsay K, Perrillo RP, and Albrecht J (1994): Quantitative detection of hepatitis C virus RNA with a solid-phase signal amplification method: definition of optimal conditions for specimen collection and clinical application in interferon-treated patients. *Hepatology* 19:1337–41.

Deschamps J, Meijlink F, and Verma IM (1985): Identification of a trancriptional enhancer element upstream from the proto-oncogene fos. *Science* 230:1174–7.

Eastman PS, Urdea M, Besemer D, Stempien M, and Kolberg J (1995): Comparison of selective polymerase chain reaction primers and differential probe hybridization of polymerase chain reaction products for determination of relative amounts of codon 215 mutant and wild-type HIV-1 populations. *J Acquir Immune Defic Syndr Hum Retrovirol* 9:264–73.

Fischbach M, Sabbioni E, and Bromley P (1993): Induction of the human growth hormone gene placed under human hsp70 promoter control in mouse cells: a quantitative indicator of metal toxicity. *Cell Biol Toxicol* 9:177–88.

Flood J, Drew WL, Miner R, Jekic-McMullen D, Shen L-P, Kolberg J, Garvey J, Follansbee S, and Poscher M: Detection of cytomegalovirus (CMV) polyradiculopathy and documentation of in vivo anti-CMV activity in cerebrospinal fluid using branched DNA signal amplification and CMV antigen assays (submitted).

Hendricks DA, Stowe BS, Hoo BS, Kolberg J, Irvine BS, Neuwald PD, Urdea MS, and Perillo RP (1995): Quantitation of HBV DNA in human serum using a branched DNA (bDNA) signal amplification assay. *Am J Clin Pathol* 104:537–46.

Ho DD, Neumann AU, Perelson AS, Chen W, Leonard JM, and Markowitz M (1995): Rapid turnover of plasma virions and CD4 lymphocytes in HIV-1 infection. *Nature* 373:123–6.

Hollander MC and Fornance AJ (1989): Induction of fos RNA by DNA-damaging agents. *Cancer Res* 49:1687–92.

Hosotsubo H et al. (1994): Quantitation of serum HBV DNA by branched DNA probe assay with chemiluminescence detection. *Med Pharm* 31:743–51.

Kapke GF, Watson G, Sheffler S, Hunt D, and Frederick C: Comparison of the Chiron Quantiplex branched DNA (bDNA) assay and the Abbott Genostics solution hybridization assay for quantification of hepatitis B viral DNA. *J Virol Hepatitis* (in press).

Kern D, Collins M, Fultz T, Detmer J, Hamren S, Peterkin JJ, Sheridan P, Urdea M, White R, Yeghiazurian T, and Todd J (1996): An enhanced sensitivity branched DNA assay for the quantification of human immunodeficiency virus type 1 RNA in plasma. *J Clin Microbiol* 34:3196–202.

Lalezari JP, Drew WL, Glutzer E, James C, Miner D, Flaherty J, Fisher PE, Cundy K, Hannigan J, Martin JC, and Jaffe HS (1995): (S)-1-[3-hydroxy-2-(phosphonylmethoxy)propyl]cytosine (cidofovir): results of a phase I/II study of a novel antiviral nucleotide analogue. *J Infect Dis* 171:788–96.

Lau JYN, Mizokami M, Ohno T, Diamond DA, Kniffen J, and Davis GL (1993): Discrepancy between biochemical and virological responses to interferon-a in chronic hepatitis C. *Lancet* 342:1208–9.

Lin HJ, Myers LE, Yen LB, Hollinger FB, Henrard D, Hooper CJ, Kokka R, Kwok S, Rasheed S, Vahey M, Winters MA, McQuay LJ, Nara PL, Reichelderfer P, Coombs RW, and Jackson JB (1994): Multicenter evaluation of quantification methods for plasma human immunodeficiency virus type 1 RNA. *J Infect Dis* 170:553–62.

MacGregor JT, Farr S, Tucker JD, Heddle JA, Tice RR, and Turteltaub KW (1995): New molecular endpoints and methods for routine toxicity testing. *Fundamental Appl Toxicol* 26:156–73.

Mellors JW, Rinaldo CRJ, Gupta P, White RM, Todd JA, and Kingsley LA (1996): Prognosis in HIV-1 infection predicted by the quantity of virus in plasma. *Science* 272:1167–70.

Mullis KB and Faloona FA (1987): Specific synthesis of DNA in vitro via a polymerase-catalyzed chain reaction. *Meth Enzymol* 155:335–50.

Orito E, Mizokami M, Suzuki K, Ohba K, Ohno T, Mori M, Hayashi K, Kato K, Iino S, and Lau JYN (1995): Loss of serum HCV RNA at week 4 of interferon-a therapy is associated with more favorable long-term response in patients with chronic hepatitis C. *J Med Virol* 46:109–15.

Perelson AS, Neuman AU, Markowitz M, Leonard JM, and Ho DD (1996): HIV-1 dynamics in vivo: virion clearance rate, infected cell life-span, and viral generation time. *Science* 271:1582–6.

Shen L-P, Sheridan P, Cao W, Dailey PJ, Urdea MS, and Kolberg J: Quantification of cytokine mRNA in peripheral blood mononuclear cells using branched DNA (bDNA) technology (in preparation).

Terrault N, Wilber J, and Feinman SV (1994): Response to interferon therapy predicted by changes in serum HCV-RNA titer early in treatment. *In* Program and Abstracts of the Digestive Diseases Meeting, New Orleans, LA.

Todd J, Pachl C, White R, Yeghiazarian T, Johnson P, Taylor B, Holodniy M, Kern D, Hamren S, Chernoff D, and Urdea M (1995): Performance characteristics for the quantitation of plasma HIV-1 RNA using branched DNA signal amplification technology. *J Acquir Immune Defic Syndr Hum Retrovirol* 10 (supplement 2):S35–S44.

Toyoda H, Nakano S, Kumada T, Takeda I, Sugiyama K, Osada T, Kiriyama S, Orito E, and Mizokami M (1996): Comparison of serum hepatitis C virus RNA concentration by branched DNA probe assay with competitive reverse transcription polymerase chain reaction as a predictor of response to interferon-alpha therapy in chronic hepatitis C patients. *J Med Virol* 48:354–9.

Wang J, Shen L-P, Najafi H, Kolberg JA, Matschinsky FM, Urdea M, and German M (1997): Regulation of insulin preRNA splicing by glucose. *Proc*

*Natl Acad Sci USA* (in press).

Watanabe Y, Hino K et al. (1993): Detection of HBV DNA by bDNA probe assay and evaluation of its clinical utility. *Med Pharm* 30:1217–26.

Wei X, Ghosh SK, Taylor ME, Johnson VA, Emini EA, Deutsch P, Lifson JD, Bonhoeffer S, Nowak MA, Hahn BH, Saag MS, and Shaw GM (1995): Viral dynamics in human immunodeficiency virus type 1 infection. *Nature* 373:117–26.

Yamada G, Takatani M, Kishi F, Takahashi M, Doi T, Tsuji T, Shin S, Tanno M, Urdea MS, and Kolberg JA (1995): Efficacy of interferon alfa therapy in chronic hepatitis C patients depends primarily on hepatitis C virus RNA level. *Hepatology* 22:1351–4.

Zaaijer HL, ter Borg F, Cuypers HTM, Hermus MCAH, and Lelie PN (1994): Comparison of methods for detection of hepatitis B virus DNA. *J Clin Microbiol* 32:2088–91.

Zhao J, Araki N, and Nishimoto SK (1995): Quantitation of matrix Gla protein mRNA by competitive polymerase chain reaction using glyceraldehyde-3-phosphate dehydrogenase as an internal control. *Gene* 155:159–65.

# Quantification of Plasmid DNA Expression in Vivo

Marston Manthorpe, Jukka Hartikka, H. Lee Vahlsing,
and Michael Sawdey

## Introduction

Gene therapy can be generally defined as the *delivery* of *polynucleotides* into specific cells for the *benefit* of the organism. The aim of *delivery* is to target functional polynucleotides to specific tissues or cells. The *polynucleotides* may be DNA or RNA, sense or antisense, single-, double-, or triple stranded, and can be packaged within viral vectors, formulated together with protein, lipid, carbohydrate, or synthetic molecules, or administered in purified form. Potential *benefits* of gene therapy include the generation of immune responses against infectious pathogens or cancer cells, production of therapeutic proteins, and correction of defective or deficient gene expression.

A simple gene therapy technique is to inject solutions of purified DNA molecules directly into target tissues. Despite the nearly ubiquitous presence of deoxyribonucleases, the injection of a saline solution containing purified plasmid DNA into tissue results in uptake of the DNA by resident cells and persistent long-term expression of the encoded genes (Wolff et al., 1990; Hartikka et al., 1996a). Thus, unlike conventional protein-based drug therapies, plasmid-DNA-based gene therapy allows individual cells to be converted into minifactories that produce a desired protein continuously. Unlike retroviral or adenoviral vectors used for gene therapy, plasmid DNA vectors do not elicit antivector immune responses that can prevent effective vector readministration. In rare instances, integration of viral gene therapy vectors into host cell DNA may result in pathogenic insertional mutagenesis, dysregulation of host cell genes, and/or the loss of vector gene expression. However, plasmid DNA within cells remains episomal (Danko and Wolff, 1994).

The success of plasmid-DNA-based gene therapy relies on the ability to develop potent, cell-specific, and regulable expression vectors. The development of improved vectors, in turn, requires the ability to precisely quantify expression levels elicited by different vectors of known sequence. Quantitative dose-response and time-course experiments can be performed to assess the extent to which DNA sequence modifications affect the dose-dependency and kinetics of gene expression. In addition, quantitative assays can be employed to monitor the effects of different tissue injection methods (Manthorpe et al., 1993), examine the effects of transfection-enhancing agents coinjected with the DNA (Wells, 1993), and develop plasmid DNA vectors whose expression can be regulated by systemic administration of small molecules (Liang et al., 1996a).

Assays used to compare plasmid DNA expression levels include in vitro transfection of cultured cells and in vivo transfection of various tissues. Screening vectors in vivo is more costly, more time-consuming, and more likely to produce highly variable results between individual animals or tissues. Nonetheless, it is our experience that relative expression levels measured in vitro do not correlate well with those measured in vivo. Thus, there is a need to develop convenient, reliable, high-throughput assays that measure DNA vector expression levels in vivo.

Over the past 4 years, we have extensively used quantitative in vivo assays to measure plasmid DNA expression levels in various tissues. In this review, we will describe how the present methodology was adopted, on the basis of considerations of reporter gene selection, DNA preparation, tissue selection, DNA delivery, tissue extraction, assay of transgene products, and data analysis. Following the methodology section, several examples will be presented showing how the assays were applied to quantify improvements in plasmid DNA expression in vivo.

## Methodology

### *Reporter gene selection*

In the first demonstration that injected plasmid DNA can be taken up and expressed within a tissue, Wolff et al. (1990) injected plasmid DNAs encoding bacterial $\beta$-galactosidase (LacZ), chloramphenicol acetyltransferase (CAT), or firefly luciferase into mouse muscle. The LacZ reporter was used for the histochemical demonstration of $\beta$-galactosidase gene product within individual myofiber cells. While it is well suited to this purpose (i.e., histochemical identification of transfected cells), LacZ is not the reporter of choice for quantitative assays, because of the presence of high levels of endogenous $\beta$-galactosidase activity in mammalian tissues and cells (Marsh, 1994; Hendrikx et al., 1994). Of the remaining two reporters (luciferase and

CAT), both exhibit minimal background activity in mammalian tissues. However, in our judgment, luciferase is the preferred reporter for the following reasons:

1) Convenience. Luciferase assays can currently be performed in microplate format utilizing a fully automated luminometer. A single plate with 96 samples requires less than an hour to set up and read. In contrast, manually performed CAT assays (Sankaran, 1992) require 2–4 hours for substrate addition, incubation, and scintillation-counting steps. Automated CAT assays utilizing robotized laboratory workstations have been described (Chauchereau et al., 1990). However, robotization does not reduce the length of time (typically 3 hr) required for sample incubations. Commercially available CAT ELISAs employing colorimetric detection (Boehringer Mannheim, Indianapolis, IN) may require 4 hr or more to complete.

2) Cost. In general, luciferase assay reagents cost less than CAT assay reagents. Moreover, luciferase standard curves for in vivo assays typically exhibit a linear range spanning 4 logs, while those of CAT span 1–2 logs. Thus, individual CAT samples not falling within the linear range of the standard curve must be reassayed until the proper dilution is found, incurring additional cost. An indirect cost is also present in the form of the radioactive waste generated from scintillation-based CAT assays, which requires proper containment and disposal.

3) Half-life of the enzyme. In cultured HepG2 cells, luciferase has a reported half-life of 3 hr while CAT has a half-life of 50 hr (Thompson et al., 1991). The relative half-lives of luciferase and CAT may also be similar in vivo, since absolute levels of CAT expression from equivalent vectors are five- to tenfold higher and since the time course of CAT expression is more protracted than that of luciferase. In practical terms, this means that CAT may be the more useful reporter for detecting very low or trace levels of gene expression in poorly transfected tissues. However, luciferase activity will better reflect real-time changes in gene transcription and messenger RNA levels, and is therefore preferred for measuring expression kinetics.

### *DNA preparation*

Large-batch preparations of DNA are recommended, since repeated experiments can be performed with the same batch. In addition, highly purified DNA free of endotoxin and other bacterial contaminants is required for injection. We employ standard shaker flask cultures in freshly prepared Terrific Broth under antibiotic selection with Kanamycin (for details on the growth and propagation of plasmids in *E. coli,* see Sambrook et al., 1989). Plasmid yields are optimized by the use of a modified pUC-derived plasmid backbone that generates high plasmid copy number in *E. coli* (Montgomery et al., 1993). With these refinements, yields of 3–6 mg per liter of culture are obtained. For DNA purification, a modified alkaline lysis procedure (Horn et al., 1995) is utilized to prepare cleared lysates, followed by double CsCl-ethidium bromide gradient ultracentrifugation (Hartikka et al., 1996a) and ethanol precipitation. As an alternative to ultracentrifugation, commercially available column-based purification procedures may be employed. However, the latter can result in high concentrations of endotoxin and other low molecular weight contaminants in the final preparation. Following ethanol precipitation, the DNA is resuspended at 4°C over-night in USP saline (McGaw, Irvine, CA) or distilled water (Gibco/BRL, Bethesda, MD) and the concentration of DNA is determined spectrophotometrically. Purity is confirmed by 260 nm/280 nm absorbance ratios of 1.8 or higher. For quality control, each batch of DNA is assayed for the presence of endotoxin (Limulus Amebocyte Lysate assay, Associates of Cape Cod, Woods Hole, MA) and protein (bicinchoninic acid assay, Pierce Chem. Co., Rockford, IL), and is free of visible bacterial RNA or DNA in ethidium-bromide-stained agarose gels. Electrophoretic analysis is also used to evaluate the percentage of supercoiled plasmid (usually >95%) and to confirm plasmid identity by digestion with at least three different restriction endonucleases.

### *Tissue selection*

In the original report describing direct intramuscular injection of plasmid DNA, Wolff et al. (1990) injected murine quadriceps. Quadriceps muscles have since been preferred because of their relatively large size and reproducibility of transfection (Wolff and Lederburg, 1995). Other muscles that have been transfected by direct injection of plasmid DNA include tibialis anterior (Wells, 1993), soleus (Vitadello et al., 1994), gastrocnemius (Sahenk et al., 1993), tongue (Prigozy et al., 1993), and heart (Ascadi et al., 1991).

Nonmuscle tissues injected with plasmid DNA alone generally exhibit much lower expression levels. Nonetheless, since these tissues represent attractive targets for gene therapy, they have been used to evaluate gene expression from various vectors and to evaluate transfection-enhancing agents. Target tissues for transfection include lung (Tsan et al., 1995; Wheeler et al., 1996), liver (Hickman et al., 1994), thymus (DeMatteo et al., 1995), tumor (Nabel et al., 1993; Plautz et al., 1993), dermis (Raz et al., 1994), thyroid (Sikes et al., 1994), brain (Holt et al., 1990), articular joints (Yovandich et al., 1995), and endothelium (Zhu et al., 1993; Nabel and Nabel, 1993). All of these tissues are readily amenable to quantitative analysis of gene expression using the methodologies described below.

## *DNA delivery*

In order to accommodate large numbers of injections, the injection procedure must be convenient and rapid. Convenience can be increased by choosing readily accessible tissues and minimizing the need for procedures such as anesthesia, animal preparation (e.g., shaving), and surgery. As described below, we have developed convenient nonsurgical methods for intramuscular, intracardiac, subcutaneous, intrahepatic, and intratumoral injections.

For intramuscular injections of mice, the quadriceps muscle group represents a readily accessible and well-characterized target. To refine the technique for injection, the anatomy of the mouse quadriceps was examined in detail. The largest of the four quadriceps muscles is the rectus femoris, which is associated with three vastus muscles (v. lateralis, v. medialis, and v. intermedialis).

Indiscriminate injection of the quadriceps results in transfection of one or more of these muscles. To maximize transfection efficiency, a method was developed to target the rectus femoris muscle (Manthorpe et al., 1993). Salient aspects of the injection method are diagramed in Figure 1A. The thigh area of 4–16 week old, awake, restrained mice is first wetted with 70% isopropanol to visualize the quadriceps. A 3cc syringe fitted with a 28 gauge 1/2 needle (Becton Dickinson, Franklin Lakes, NJ, Cat. # BD9430) and a plastic collar (see below) is inserted into the distal one-third of the quadriceps, immediately proximal to the patella, at an angle of approximately 60° relative to the longitudinal axis. Fifty $\mu$l of the solution for injection is then delivered over a period of 1–2 sec. Injection depth is limited to 2 mm (the depth of the rectus femoris muscle) by gluing a plastic collar derived from a standard micropipette tip to the base of the syringe, so that the needle tip protrudes 2 mm past the end of the collar. A typical experiment uses 100 muscles (50 mice), where 10 muscles (five mice per cage) are injected with each test plasmid. With practice, a two-person team (one to position and restrain the animal, the other to swab and inject the muscle) can inject both quadriceps of a mouse in less than 20 sec. At this level of efficiency, 50 mice can be injected in less than 45 min.

Transfection of other tissues by direct injection of awake mice is carried out as depicted in Figure 1 (panels B–G). The tibialis anterior muscle is injected with 50 $\mu$l solution into the proximal two-thirds of the muscle (panel B). Heart muscles can be quickly and accurately transfected with 100 $\mu$l DNA solution injected adjacent to the sternum into the left ventricular apex (panel C). The liver can be injected by inserting the needle

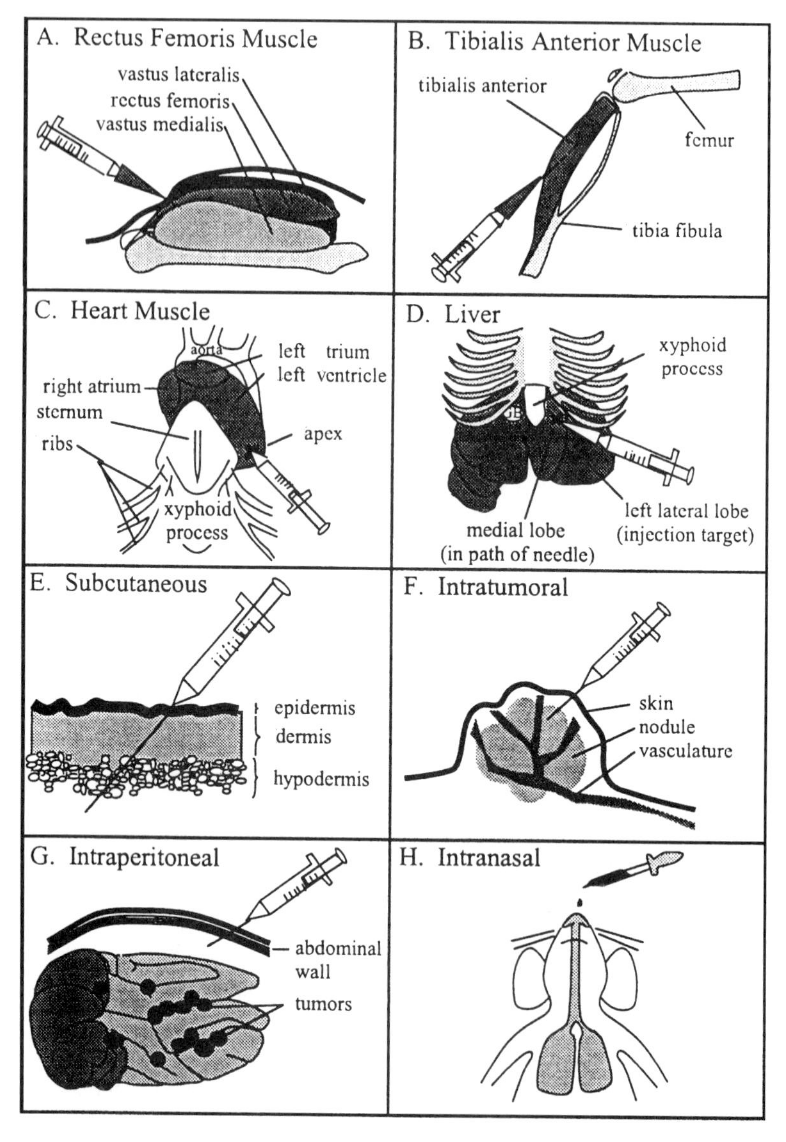


FIGURE 1. Methods for injection of mouse tissues. Awake, restrained mice can be quickly injected with plasmid DNA solutions into A. rectus femoris muscle, B. tibialis anterior muscle, C. heart muscle, D. liver, E. subcutaneous tissue, F. subcutaneous tumor, and G. intraperitoneal tumor. H. Lightly anesthetized mice can be intubated intranasally to introduce DNA into the lung.

adjacent to the xyphoid process, and penetrating to a depth that traverses the overlying medial lobe to target the left lateral lobe (panel D). Skin can be easily transfected by direct injection of DNA into the subcutaneous tissue (panel E). Intradermal tumors can be transfected by direct injection into individual tumor nodules (panel F). Intraperitoneal tumors are transfected by injection of DNA solution into the peritoneal cavity (panel G). Lungs are transfected with 100 $\mu$l DNA solution administered intranasally in mice lightly anesthetized with Metophane (Whitman Moore, Inc., Mundelien, IL; panel H).

### *Tissue extraction*

In order to carry out extractions for large numbers of tissues in a timely manner, a procedure was developed that enabled rapid processing and extraction of tissue samples with minimal loss of reporter activity. For this purpose, control experiments were performed with uninjected tissue samples spiked with a known amount of purified luciferase. Different tissue-processing methods were attempted, including (1) homogenization in the presence or absence of detergent with tight-fitting Teflon™ or glass pestles, (2) blending with rotating stainless steel blade devices, (3) vortexing samples with glass or plastic beads, and (4) extrusion of tissues through narrow orifices. All of these methods resulted in recovery of less than 75% of the added luciferase, which is possibly related to the generation of foam and heat that inactivated enzyme activity. Better recoveries were obtained by pulverizing frozen tissues, using a standard porcelain mortar and pestle. However, this tedious method was not suitable for processing large numbers of samples. Thus, the complete extraction procedure for mouse tissues was modified as depicted in Figure 2.

Animals are euthanized at various times postinjection with an overdose of sodium pentobarbital (Euthanasia-5, H. Schein Inc., Port Washington, NY; 4.5 mg/mouse i.p.), and fresh tissue is collected and immediately frozen on dry ice in 1.5 ml microcentrifuge tubes. Tissue samples can be stored at –78°C until processing. Frozen tissue pieces (<500 mg wet weight) are transferred into 2 ml microcentrifuge tubes containing 0.4 ml of frozen lysis buffer (Cell Culture Lysis Reagent, Promega, Madison, WI), and the tubes are inserted into an aluminum heat block partially immersed in a liquid nitrogen bath. The frozen tissue is ground into a fine powder using a reversible electric masonry drill (SKIL Xtra-tool 600, SKIL Corp., Chicago, IL) mounted onto a drill press. The drill bit size (19/64″) is such that it just fits into the microcentrifuge tube. The bits are prechilled in liquid nitrogen to prevent sample adherence, and must be driven in thereverse direction to prevent ground tissue powder from rising through the bit grooves. To

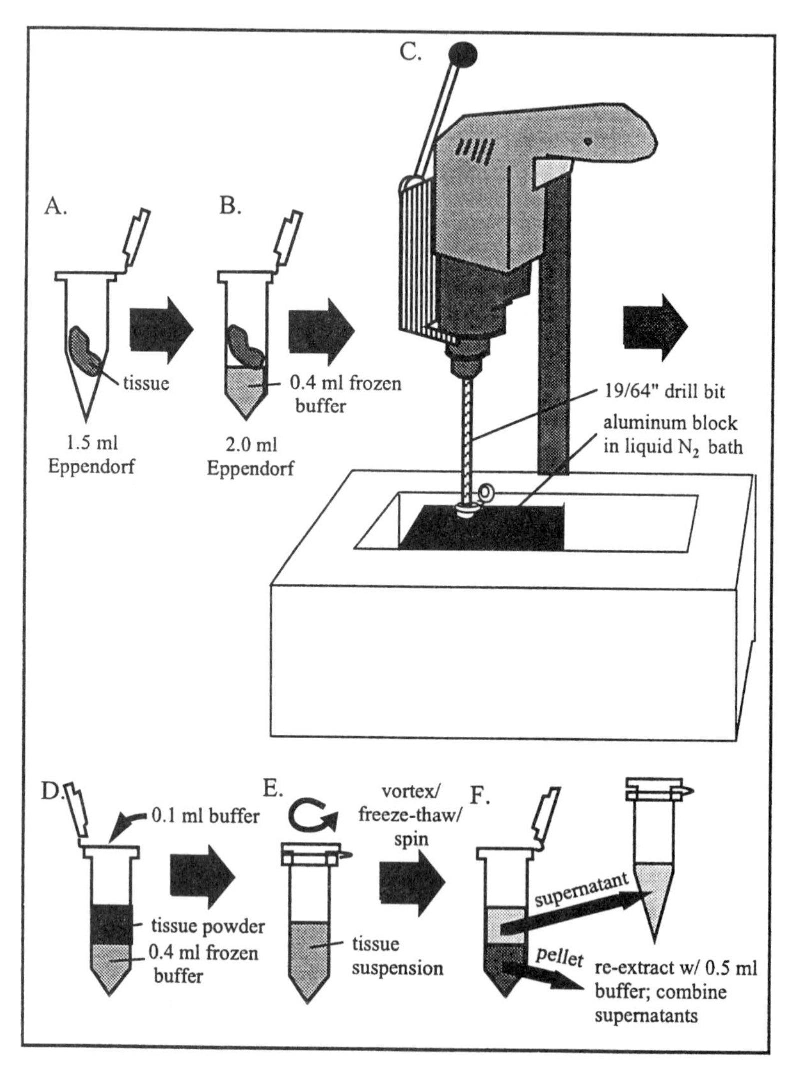


FIGURE 2. Rapid procedure for extraction of plasmid-DNA-encoded gene products from mouse tissues: A. tissue is rapidly collected and frozen; B. tissue is removed to a tube containing 0.4 ml frozen lysis buffer; C. tissue is pulverized with a drill bit run in reverse direction; D. 0.1 ml buffer is added to thaw tissue powder; E. tissue powder is extracted by vortexing; F. supernatant is removed and combined with a second 0.5 ml extraction.

minimize cross-contamination, bits are cleaned between samples and changed between sets of ten samples.

With practice, 50 tissue pieces can be pulverized into powder in 1 hr using the drilling procedure. Care must be taken to avoid cracking the tubes as the result of drill misalignment or excessive downward pressure. However, in the rare event that a tube does crack, the powder and remaining tissue can be transferred to another tube for completion of grinding. Sample powders not extracted immediately after grinding can be stored frozen at –78°C without loss of luciferase activity.

For extraction, frozen muscle powders are thawed by adding 0.1 ml lysis buffer. The samples are vortexed at room temperature for 15 min using a multitube platform, freeze-thawed three times using alternating liquid nitrogen and 23°C water baths, and centrifuged 3 min at 10,000 × g. (Note: luciferase activity is rapidly lost if samples are thawed at temperatures exceeding 37°C.) The supernatant is transferred to a separate 1.5 ml tube kept at 4°C, another 0.5 ml lysis buffer is added to the pellet, and the extraction process is repeated without freeze-thawing. The second supernatant is combined with the first, and the extract is stored at –78°C until assay.

In control experiments, this procedure allowed the recovery of <95% of a luciferase aliquot added to uninjected muscle samples prior to grinding and extraction. We have employed this same procedure to successfully process and extract tumor, lung, liver, heart, and skin samples from plasmid-DNA-injected animals for analysis of luciferase activity. While the extraction procedure described above has been optimized for luciferase, it can be adapted for use with other reporters simply by substituting an optimal lysis buffer. For example, for extraction of CAT-expressing tissues, Reporter Lysis Buffer (Promega, Madison, WI) can be employed. For other reporters or gene products, the investigator is advised to first perform preliminary spiking experiments to assess protein recovery and stability.

### *Assay of transgene products*

Luciferase activity is assayed using a 96-well microplate luminometer (Model ML2250, Dynatech, Chantilly, VA). Tissue extract (10–20 μl) is placed in each microwell and the plate inserted into the luminometer. In a fully automated procedure, 100 μl of luciferase substrate is added to each well in succession, and sample relative light units (RLU) are recorded several times within 1–2 sec after substrate addition. The luciferase content of the samples is calculated from peak RLUs, using a standard curve of purified firefly luciferase diluted in pooled extract from uninjected tissues. The specific activity of the commercial firefly luciferase enzyme reagent (Analytical Luminescence Labs, Ann Arbor, MI, Cat. No. 2400) is reported by the manufacturer to be $1.69 \times 10^{13}$ RLU/mg protein. For a given extract, the luciferase assay is highly reproducible, yielding <5% variation in RLUs within triplicate samples. In control experiments, representative muscle extracts containing high, medium and low levels of luciferase activity have been stored at −78°C in replicate aliquots and periodically assayed over a three-year interval, yielding <10% variation in RLUs.

CAT enzymatic activity is quantified using $^3$H-acetyl-CoA and extraction of reaction product directly into scintillation fluid as described(Sankaran, 1992). Reactions are incubated at 37°C for 3 hr prior to extraction and counting. To minimize the need to reassay samples exceeding the linear range of the standard curve, multiple dilutions of unknown samples can be assayed. As with luciferase activity measurements, a standard curve is performed in the presence of tissue extract and run concurrently with the test samples, using a commercially available CAT standard (Sigma, St. Louis, MO).

### *Data analysis*

Statistical comparisons are performed using the nonparametric Mann-Whitney rank sum test (SigmaStat version 2.0, Jandel Scientific Software, San Rafael, CA). This rank-order test is preferred because it makes no assumptions regarding the type of distribution of values around the mean (e.g., normal, log-normal, etc.), and it is valid for use in comparing two independent groups that contain the same or different numbers of values. Experimental group averages are considered significantly different fromcontrol group averages if the $P$ value is less |than 0.05.

## Applications

The above methodology has been used extensively to evaluate various parameters of expression after direct injection of plasmid DNA into mouse tissues. These parameters include expression potency and persistence, assessment of vector improvements, analysis of expression variability in a single tissue or in multiple tissues, the effect of transfection-inhibiting or -enhancing agents, and regulable vector expression.

### *Expression potency and persistence*

The potency of plasmid DNA vectors in vivo can best be determined by performing dose-response experiments. In the experiment shown in Figure

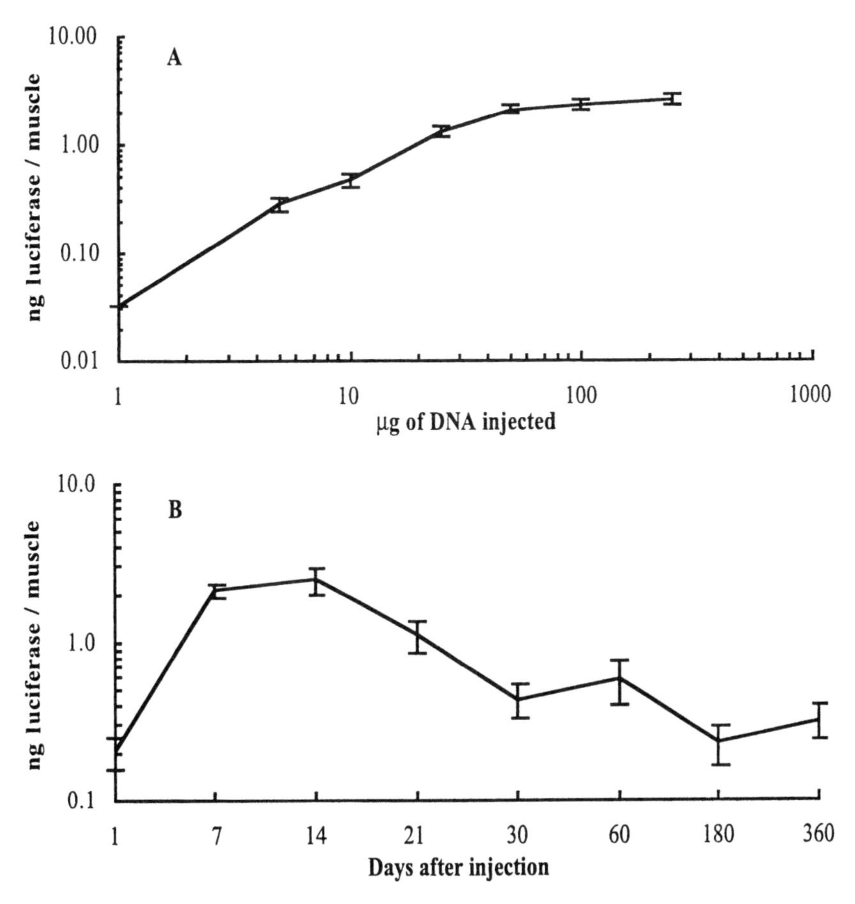


FIGURE 3. Dose response and time course of expression of nRSVLux DNA injected into mouse rectus femoris. A. Dose response at 7 days postinjection of 50 $\mu$l saline containing the indicated dose of DNA. Values are averages ± standard errors of the following number of muscles per dose: 1 $\mu$g (60), 5 $\mu$g (58), 10 $\mu$g (68), 25 $\mu$g (87), 50 $\mu$g (198), 100 $\mu$g (69), and 250 $\mu$g (49). B. Time course of expression up to 360 days postinjection of 50 $\mu$g DNA. Values are averages ± standard errors of the following number of muscles per time point: 1 d (40), 7 d (219), 14 d (40), 21 d (30), 30 d (40), 60 d (49), 180 d (69), and 360 d (78).

3A, mouse rectus femoris muscles were injected with 50 $\mu$l of saline containing various amounts of nRSVLux plasmid DNA (see Figure 4), and the muscles were processed for luciferase assays as described.

Luciferase expression rose from 0.03 ng luciferase/muscle, when 1 $\mu$g of DNA was injected, to 2.0–2.5 $\mu$g/muscle when 50–250 $\mu$g was injected. Thus,

| Plasmid | Plasmid Map | | | | | | ng Luciferase Muscle Avg ± SE | Med | Max |
|---|---|---|---|---|---|---|---|---|---|
| nRSVLux | pUC | A | RSV | | Lux | SV40 | 2.2 ± 0.2 | 1.4 | 18 |
| VR1204 | pUC | A | CMV | | Lux | SV40 | 5.2 ± 0.6 | 2.2 | 54 |
| VR1205 | pUC-O | A | CMV | int | Lux | SV40 | 6.5 ± 0.3 | 3.3 | 72 |
| VR1210 | pUC-O | K | CMV | int | Lux | BGH | 53 ± 4.4 | 31 | 315 |
| VR1216 | pUC-O | K | CMV | int | Lux | BGH | 82 ± 5.9 | 63 | 529 |
| VR1223 | pUC-O* | K | CMV | int | Luc+ | BGH | 175 ± 16 | 110 | 1720 |
| VR1255 | pUC-O* | K | CMV | int | Luc+ | RBG | 301 ± 30 | 250 | 1290 |

FIGURE 4. Comparison of intramuscular expression of different plasmid DNAs encoding firefly luciferase. Various elements of the nRSVLux plasmid were modified as depicted in the plasmid maps to yield the plasmids labeled VR1204 to VR1255. Fifty $\mu$g DNA in 50$\mu$l saline was injected into mouse rectus femoris muscle. The muscles were collected at 7 days postinjection and assayed for luciferase activity. Data are presented as average ± SE [$n$ = 259 (nRSVLux); 194 (VR1204); 851 (VR1205); 181 (VR1210); 201 (VR1216); 183 (VR1223); 78 (VR1255)]. Abbreviations: A, $\beta$-lactamase (ampicillin resistance) gene; BGH, bovine growth hormone terminator; CMV, human cytomegalovirus immediate early 1 promoter/enhancer; int, CMV intron A; K, aminoglycoside 3-phosphotransferase (kanamycin resistance) gene; Lux, peroxisomal luciferase; Luc+, modified (cytoplasmic) luciferase; RBG, minimal rabbit $\beta$-globin terminator; pUC, pUC18 plasmid backbone; pUC-O, pUC18 without SV40 origin of replication; pUC-O*, modified pUC-O plasmid backbone; RSV, Rous sarcoma virus promoter/enhancer; SV40, SV40 early region terminator; Med, median value; Max, maximum value. (Adapted from Hartikka et al., 1996a.)

expression rises linearly with doses up to 50$\mu$g. The plateau in expression at doses greater than 50$\mu$g suggests that factors governing the uptake or expression of plasmid DNA may be saturated at these doses.

To measure persistence of expression, time-course experiments can be performed. In the experiment shown in Figure 3B, mouse rectus femoris muscles were injected with 50$\mu$l of saline containing 50$\mu$g of nRSVLux DNA and harvested at the indicated times postinjection. Luciferase expression rose to a peak at about 2.5 ng/muscle by 7–14 days postinjection, and then declined to about 0.2–0.6 ng/muscle for the remaining time points. The probable cause for this decline in expression is that the mouse develops a cellular immune response against the foreign transgene product (i.e., insect luciferase). This interpretation is supported by the finding that luciferase expression persists at high levels in immunocompromised mice (Hartikka et al., 1996b), and by the finding that injection of plasmid DNA encoding murine "self" proteins results in persistent long-term expression in immunocompetent mice (Margalith et al., 1996; Tripathy et al., 1996).

*Assessment of vector improvements*

Performing dose responses and time courses for each test plasmid is impractical when assessing vector improvements. Since the dose-response and time-course analyses shown in Figure 3 indicated that optimal expression in mouse rectus femoris muscle was obtained by injecting 50 $\mu$g DNA in 50 $\mu$l saline and measuring expression at 7 days postinjection, we used these conditions to compare expression of test plasmids with nRSVLux. The results, as well as a simplified plasmid map for each DNA, are shown in Figure 4.

Incremental changes in vector components such as promoter/enhancers, transcriptional terminators, plasmid backbone elements, and the luciferase gene itself increased expression in muscle. The Luc+ gene (Promega, Madison, WI) has been modified in several ways to enhance its utility as a reporter gene. However, the primary reason for the increased expression of Luc+ in muscle appears to be the removal of the peroxisomal targeting signal (Hartikka et al., 1996a). The Luc+ gene also appears to confer higher expression levels in lung and tumor (data not shown). The VR1223 and VR1255 vectors, representing the cumulative effect of incremental vector improvements, expressed 80- and 137-fold more luciferase, respectively, than the original nRSVLux vector. Maximum expression values from both VR1223 and VR1255 exceeded 1 $\mu$g per muscle.

*Analysis of expression variability*

The data presented in Figure 4 demonstrate median expression values that are consistently lower than the average (mean) expression values, suggesting a disproportionate number of low expression values. Therefore, a frequency distribution analysis was carried out on the data from two vectors, nRSVLux and VR1255. The results are shown in Figure 5.

The data indicate that the vector improvements embodied by VR1255 generally assure that positive expression values will be obtained. In fact, almost 90% of expression values were above 64 ng/muscle. In contrast, the less potent DNA vector, nRSVLux, yielded no expression values above

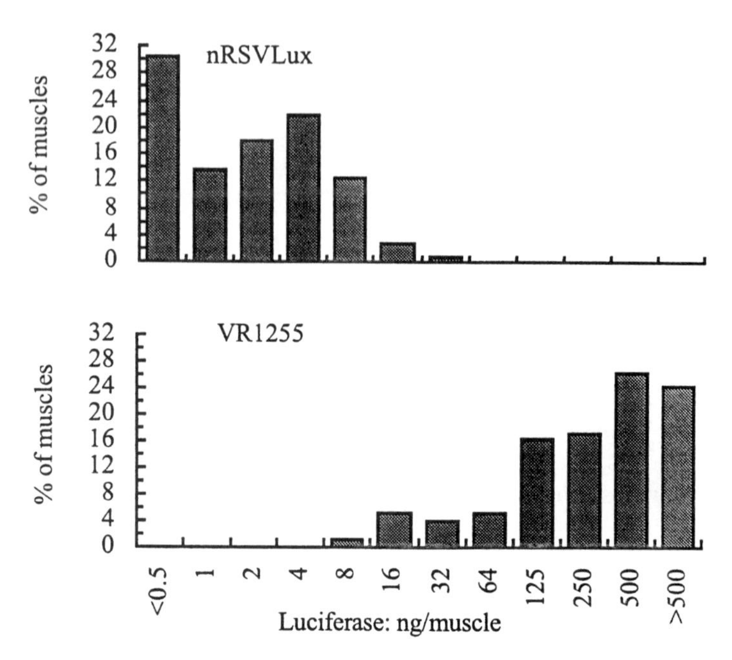


FIGURE 5. Distribution of intramuscular expression values for nRSVLux and VR1255. The expression values for individual muscles were placed into bins ranging from <0.5 ng/muscle to >500 ng/muscle as indicated on the abscissa. The percentage of the total number of values (nRSVLux = 259 values; VR1255 = 78 values) is shown on the ordinate.

32 ng/muscle, and 95% of the expression levels were below 8 ng/muscle. Thus, the VR1255 vector should provide increased efficacy in genetic vaccination, therapeutic protein secretion, and/or other gene therapy applications.

### *Analysis of expression in multiple tissues*

Tissues other than rectus femoris muscle are also transfected by direct plasmid DNA injection. In the experiment shown in Table I, selected mouse tissues were injected with VR1223 plasmid DNA (see Figure 4), and tissue extracts were assayed for luciferase enzyme activity.

In these experiments, VR1223 expressed 175,000 pg luciferase/muscle in rectus femoris muscle. Surprisingly, eightfold more luciferase (1,400,000 pg muscle) was expressed in tibialis anterior muscle. Thus, different mouse skeletal muscles express considerably different levels of transgene product

TABLE I. Luciferase expression in tissues in vivo.

| TISSUE | INJECTION SITE | $\mu$g VR1223[1] | DAY AFTER INJ. | pg LUX PER UNIT | $n$ | UNIT |
|---|---|---|---|---|---|---|
| Skeletal muscle | Tibialis anterior | 50 | 7 | 1,400,000 | 20 | Tibialis ant. |
| Skeletal muscle | Rectus femoris | 50 | 7 | 175,000 | 851 | Quadriceps |
| Tumor tumor (Renca) | Subcutaneous (tumor nodes) | 50 | 2 | 5,600 | 20 | Entire |
| Liver | Medial lobe | 20 | 2 | 1,100 | 20 | Entire lobe |
| Skin | Subcutaneous | 50 | 2 | 116 | 20 | 1 $cm^2$ |
| Lung | Intranasal | 130 | 3 | 17 | 40 | Entire lung |

[1] The indicated amounts of VR1223 were injected into the tissues shown. The tissues were collected at the indicated times postinjection and assayed for luciferase activity.

after injection of the same amount of a given plasmid. The basis for this differential expression may be related to muscle physiology or anatomy, to the nature of the cell types transfected, or to physical aspects of the injection procedure. In any case, such results encourage a more detailed comparison of plasmid DNA expression among different muscles.

The data shown in Table I represent peak time points of expression determined empirically for each tissue. Expression levels in lung, skin, liver, and tumor were considerably lower than in skeletal muscle. Intranodal injection into subcutaneous renal cell carcinoma (Renca) tumors elicited 5600 pg luciferase/tumor, 250-fold lower than tibialis muscle. Liver, at 1100 pg/lobe, expressed about 1000-fold lower than muscle, and skin and lung expressed luciferase at very low levels (116 pg/skin piece and 17 pg/lung). The reason that skeletal muscle is so readily transfected by plasmid DNA is unclear. It may reflect uptake mechanisms mediated by structures unique to or prevalent in skeletal muscle, such as the T-tubule system (Wolff et al., 1990) or caveolae (Wolff et al., 1992), respectively.

### *Analysis of transfection-enhancing agents*

Several laboratories have reported that intramuscular injection of the anesthetic bupivacain (BPVC) 5 days before plasmid DNA injection will en-

TABLE II. Bupivacaine (BPVC) inhibits muscle transfection.

| Vector | μg DNA Injected | No BPVC | Inject BPVC 5 Days before DNA Injection | Coinject BPVC with DNA |
|---|---|---|---|---|
| nRSVLux | 10 | 0.75 (37) | 1.92 (22) | 0.014 (22) |
| VR1223 | 50 | 168 (19) | 201 (14) | 31 (14) |

ng luciferase/muscle (n).

hance expression of encoded reporter genes (Thomason and Booth, 1990; Danko et al., 1993a; Vitadello et al., 1994; Levy et al., 1996). BPVC is known to cause degeneration of myofibers followed by regeneration of new muscle via satellite cell proliferation, myotube formation, and myotube fusion with myofiber stumps. The increase in transfection may be due to preferential transfection of proliferating satellite cells or fusing myotubes, to protective residence of plasmid in basal lamina tubes left intact after myofiber degeneration, or to other phenomena. We have repeated these BPVC-injection experiments using our quantitative methods. Representative results are shown in Table II. We first used the most frequently reported BPVC treatment conditions (10 μg plasmid in 50 μl saline injected i.m. 5 days after i.m. injection of 100 μl 0.75% BPVC). Muscles were collected 7 days after DNA injection, the time of peak expression (see Figure 3). The results using nRSVLux (see map in Figure 4) showed a 2.6-fold BPVC-induced increase (0.75 to 1.92 ng/muscle). However, using the more potent luciferase plasmid, VR1223 (see Figure 4), BPVC had little effect on expression (168 to 201 ng luciferase/muscle). BPVC injections at 1, 3, 7, or 10 days before DNA injection also had little effect on VR1223 expression (data not shown). In contrast, coinjection of BPVC with either nRSVLux or VR1223 resulted in a dramatic decrease in luciferase expression to 1.9% (nRSVLux) or 18% (VR1223) of the non-BPVC control values. Thus, BPVC preinjection increases the expression of suboptimal plasmids such as nRSVLux, but has little effect on highly potent plasmids such as VR1223, while BPVC coinjection is strongly detrimental, regardless of the vector used.

Given the inability to achieve high-level transfection of nonmuscle tissues, many investigators have attempted to identify substances that enhance the delivery of DNA into cells (Vlassov et al., 1994; Schreier, 1994; Miller and Vile, 1995; Cooper, 1996). One such category of substances is termed cytofectins (Felgner et al., 1995). The cytofectin DMRIE:DOPE, consisting of a cationic lipid (dimyristoyl Rosenthal inhibitor ether, or

DMRIE) associated in a 50:50 molar ratio with a neutral lipid (dioleoyl phosphatidyethanolamine, or DOPE), is currently being tested in clinical trials (Vogelzang et al., 1994; Hersh et al., 1994). Cytofectins can impart a net positive charge to DNA-lipid complexes, possibly facilitating their contact with negatively charged cell surfaces. In the experiment shown in Figure 6, plasmid DNA complexed with DMRIE:DOPE was injected into selected mouse tissues, and the effect on reporter gene expression was determined.

DMRIE:DOPE enhanced transfection of subcutaneous or intraperitoneal (melanoma) tumors 5.3-fold and 62-fold, respectively, relative to plasmid DNA alone. Liver and skin transfection were only marginally enhanced (1.3- and 1.1-fold), while transfection of intradermal (Renca) tumors and lung was inhibited. However, other cytofectins have recently been identified that enhance lung transfection approximately 100-fold relative to plasmid DNA alone (Wheeler et al., 1996).

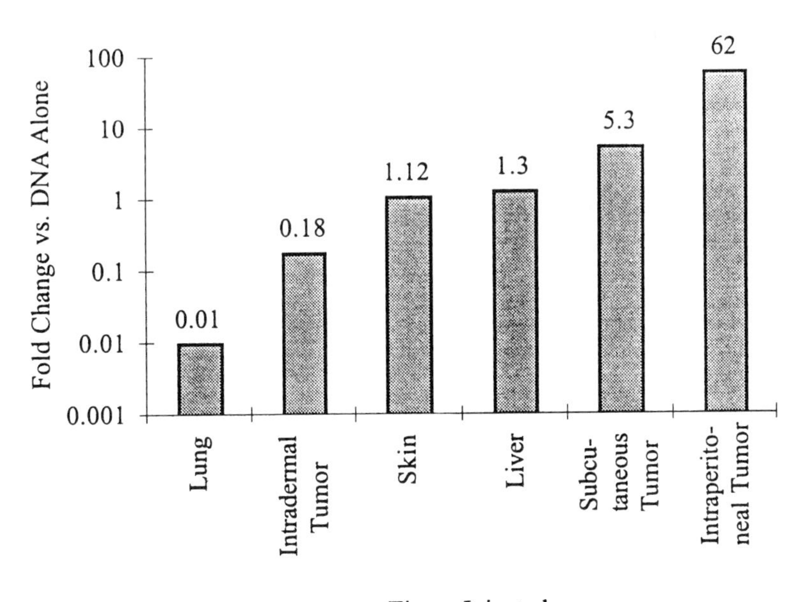


FIGURE 6. Effect of DMRIE:DOPE on plasmid DNA expression in different tissues. DMRIE:DOPE was complexed with plasmid DNA at a molar ratio of 4:1 (DNA:DMRIE) and delivered to the indicated tissues. Plasmid DNA alone was delivered separately as a control. Fold change values derive from $n$ = 15 (lung), 10 (intradermal tumor), 20 (skin), 5 (liver), 10 (subcutaneous tumor), and 21 (peritoneal tumor).

*Regulable vector expression*

For certain gene therapy applications (e.g., hormonal replacement therapy), it would be desirable to regulate gene expression from injected plasmids. A candidate vector system for this purpose has been developed (Liang et al., 1996b), based on the chimeric tetracycline-responsive transcriptional activator originally described by Gossen and Bujard (1992). The details of the system are depicted in Figure 7. Briefly, VR1502 constitutively expresses the tet-VP16 transactivator protein. In the absence of tetracycline, the transactivator binds avidly to the tet operator sequence (tetO) and either stimulates or inhibits gene transcription, depending on the positioning of the tetO sequence with respect to the promoter (TATA). VR1250, which expresses luciferase, contains seven copies of the tetO sequence in the stimulatory position upstream of the promoter. VR1340, which expresses CAT, contains four copies of the tet operator sequence in the inhibitory position downstream of the promoter.

Selected combinations of these plasmids were coinjected into mouse rectus femoris muscle (Figure 7B). When VR1250 and VR1340 were coinjected, CAT expression was high and luciferase expression was low, reflecting the relative strength of the promoter/enhancers present in these vectors. When VR1250, VR1340, and VR1502 were coinjected, and tetracycline was absent, CAT expression was downregulated by binding of the tet transactivator downstream of the CMV promoter/enhancer in VR1340, where the transactivator functions to block transcriptional elongation. In contrast, luciferase expression was increased dramatically under these circumstances, as the result of tet-VP16 binding and transcriptional activation of the minimal promoter (TATA) of VR1250. The systemic administration of tetracycline, by causing dissociation of the tet transactivator from the DNA, resulted in the same expression profile observed in the absence of VR1502. These experiments provided an initial proof that expression from intramuscularly injected plasmid DNAs can be regulated in a positive or negative fashion by systemic administration of tetracycline, a clinically approved antibiotic drug that can be given orally.

## Summary and Conclusions

A quantitative procedure has been developed for measuring reporter gene expression in tissues directly injected with plasmid DNA. This procedure is conducive to high-throughput, in vivo analyses. Several applications have been illustrated, including the rapid evaluation of plasmid DNA vector improvements, identification of transfection-inhibiting or -enhancing

A. Plasmids Used

Plasmid: Maps

| Vector Name | Enhancer | Promoter | 5' Leader | Intron | Gene | Terminator |
|---|---|---|---|---|---|---|
| VR1250: | $(tetO)_7$ | CMV | CMV | CMV | Lux | BGH |
| VR1340: | CMV | CMV | $(tetO)_4$ | CMV | CAT | BGH |
| VR1502: | CMV | CMV | CMV | (none) | tet-VP16 | SV40 |

B. Relative Luciferase/CAT Activity in Mouse Skeletal Muscle

| | -VR1502 | +VR1502 | +VR1502 +Tc |
|---|---|---|---|
| Lux | 0.2 | 100 | 3 |
| CAT | 100 | 10 | 100 |

FIGURE 7. Reciprocal regulation of luciferase and CAT expression after intramuscular injection of plasmid DNA. Mouse rectus femoris muscles were coinjected with 50 $\mu$g of the indicated DNA(s), and the muscles were assayed for reporter enzyme activity at 7 days postinjection. A. The tetracycline operator sequence (tetO)-containing vectors, VR1250 and VR1340, were injected in the presence or absence of the VR1502 vector encoding the tet-VP16 fusion protein that binds to the (tetO) sequence. B. Systemic administration of tetracycline (Tc) down-regulates VR1250 and up-regulates VR1340 expression. For abbreviations, see Figure 4. Values derive from $n = 20$ muscles per condition (with and without tetracycline).

agents, analysis of expression in multiple tissues, discovery of more potent vectors that confer persistent long-term expression, and development of vectors where expression can be regulated.

The methodologies illustrated in this review chapter for convenient quantification of *reporter gene* expression may find increasing utility in studies of the expression of *functional therapeutic genes*. However, the reporter genes discussed (luciferase, CAT) are often used because they are retained intracellularly, which allows an accurate assessment of their total expression within a given tissue. In contrast, therapeutic plasmids may encode proteins that 1) remain associated with the cell intracellularly, 2) are presented extracellularly via the cell membrane, 3) are secreted into the local tissue environment, or 4) are released into the blood. Each of these protein destinations presents the researcher with a particular set of problems for analytical quantification. For therapeutic applications, quantification of the level of transgene *expression* is important *per se* but most important is the quantification of the level of *function* restored by transgene expression. Thus, investigators often use the ideal approach of correlating the level of expressed protein with its biological activity. Selected examples follow.

1) Quantification of proteins that remain associated with the cell intracellularly. The muscular dystrophic mouse mdx expresses a defective dystrophin protein whose malfunction causes chronic myofiber degeneration and regeneration. Intramuscular injection of plasmid DNA encoding luciferase in mdx mice resulted in loss of luciferase expression by 2 months (Danko et al., 1993a). However, coinjection of luciferase DNA with DNA encoding normal mouse dystrophin resulted in stable expression of luciferase in the muscle. Such coquantification suggests that the plasmid-derived normal dystrophin protein protects mdx mouse myofibers from degeneration.

2) Quantification of proteins that are presented extracellularly via the cell membrane. Perhaps the most studied therapeutic application of direct plasmid DNA injection is vaccination against infectious diseases or cancer (Brown et al., 1996; Ulmer et al., 1996; Conry et al., 1996). Many cell types transfected with plasmid DNA expressing a foreign protein will proteolytically cleave the protein into short polypeptides and present these peptides at the cell surface associated with the class I major histocompatibility complex. Such peptide presentation can elicit a systemic cytotoxic T-lymphocytic immune response against any cell expressing the foreign protein. For example, intramuscular injection of plasmid DNAs encoding viral, mycobacterial, bacterial, protozoan, or platyhelminthian proteins into animals has afforded systemic protection to the animal upon challenge with a lethal dose of the live pathogen (Ulmer et al., 1996). In all these examples, rather than quantifying the levels of expressed protein antigen, the investigators quantified animal infection or survival, which indirectly reflects efficiency of immune response generation.

3) Quantification of proteins that are secreted into the local tissue environment. Plasmid DNAs encoding IL-2 (Vile and Hart, 1994; Parker et al., 1996; Saffran et al., 1996) have been directly injected into established mouse melanoma or renal cell tumors. Biologically active IL-2 was quantified from cell cultures derived from injected tumors. In one case (Saffran et al., 1996) renal cell carcinoma tumors were completely cured after IL-2

plasmid DNA injections, and the cured animals generated quantifiable tumor-specific cytolytic immune responses. In this example, a defined level of plasmid DNA expression in tumors correlated with plasmid-induced antitumor efficacy.

4) Quantification of proteins that are released into the blood. Intramuscular injection of plasmid DNAs encoding interleukin-2 (Raz et al., 1995), erythropoetin (Tripathy et al., 1996), or leptin (Liang et al., 1996b) results in secretion of these proteins into the blood. In the case of interleukin-2, the proteins were measured directly via ELISA of blood samples. The presence of erythropoetin or leptin in the blood of mice was inferred from quantification of, respectively, sustained high levels of hematocrit or weight loss plus hypoglycemia. These examples illustrate that intramuscular injection of plasmid DNA can produce quantifiable systemic responses to gene products.

The above are but a few examples of the utility of quantifying gene products or biological responses after injection of plasmid DNA. New therapeutic applications of direct plasmid DNA gene therapy are being developed at a rapid pace in diverse areas such as DNA vaccination, immunomodulation, systemic protein delivery, antitumor therapy, and correction of genetic disorders. Each specific application of plasmid DNA-based gene therapy will require convenient methods for correlating transgene expression levels with biological effects. This chapter has attempted to detail some methods used to quantify levels of expression currently achievable by direct plasmid DNA delivery.

## REFERENCES

Acsadi G, Dickson G, Love DR, Jani A, Walsh FS, Gurusinghe A, Wolff JA, and Davies KE (1991): Human dystrophin expression in mdx mice after intramuscular injection of DNA constructs. *Nature* 352:815–18.

Brown F, Norrby E, Burton D, and Mekalanos J (1996): *Vaccines96: Molecular approaches to the control of infectious diseases.* Cold Spring Harbor Laboratory Press, Plainview, New York.

Chauchereau A, Astinotti D, and Bouton M (1990): Automation of a chloramphenicol acetyltransferase assay. *Anal Biochem* 188:310–16.

Conry RM, LoBuglio AF, and Curiel DT (1996): Polynucleotide-mediated immunization therapy of cancer. *Semin Oncol* 23:135–47.

Cooper MJ (1996): Noninfectious gene transfer and expression systems for cancer gene therapy. *Semin Oncol* 23:172–87.

Danko I, Fritz, Jiao K, Hogan K, Latendresse JS, and Wolff JA (1993a): Pharmacological enhancement of in vivo foreign gene expression in muscle. *Gene Ther* 1:114–21.

Danko I, Fritz JD, Latendresse JS, Herweijer H, Schultz E, and Wolff JA (1993b): Dystrophin expression improves myofiber survival in mdx muscle following intramuscular plasmid DNA injection. *Hum Mol Genet* 2:2055–60.

Danko I and Wolff JA (1994): Direct gene transfer into muscle. *Vaccine* 12:1499–502.

DeMatteo RP, Raper SE, Ahn M, Fisher KJ, Burke C, Radu A, Widera G, Claytor BR, Barker CF, and Markmann JF (1995): Gene transfer to the thymus. A means of abrogating the immune response to recombinant adenovirus. *Ann Surg* 222:229–42.

Felgner PL, Tsai YJ, Sukhu L, Wheeler CJ, Manthorpe M, Marshall J, and Cheng SH (1995): Improved cationic lipid formulations for in vivo gene therapy. *Ann NY Acad Sci* 772:126–39.

Gossen M and Bujard H (1992): Tight control of gene expression in mammalian cells by tetracycline-responsive promoters. *Proc Natl Acad Sci USA* 89:5547–51.

Hartikka J, Sawdey M, Cornefert-Jensen F, Margalith M, Barnhart K, Nolasco M, Vahlsing HL, Meek J, Marquet M, Hobart P, Norman J, and Manthorpe M (1996a): An improved plasmid DNA expression vector for direct injection into skeletal muscle. *Hum Gene Ther* 7:1205–17.

Hartikka J, Cornefert-Jensen F, Doh S, Margalith M, Liang X, Gromkowski S, Norman J, Hobart P, and Manthorpe, M (1996b): Transient expression of plasmid DNA in mouse skeletal muscle is due to a host immune response against foreign reporter proteins. *Cold Spr Harb Symp* (in press).

Hendrikx PJ, Martens AC, Visser JW, and Hagenbeek A (1994): Differential suppression of background mammalian lysosomal beta-galactosidase increases the detection sensitivity of LacZ-marked leukemic cells. *Anal Biochem*

222:456–60.

Hersh EM, Akporiaye E, Harris D, Stopeck AT, Unger EC, Warneke JA, and Kradjian SA (1994): Phase I study of immunotherapy of malignant melanoma by direct gene transfer. *Hum Gene Ther* 5:1371–84.

Hickman MA, Malone RW, Lehmann-Bruinsma K, Sih TR, Knoell D, Szoka FC, Walzem R, Carlson DM, and Powell JS (1994): Gene expression following direct injection of DNA into liver. *Hum Gene Ther* 5:1477–83.

Holt CE, Garlick N, and Cornel E (1990): Lipofection of cDNAs in the embryonic vertebrate central nervous system. *Neuron* 4:203–14.

Horn N, Meek J, Budahazi G, and Marquet M (1995): Cancer gene therapy using plasmid DNA: purification of DNA for human clinical trials. *Hum Gene Ther* 6:565–73.

Levy (1996): Characterization of plasmid DNA transfer into mouse skeletal muscle: evaluation of uptake mechanism, expression and secretion of gene products into blood. *Gene Ther* 3:201–11.

Liang X, Hartikka J, Sukhu L, Manthorpe M, and Hobart P (1996a): Novel, high expressing, and antibiotic controlled plasmid vectors designed for use in gene therapy. *Gene Ther* 3:350–6.

Liang X, Carner K, Doh SG, and Hobart P (1996b) Weight regulation in *ob/ob* mice by intramuscular injection of plasmid DNA expressing leptin. *Cold Spr Harb Sym Gene Ther* (in press).

Manthorpe M, Cornefert-Jensen F, Hartikka J, Felgner J, Rundell A, Margalith M, and Dwarki VJ (1993): Gene therapy by intramuscular injection of plasmid DNA: studies on firefly luciferase gene expression in mice. *Hum Gene Ther* 4:419–31.

Margalith M, Tripathy S, Leiden J, and Hobart P (1996): Sustained delivery of a biologically active protein by repeat intramuscular injection of plasmid DNA. *Cold Spr Harb Symp* (in press).

Marsh J (1994): Kinetic determination of cellular LacZ expression. *Genet Anal Tech Appl* 11:20–3.

Miller N and Vile R (1995): Targeted vectors for gene therapy. *Faseb J* 9:190–9.

Montgomery DL, Shiver JW, Leander KR, Perry HC, Friedman A, Martinez D,

Ulmer JB, Donnelly JJ, and Liu MA (1993): Heterologous and homologous protection against influenza A by DNA vaccination: optimization of DNA vectors. *DNA Cell Biol* 12:777–83.

Nabel GJ, Nabel EG, Yang ZY, Fox BA, Plautz GE, Gao X, Huang L, Shu S, Gordon D, and Chang AE (1993): Direct gene transfer with DNA-liposome complexes in melanoma: expression, biologic activity, and lack of toxicity in humans. *Proc Natl Acad Sci USA* 90:11307–11.

Nabel EG and Nabel GJ (1993): Direct gene transfer: basic studies and human therapies. *Thromb Haemostasis* 70:202–3.

Parker SP, Khatibi S, Margalith M, Anderson D, Yamkauckas M, Gromkowski SH, Latimer T, Lew D, Marquet M, Manthorpe M, Hobart P, Hersh E, Stopeck AT, and Norman J (1996): Plasmid DNA gene therapy: studies with the human interleukin-2 gene in tumor cells in vitro and in the murine B16 melanoma model in vivo. *Cancer Gene Ther* 3:175–85.

Plautz GE, Yang ZY, Wu BY, Gao X, Huang L, and Nabel GJ (1993): Immunotherapy of malignancy by in vivo gene transfer into tumors [see comments]. *Proc Natl Acad Sci USA* 90:4645–9.

Prigozy T, Dalrymple K, Kedes L, and Shuler C (1993): Direct DNA injection into mouse tongue muscle for analysis of promoter function in vivo. *Som Cell Mol Genet* 19:111–22.

Raz E, Carson DA, Parker SE, Parr TB, Abai AM, Aichinger G, Gromkowski SH, Singh M, Lew D, and Yankauckas MA (1994): I ntradermal gene immunization: the possible role of DNA uptake in the induction of cellular immunity to viruses. *Proc Natl Acad Sci USA* 91:9519–23.

Raz E, Watanabe A, Baird SM, Eisenberg RA, Parr TB, Lotz M, Kipps TJ, and Carson DA (1993): Systemic immunological effects of cytokine genes injected into skeletal muscle. *Proc Natl Acad Sci USA* 90:4523–7.

Saffran DC, Yankauckas M, Anderson D, Khatibi S, Abai A, Margalith M, Barnhart K, Manthorpe M, Parker S, and Norman J (1996): Direct injection of plasmid DNA containing the IL-2 gene results in immune rejection of established tumors in mice. *Cold Spr Harb Symp Gene Ther* (in press).

Sahenk Z, Seharaseyon J, Mendell JR, and Burghes AH (1993): Gene delivery to spinal motor neurons. *Brain Res* 606:126–9.

Sambrook J, Fritsch EF, and Maniatis T (1989): *Molecular Cloning: A Laboratory Manual*, 2nd Edition. Cold Spring Harbor Laboratory Press, Plainview, New York.

Sankaran L (1992): A simple quantitative assay for chloramphenicol acetyltransferase by direct extraction of the labeled product into scintillation cocktail. *Anal Biochem* 200:180–6.

Schreier H (1994): The new frontier: gene and oligonucleotide therapy. *Pharm Acta Helv* 68:145–59.

Sikes ML, O'Malley BW, Jr, Finegold MJ, and Ledley FD (1994): In vivo gene transfer into rabbit thyroid follicular cells by direct DNA injection. *Hum Gene Ther* 5:837–44.

Thomason DB and Booth FW (1990): Stable incorporation of a bacterial gene into adult rat skeletal muscle in vivo. *Am J Physiol* 258:C578–81.

Thompson JF, Hayes LS, and Lloyd DB (1991): Modulation of firefly luciferase stability and impact on studies of gene regulation. *Gene* 103:171–7.

Tripathy SK, Svensson EC, Black HB, Glodwasser E, Margalith M, Hobart PM, and Leiden JM (1996): Long-term expression of erythropoetin in the systemic circulation of mice after intramuscular injection of a plasmid DNA vector. *Proc Natl Acad Sci USA* 93 (in press).

Tsan MF, White JE, and Shepard B (1995): Lung-specific direct in vivo gene transfer with recombinant plasmid DNA. *Am J Physiol* 268:L1052–6.

Ulmer JB, Sadoff JC, and Liu M (1996): DNA vaccines. *Curr Opin Immunol* 8:531–6.

Vile RG and Hart IR (1993): Use of tissue-specific expression of the herpes simplex virus thymidine kinase gene to inhibit growth of established murine melanomas following direct intratumoral injection of DNA. *Cancer Res* 53:3860–4.

Vitadello M, Schiaffino MV, Picard A, Scarpa M, and Schiaffino S (1994): Gene transfer in regenerating muscle. *Hum Gene Ther* 5:11–18.

Vlassov VV, Balakireva LA, and Yakubov LA (1994): Transport of oligonucleotides across natural and model membranes. *Biochim Biophys Acta* 1197:95–108.

Vogelzang NJ, Lestingi TM, Sudakoff G, and Kradjian SA (1994): Phase I study of immunotherapy of metastatic renal cell carcinoma by direct gene transfer into metastatic lesions. *Hum Gene Ther* 5:1357–70.

Wells DJ (1993): Improved gene transfer by direct plasmid injection associated with regeneration in mouse skeletal muscle. *Febs Lett* 332:179–82.

Wheeler CJ, Felgner PL, Tsai YT, Marshall J, Sukhu L, Doh G, Hartikka JH, Nietupski J, Manthorpe M, Nichols M, Plewe M, Liang X, Norman J, Smith A, and Cheng SH (1996): A novel cationic lipid greatly enhances plasmid DNA delivery and expression in mouse lung. *Proc Natl Acad Sci USA* (in press).

Wolff JA, Lederburg J (1995): An early history of gene transfer and therapy. *Hum Gene Ther* 5:469–80.

Wolff JA, Malone RW, Williams P, Chong W, Acsadi G, Jani A, and Felgner PL (1990): Direct gene transfer into mouse muscle in vivo. *Science* 247:1465–8.

Wolff JA, Dowty ME, Jiao S, Repetto G, Berg BK, Ludtke JJ, Williams P, and Slautterback DB (1992): Expression of naked plasmids by cultured myotubes and entry of plasmids into T tubules and caveolae of mammalian skeletal muscle. *J Cell Sci* 103:1249–59.

Yovandich J, O'Malley B, Jr, Sikes M, and Ledley FD (1995): Gene transfer to synovial cells by intra-articular administration of plasmid DNA. *Hum Gene Ther* 6:603–10.

Zhu N, Liggitt D, Liu Y, and Debs R (1993): Systemic gene expression after intravenous DNA delivery into adult mice. *Science* 261:209–11.

# Index